Auto-Segmentation for Radiation Oncology

Series in Medical Physics and Biomedical Engineering

Series Editors
Kwan-Hoong Ng, E. Russell Ritenour, and Slavik Tabakov

Recent books in the series:

Introduction to Megavoltage X-Ray Dose Computation Algorithms
Jerry Battista

Problems and Solutions in Medical Physics
Nuclear Medicine Physics
Kwan Hoong Ng, Chai Hong Yeong, Alan Christopher Perkins

The Physics of CT Dosimetry
CTDI and Beyond
Robert L. Dixon

Advanced Radiation Protection Dosimetry
Shaheen Dewji, Nolan E. Hertel

On-Treatment Verification Imaging A Study Guide for IGRT
Mike Kirby, Kerrie-Anne Calder

Modelling Radiotherapy Side Effects Practical Applications for Planning Optimisation
Tiziana Rancati, Claudio Fiorino

Proton Therapy Physics, Second Edition
Harald Paganetti (Ed)

e-Learning in Medical Physics and Engineering
Building Educational Modules with Moodle
Vassilka Tabakova

Diagnostic Radiology Physics with MATLAB®
A Problem-Solving Approach
Johan Helmenkamp, Robert Bujila, Gavin Poludniowski (Eds)

Auto-Segmentation for Radiation Oncology
State of the Art
Jinzhong Yang, Gregory C. Sharp, and Mark J. Gooding

For more information about this series, please visit: https://www.routledge.com/Series-in-Medical-Physics-and-Biomedical-Engineering/book-series/CHMEPHBIOENG

About the Series

The *Series in Medical Physics and Biomedical Engineering* describes the applications of physical sciences, engineering, and mathematics in medicine and clinical research.

The series seeks (but is not restricted to) publications in the following topics:

- Artificial organs
- Assistive technology
- Bioinformatics
- Bioinstrumentation
- Biomaterials
- Biomechanics
- Biomedical engineering
- Clinical engineering
- Imaging
- Implants
- Medical computing and mathematics
- Medical/surgical devices

- Patient monitoring
- Physiological measurement
- Prosthetics
- Radiation protection, health physics, and dosimetry
- Regulatory issues
- Rehabilitation engineering
- Sports medicine
- Systems physiology
- Telemedicine
- Tissue engineering
- Treatment

The *Series in Medical Physics and Biomedical Engineering* is an international series that meets the need for up-to-date texts in this rapidly developing field. Books in the series range in level from introductory graduate textbooks and practical handbooks to more advanced expositions of current research.

The *Series in Medical Physics and Biomedical Engineering* is the official book series of the International Organization for Medical Physics.

THE INTERNATIONAL ORGANIZATION FOR MEDICAL PHYSICS

The International Organization for Medical Physics (IOMP) represents over 18,000 medical physicists worldwide and has a membership of 80 national and 6 regional organizations, together with a number of corporate members. Individual medical physicists of all national member organisations are also automatically members.

The mission of IOMP is to advance medical physics practice worldwide by disseminating scientific and technical information, fostering the educational and professional development of medical physics and promoting the highest quality medical physics services for patients.

A World Congress on Medical Physics and Biomedical Engineering is held every three years in cooperation with International Federation for Medical and Biological Engineering (IFMBE) and International Union for Physics and Engineering Sciences in Medicine (IUPESM). A regionally based international conference, the International Congress of Medical Physics (ICMP) is held between world congresses. IOMP also sponsors international conferences, workshops and courses.

The IOMP has several programmes to assist medical physicists in developing countries. The joint IOMP Library Programme supports 75 active libraries in 43 developing countries, and the Used Equipment Programme coordinates equipment donations. The Travel Assistance Programme provides a limited number of grants to enable physicists to attend the world congresses.

IOMP co-sponsors the *Journal of Applied Clinical Medical Physics.* The IOMP publishes, twice a year, an electronic bulletin, *Medical Physics World.* IOMP also publishes e-Zine, an electronic news letter about six times a year. IOMP has an agreement with Taylor & Francis for the publication of the *Medical Physics and Biomedical Engineering* series of textbooks. IOMP members receive a discount.

IOMP collaborates with international organizations, such as the World Health Organisations (WHO), the International Atomic Energy Agency (IAEA) and other international professional bodies such as the International Radiation Protection Association (IRPA) and the International Commission on Radiological Protection (ICRP), to promote the development of medical physics and the safe use of radiation and medical devices.

Guidance on education, training and professional development of medical physicists is issued by IOMP, which is collaborating with other professional organizations in development of a professional certification system for medical physicists that can be implemented on a global basis.

The IOMP website (www.iomp.org) contains information on all the activities of the IOMP, policy statements 1 and 2 and the 'IOMP: Review and Way Forward' which outlines all the activities of IOMP and plans for the future.

Auto-Segmentation for Radiation Oncology

State of the Art

Edited by
Jinzhong Yang, Gregory C. Sharp, and Mark J. Gooding

CRC Press
Taylor & Francis Group
Boca Raton London New York

CRC Press is an imprint of the
Taylor & Francis Group, an **informa** business

First edition published [2021]
by CRC Press
6000 Broken Sound Parkway NW, Suite 300, Boca Raton, FL 33487-2742

and by CRC Press
2 Park Square, Milton Park, Abingdon, Oxon, OX14 4RN

ISBN: 978-0-367-33600-4 (hbk)
ISBN: 978-0-367-76122-6 (pbk)
ISBN: 978-0-429-32378-2 (ebk)

Typeset in Times
by Deanta Global Publishing Services, Chennai, India

Contents

PART III   Clinical Implementation Concerns

Foreword I

Auto-segmentation plays an important everyday role in radiotherapy clinics. It is an essential prerequisite step for complex treatment process that aims to identify the gross tumor volume and surrounding organs at risk in order to maximize tumor control while minimizing normal tissue toxicity, which is the holy grail of radiotherapy. With the ever-increasing sophistication of treatment planning procedures and improved delivery systems, accurate contouring becomes a major limitation to an efficient and lean radiotherapy clinic. The process of segmentation has historically been done manually, which makes it not only a laborious and time-consuming task but also prone to errors that may affect the planning quality as well as the patient's treatment outcomes. Auto-segmentation promises to alleviate this problem; however, its application in the clinic has been slow to meet its demands. Recent years have witnessed the development of more accurate auto-segmentation tools using advanced machine and deep learning techniques that aim at transforming how these tools can be applied reliably in the clinic, but with key questions about their nature and proper implementation remaining open. These questions are thoughtfully answered by the new textbook by Gooding, Sharp, and Yang.

This delightful textbook *Auto-Segmentation for Radiation Oncology: State of the Art* takes the interested reader on a three-part journey from traditional multi-atlas auto-segmentation techniques to more modern deep learning approaches and their implementation in the clinic. In Part I, atlas segmentation methods are discussed: their selection process, combination with deformable registration, and eventually their evaluation. In Part II, deep learning auto-segmentation is presented, starting from architectural designs, comparisons of 2D and 3D approaches with the common U-net architecture, organ-specific versus multi-class segmentation, effect of loss function and augmentation, ending with pitfalls, in each case using publicly available challenge datasets as examples. Part III addresses practical issues related to clinical commissioning and challenges from data curation to evaluation.

This is a must-have textbook for anyone starting on auto-segmentation in radiotherapy or interested in the topic for academic or professional reasons. Finally, I would like to commend the authors on their achievement, bringing a team from industry and academia to address this important topic in radiotherapy and making it feasible for the reader.

Issam El-Naqa
Chair of Machine Learning
H. Lee Moffitt Cancer Center

Foreword II

Artificial intelligence (AI) commonly refers to the computational technologies that enable machines to perform tasks requiring human intelligence. In the past several years, a new rising tide of AI has been changing the world in many aspects, including healthcare, which is mainly attributable to the availability of big data in the internet era, the easy access to GPU-based high-performance computing, and the emergence of deep learning (DL) algorithms. Given its broad and exponentially growing applications in medicine, AI has the potential to fundamentally transform the way medicine is practiced.

Radiation oncology is a crucial pillar of cancer care and is indicated for more than 50% of cancer patients in most places. Over the past few decades, technological advances have made radiotherapy increasingly complex and precise, which has resulted in a near complete reliance on human–machine interactions including both software and hardware. In radiotherapy, the patient treatment workflow includes many complex tasks like patient image acquisition, segmentation of tumor target volumes and organs at risk (OARs), treatment planning, patient positioning and immobilization, treatment delivery, and post-treatment follow-up. Many of these tasks still require time-consuming manual input, despite technological advances. The technology-intensive nature of radiation oncology and its reliance on digital data and human–machine interactions actually make AI particularly suited to improve the accuracy and efficiency of its clinical workflow. One good example is DL-based auto-segmentation.

Target and OAR segmentation is particularly important for radiotherapy since precise and quantitative planning and assessment of the radiation dose needs to be delivered to the target and each OAR. Auto-segmentation has been a major research topic over the last several decades. Various technologies have been developed and tested, ranging from intensity thresholding and region growing, to active contours and level-sets, to probability-based and prior knowledge-based techniques. Among these, atlas-based auto-segmentation has become popular in recent years and has been implemented commercially and clinically. However, its success is still quite limited, according to feedback from clinicians and the extent of its deployment in clinical practice. The main complaint has been the lack of accuracy for some challenging organs, which requires tedious and time-consuming manual correction of the auto-segmentation results.

In the last few years, DL-based auto-segmentation has shifted the paradigm in this important area of medical imaging and radiation oncology. It has achieved a performance better than any traditional methods and, on many tasks, comparable to human experts. Much work has been done on this topic, which has led to extensive publications and some commercial products. The three editors of this book are among the leaders of those research and commercialization efforts. At the 2017 AAPM Annual Conference, the book's editors, together with other colleagues, organized a grand challenge on thoracic auto-segmentation, which was a great success at the time. From this grand challenge, much has been learned about atlas-based and DL-based auto-segmentation for radiation oncology, which serves as a solid foundation for this book.

Given the ongoing heated research efforts and the emerging clinical implementation, now is the perfect time for this book. The book is well organized and balanced with three major sections: atlas-based technologies, DL-based technologies, and clinical implementation issues. The editors are very experienced and can ensure the quality of the book. Anyone who is interested in research in and the clinical use of image auto-segmentation for radiation oncology should read this book.

Steve Jiang
Vice Chair, Department of Radiation Oncology
Director, Medical Artificial Intelligence and Automation Laboratory
University of Texas Southwestern Medical Center

Editors

Jinzhong Yang earned his BS and MS degrees in Electrical Engineering from the University of Science and Technology of China, in 1998 and 2001, and his PhD degree in Electrical Engineering from Lehigh University in 2006. In July 2008, Dr Yang joined the University of Texas MD Anderson Cancer Center as a Senior Computational Scientist, and since January 2015 he has been an Assistant Professor of Radiation Physics. Dr Yang is a board-certified medical physicist. His research interest focuses on deformable image registration and image segmentation for radiation treatment planning and image-guided adaptive radiotherapy, radiomics for radiation treatment outcome modeling and prediction, and novel imaging methodologies and applications in radiotherapy.

Greg Sharp earned a PhD in Computer Science and Engineering from the University of Michigan and is currently Associate Professor in Radiation Oncology at Massachusetts General Hospital and Harvard Medical School. His primary research interests are in medical image processing and image-guided radiation therapy, where he is active in the open source software community.

Mark Gooding earned his MEng in Engineering Science in 2000 and DPhil in Medical Imaging in 2004, both from the University of Oxford. He was employed as a postdoctoral researcher both in university and hospital settings, where his focus was largely around the use of 3D ultrasound segmentation in women's health. In 2009, he joined Mirada Medical Ltd, motivated by a desire to see technical innovation translated into clinical practice. While there, he has worked on a broad spectrum of clinical applications, developing algorithms and products for both diagnostic and therapeutic purposes. If given a free choice of research topic, his passion is for improving image segmentation, but in practice he is keen to address any technical challenge. Dr Gooding now leads the research team at Mirada, where in addition to the commercial work he continues to collaborate both clinically and academically.

Contributors

Payton Bruckmeier
Department of Physics and Astronomy
Louisiana State University
Baton Rouge, LA, USA

Carlos E. Cardenas
Department of Radiation Physics
The University of Texas MD Anderson
    Cancer Center
Houston, TX, USA

Ken Chang
Athinoula A. Martinos Center for
    Biomedical Imaging
Department of Radiology
Massachusetts General Hospital,
    Boston, MA, USA

Quan Chen
Department of Radiation Medicine
University of Kentucky
Lexington, KY, USA

and

Carina Medical LLC
Lexington, KY, USA

Laurence Court
Department of Radiation Physics
The University of Texas MD Anderson
    Cancer Center
Houston, TX, USA

Walter J. Curran
Department of Radiation Oncology
Emory University
Atlanta, GA, USA

and

Winship Cancer Institute
Emory University
Atlanta, GA, USA

Raymond Fang
Medical Scientist Training Program
Northwestern University Feinberg
    School of Medicine
Chicago, IL, USA

Xi Fang
Department of Biomedical Engineering
Rensselaer Polytechnic Institute
Troy, NY, USA

Xue Feng
Department of Biomedical Engineering
University of Virginia
Charlottesville, VA, USA

and

Carina Medical LLC
Lexington, KY, USA

Yabo Fu
Department of Radiation Oncology
Emory University
Atlanta, GA, USA

Mishka Gidwani
Athinoula A. Martinos Center for
    Biomedical Imaging
Department of Radiology
Massachusetts General Hospital,
    Charlestown, MA, USA

Mark J. Gooding
Mirada Medical Ltd
Oxford, UK

Dongdong Gu
Hunan University
Changsha, China

Thomas Guerrero
Beaumont Artificial Intelligence Lab
Beaumont Research Institute
Royal Oak, MI, USA

and

Department of Radiation Oncology
Beaumont Health System
Royal Oak, MI, USA

Jayashree Kalpathy-Cramer
Department of Radiology
Massachusetts General Hospital
Boston, MA, USA

and

MGH and BWH Center for Clinical Data
   Science
Boston, MA, USA

Nagarajan Kandasamy
Department of Electrical and Computer
   Engineering Drexel University
Philadelphia, PA, USA

Yang Lei
Department of Radiation Oncology
Emory University
Atlanta, GA, USA

Matthew D. Li
Athinoula A. Martinos Center for Biomedical
   Imaging
Department of Radiology
Massachusetts General Hospital
Boston, MA, USA

Tian Liu
Department of Radiation Oncology
Emory University
Atlanta, GA, USA

and

Winship Cancer Institute
Emory University
Atlanta, GA, USA

Jay B. Patel
Athinoula A. Martinos Center for Biomedical
   Imaging
Department of Radiology
Massachusetts General Hospital
Boston, MA USA

Xi Pei
Department of Nuclear Science and
   Engineering
School of Nuclear Science and Technology
University of Science and Technology of China
Hefei, China

and

Institute of Nuclear Medical Physics
University of Science and Technology of China
Hefei, China

Zhao Peng
Department of Nuclear Science and
   Engineering
School of Nuclear Science and Technology
University of Science and Technology of China
Hefei, China

and

Institute of Nuclear Medical Physics
University of Science and Technology of China
Hefei, China

Evan Porter
Department of Medical Physics
Wayne State University
Detroit, MI, USA

and

Beaumont Artificial Intelligence Lab
Beaumont Research Institute
Royal Oak, MI, USA

Richard L.J. Qiu
Department of Radiation Oncology
Emory University
Atlanta, GA, USA

and

Winship Cancer Institute
Emory University
Atlanta, GA, USA

James Shackleford
Department of Electrical and Computer
    Engineering
Drexel University
Philadelphia, PA, USA

Keyur Shah
Department of Electrical and Computer
    Engineering
Drexel University
Philadelphia, PA, USA

Hongming Shan
Institute of Science and Technology for Brain-
    inspired Intelligence and MOE Frontiers
    Center for Brain Science
Fudan University
Shanghai, China

and

Shanghai Center for Brain Science and Brain-
    inspired Technology
Shanghai, China

Gregory C. Sharp
Department of Radiation Oncology
Massachusetts General Hospital
Boston, MA, USA

Zaid A. Siddiqui
Beaumont Artificial Intelligence Lab
Beaumont Research Institute
Royal Oak, MI, USA

and

Department of Radiation Oncology
Beaumont Health System
Royal Oak, MI, USA

David Solis
Department of Medical Physics
Mary Bird Perkins Cancer Center
Baton Rouge, LA, USA

and

Department of Physics and Astronomy
Louisiana State University
Baton Rouge, LA, USA

Harini Veeraraghavan
Department of Medical Physics
Memorial Sloan-Kettering Cancer Center
New York, NY, USA

Ge Wang
Department of Biomedical Engineering
Rensselaer Polytechnic Institute
Troy, NY, USA

Tonghe Wang
Department of Radiation Oncology
Emory University
Atlanta, GA, USA

and

Winship Cancer Institute
Emory University
Atlanta, GA, USA

X. George Xu
Department of Nuclear Science and
    Engineering
School of Nuclear Science and Technology
University of Science and Technology of China
Hefei, China

and

Institute of Nuclear Medical Physics
University of Science and Technology of China
Hefei, China

and

Department of Radiation Oncology
The First Affiliated Hospital of University of
    Science and Technology of China
Hefei, China

Zhong Xue
Shanghai United Imaging Intelligence Co., Ltd.
Shanghai, China

Pingkun Yan
Department of Biomedical Engineering
Rensselaer Polytechnic Institute
Troy, NY, USA

Jinzhong Yang
Department of Radiation Physics
The University of Texas MD Anderson
    Cancer Center
Houston, TX, USA

Xiaofeng Yang
Department of Radiation Oncology
Emory University
Atlanta, GA, USA

and

Winship Cancer Institute
Emory University
Atlanta, GA, USA

Leonid Zamdborg
South Carolina Oncology Associates
Columbia, SC, USA

Jieping Zhou
Department of Radiation Oncology
The First Affiliated Hospital of University of
    Science and Technology of China
Hefei, China

1 Introduction to Auto-Segmentation in Radiation Oncology

Jinzhong Yang, Gregory C. Sharp, and Mark J. Gooding

CONTENTS

1.1 INTRODUCTION

In the past two decades, the advancement in radiation therapy has allowed delivery of radiation to the treatment target with optimized spatial dose distribution that minimizes radiation toxicity to the adjacent normal tissues [1, 2]. In particular, with the advent of intensity modulated radiation therapy (IMRT) [1], fast and accurate delineation of targets and concerned organs at risk (OARs) from computed tomography (CT) images is extremely important for treatment planning in order to achieve a favorable dose distribution for treatment. Traditionally, these structures are manually delineated by clinicians. This manual delineation is time-consuming, labor intensive, and often subject to inter- and intra-observer variability [3, 4]. With the technology development in medical imaging computing, as well as the availability of more and more curated image and contour data, auto-segmentation has become increasingly available and important in radiation oncology to provide fast and accurate contour delineation in recent years. More and more auto-segmentation tools have been gradually used in routine clinical practice.

1.2 EVOLUTION OF AUTO-SEGMENTATION

The improvement in the performance of auto-segmentation algorithms has evolved alongside the capability of the algorithms to use prior knowledge for new segmentation tasks [5]. In the early stages of development, limited by the computer power and the availability of segmented data, most segmentation techniques used no or little prior knowledge, relying on the developer to encode their belief in what would provide good segmentation. These methods are referred to as low-level segmentation approaches. Methods include intensity thresholding, region growing, and heuristic edge detection algorithms [6–8]. More advanced techniques were developed in an attempt to avoid heuristic approaches leading to the introduction of uncertainty models and optimization methods. Region-based techniques, such as active contours, level-sets, graph cuts, and watershed algorithms, have been used in medical imaging auto-segmentation [9–12]. Probability-based auto-segmentation techniques, such as Gaussian mixture models, clustering, k-nearest neighbor, and Bayesian classifiers, rose in popularity with the turn of the century thanks to the availability of higher computing power [13, 14]. These techniques employ a limited quantity of prior knowledge in the form of statistical information about each organ's appearance acquired from example images.

In the last two decades, a large amount of exploratory work has been invested in making better use of prior knowledge, such as shape and appearance characteristics of anatomical structures, to compensate for insufficient soft tissue contrast of CT data which prevents accurate boundary definition using low-level segmentation methods. The approaches can be grouped as (multi)atlas-based segmentation, model-based segmentation, and machine learning-based segmentation [15], taking into account prior knowledge using differing techniques and to differing extents.

Single atlas-based segmentation uses one reference image, referred to as an atlas, in which structures of interest are already segmented, as prior knowledge for new segmentation tasks [16]. The segmentation of a new patient image relies on deformable registration, finding the transformation between the atlas and the patient image to map the contours from the atlas to the patient image. Various deformable registration algorithms have been used for this purpose [17–22], with intensity-based algorithms being popular to achieve full automation. The segmentation performance largely depends on the performance of deformable registration, which, in turn, depends on the similarity of the morphology of organs of interest between the atlas and the new image. To achieve good segmentation results, varied atlas selection strategies have been proposed [23–29]. Alternatively, using an atlas that reflects the anatomy of an average patient can potentially improve segmentation performance [30, 31].

Atlas-based segmentation is also impacted by the inter-subject variability since inaccurate contouring in the atlas will be propagated to the patient image. Instead of using a single atlas, multi-atlas approaches use a number of atlases (normally around ten) as prior knowledge for segmentation of new images [32–37]. Similar to single atlas-based approaches, deformable registration is used to map atlas contours from each atlas to the patient image. Then an additional step, frequently referred as to label/contour fusion, is performed to combine the individual segmentations from each atlas to produce a final segmentation that is the best estimate of the true segmentation [29, 38–41]. Multi-atlas segmentation has been shown to minimize the effects of inter-subject variability and improve segmentation accuracy over single atlas approaches. In the past decade, multi-atlas segmentation has been one of the most effective segmentation approaches in different grand challenges [42–44]. This approach has been validated for clinical radiation oncology applications in contouring normal head and neck tissue [45], cardiac substructures [46], and brachial plexus [34], among others. Commercial implementation of multi-atlas segmentation is also available from multiple vendors [15]. Part I of this book focuses on multi-atlas segmentation, particularly considering atlas selection strategies, deformation registration choice, and the impact of label/contour fusion.

When more contoured images are available, characteristic variations in the shape or appearance of structures of interest can be used for auto-segmentation. Statistical shape models (SSM) or statistical appearance models (SAM) can model the normal range of shape or appearance using a training set of atlases. These approaches have the benefit, compared to atlas-based methods, of restricting the final segmentation results to anatomically plausible shapes described by the models [47]. Consequently, they have shown good performance where registration is poor, for example where there is limited contrast. However, model-based segmentation is less flexible to extreme anatomical variation due to the limitation of specific shapes characterized by the statistical models, particularly where the size and content of the training data are limited. In radiation oncology applications, model-based segmentation is mostly used for the segmentation of structures in the pelvic region [48–50].

To take better advantage of prior knowledge, without using data directly as in atlas-based contouring, general machine learning approaches have been used to offer greater flexibility over model-based methods (although model-based methods are effectively limited-capacity machine learning models). Machine learning approaches can aid in segmentation by learning appropriate priors of organs shapes and image context and appearance for voxel classification [51–53]. Support vector machines and tree ensemble (i.e. random forests) algorithms have shown promising results in thoracic, abdominal, and pelvic tumor and normal tissue segmentation [54–56]. These generally employ human-engineered features, usually derived from the image intensity histograms, and use large databases of patients as inputs to train the segmentation model.

Deep learning is a specific part of the broader field of machine learning where algorithms are able to learn data representations on their own. More specifically, deep learning uses artificial neural networks with multiple (two or more) hidden layers (those between input and output layers) to learn features from a dataset by modeling complex non-linear relationships. The advancement of deep learning is attributed to the availability of more curated data, the advancement in computer power (e.g. Graphics Processing Unit (GPU) applications), and efficient algorithms. Previously, deep architectures were prone to model overfitting; however, algorithmic advances over the past decade have allowed for the use of very deep architectures (100+ layers) to achieve "superhuman" performance in some tasks. Furthermore, the application of GPUs to speed up computations has allowed the field to progress rapidly.

Convolutional neural networks (CNN) are of particular interest in computer vision tasks (i.e. segmentation, detection, classification) as these learn the filters or kernels that were previously engineered for use in traditional approaches [57]. CNNs allow for the classification of each individual pixel in the image; however, this becomes computationally expensive as the same convolutions are computed several times due to the large overlap between input patches from neighboring pixels. Fully convolutional networks (FCNs), introduced by Long et al. [58], overcome the loss of spatial information resulting from the implementation of fully connected layers as final layers of classification CNNs. Most FCNs used for medical image segmentation are based on 2D or 3D variants of successful methods adapted from computer vision. Improvements in 3D convolution computation efficiency and hardware, in particular the fast increase in available GPU memory, have enabled the extension of these methods to 3D imaging. The most popular medical image segmentation FCN architecture is the U-net and its variants [59–61]. In more recent grand challenges, deep learning-based segmentation approaches have been the most powerful and dominant segmentation approaches [44, 62]. Part II of this book focuses on deep learning-based auto-segmentation, considering a range of topics including architecture design and selection, loss function choice, and data augmentation methods.

1.3 EVALUATION OF AUTO-SEGMENTATION

Although auto-segmentation algorithms have been available for more than two decades, clinical use of auto-segmentation is limited. This is partly due to the lack of an effective approach for their evaluation, and a perception that auto-segmentation is of lower quality than human segmentation. In recent years, the concept of a "grand challenge" has emerged as an unbiased and effective approach for evaluating different segmentation approaches [42–44]. In a grand challenge, the participants are invited to evaluate their algorithms using a common benchmark dataset, with the algorithm performance being scored by an impartial third party. This framework allows the different segmentation approaches to be evaluated more evenly and reduces the risk of evaluation error due to overfitting to test cases and allows direct comparison of methods. Grand challenges attract some of the best academic and industrial researchers in the field. The competition is friendly and stimulates scientific discussion among participants, potentially leading to new ideas and collaboration.

This book was inspired by the 2017 AAPM Thoracic Auto-segmentation Challenge held as an event of the 2017 Annual Meeting of American Association of Physicists in Medicine (AAPM) [44]. This grand challenge invited participants from around the globe to apply their algorithms to perform auto-segmentation of OARs from real patient CT images collected from a variety of institutions. The organs to be segmented were the esophagus, heart, lung, and spinal cord. The grand challenge consisted of two phases: an offline contest and an online contest. The offline contest was conducted in advance of the AAPM 2017 Annual Meeting. The training data consisted of planning CT scans from 36 different patients with curated contours. These were made available to the participants prior to the offline contest through The Cancer Imaging Archive (TCIA) [63]. The participants were given one month to train or refine their algorithms using the training data. An additional 12 test cases were distributed to the participants, without contours, for the offline contest. Participants were given three weeks to process these test cases with their algorithms and submit the segmentation results to the grand challenge website (http://autocontouringchallenge.org). The segmentations were then evaluated by the organizers of the grand challenge. More than 100 participants registered on the challenge website by the time the offline contest concluded, and 11 participants submitted their offline results to the contest. Seven participants from the offline contest participated in the online challenge with three remote and four on-site participants. The online contest was held at the AAPM 2017 Annual Meeting and was followed by a symposium focusing on the challenge. During the online contest, the participants had two hours to process 12 previously unseen test cases. The segmentations were evaluated by the organizers and the challenge results were announced at the symposium the day after the online competition. This grand challenge provided a unique opportunity for participants to compare their automatic segmentation algorithms with those of others from academia, industry, and government in a structured, direct way using the same datasets. All online challenge participants were invited to contribute a chapter to this book (although not all chose to do so) addressing a specific strength of their segmentation algorithms. All chapter authors were encouraged to use the same common benchmark dataset to demonstrate the aspects of the methods in a consistent manner.

1.4 BENCHMARK DATASET

The benchmark dataset used throughout this book was curated for the 2017 AAPM Thoracic Auto-segmentation Challenge. This dataset consists of CT scans with manually drawn contours for use in testing normal thoracic tissue auto-segmentation. These scans were acquired for treatment planning purposes, with each patient in the treatment planning position using cradles for immobilization. The image quality is sufficient for contouring both tumors and organs at risk for treatment planning. The dataset consists of treatment-simulation CT scans obtained before radiotherapy from 60 patients with thoracic cancer together with manually drawn OAR contours. Scans were obtained from three institutions: the MD Anderson Cancer Center (MDACC), the Memorial Sloan-Kettering Cancer Center (MSKCC), and the Stichting Maastricht Radiation Oncology (MAASTRO) clinic. Each institution provided images from 20 patients acquired according to local institutional protocol, including the mean intensity projection from four-dimensional CT (4D CT) scans, the exhale phase from 4D CT scans, and free-breathing CT scans. The datasets were divided into three groups, consisting of 36 training cases, 12 offline test cases, and 12 online test cases. Each group contained an equal distribution of cases from each institution. All CT scans covered the entire thoracic region. Scans from patients with collapsed lungs or with the esophagus terminating superior to the lower lobes of lungs were excluded. The median patient age was 70 years (range 37–92 years) at the time of CT scanning; 33 patients (55%) were men and 27 (45%) were women. All scans had a field of view of 50 cm and a reconstruction matrix of 512×512 pixels. Slice spacing varied between institutions: 1 mm (MSKCC), 2.5 mm (MDACC), and 3 mm (MAASTRO). All CT scans covered the entire thoracic region.

Manually drawn contours were collected from clinical treatment plans from all three institutions for this dataset. Contours of the following structures were included: left and right lungs, esophagus, heart, and spinal cord. The clinical contours were quality checked and edited to adhere to Radiation Therapy Oncology Group (RTOG) contouring guidelines, RTOG-1106, [64, 65] as closely as possible. The editing specifically did not remove inter-institutional variability in the original contours. The dataset was expected to include several areas of inconsistency representative of clinical practice. These anatomic structures were selected because they are important organs at risk in radiotherapy treatment planning and are of particular interest in plan evaluation. The spinal cord is a critical organ that must be protected from excessive doses to prevent permanent injury. Radiation-induced

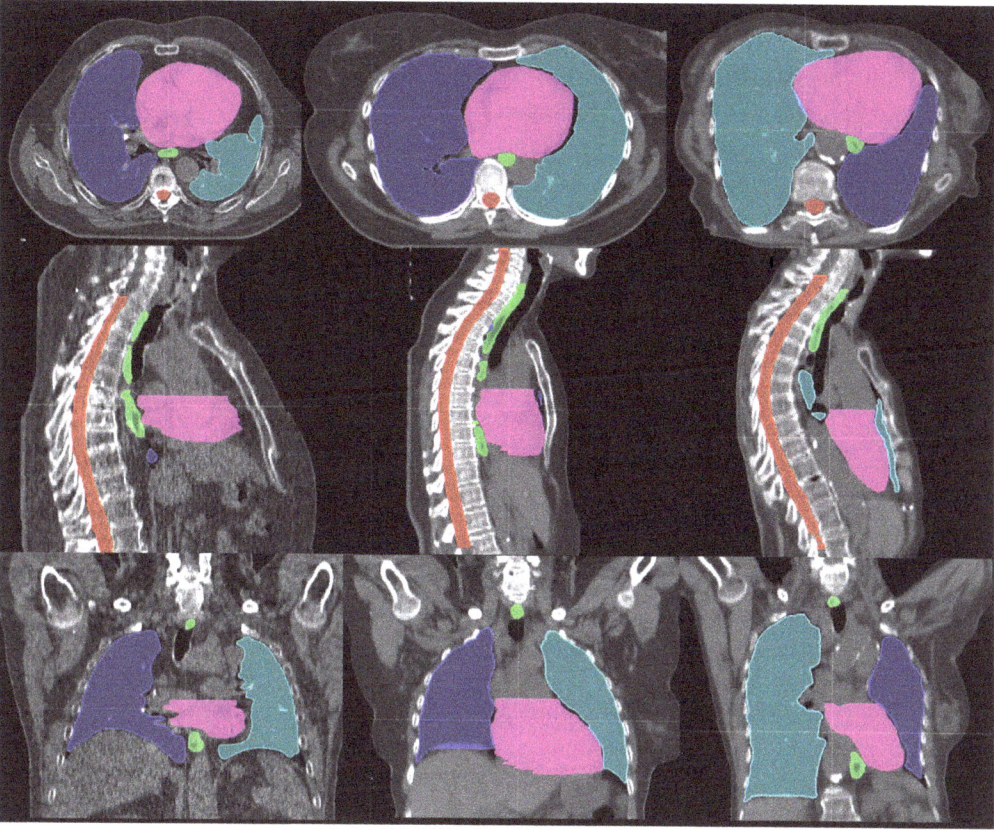

FIGURE 1.1 Axial (top row), sagittal (middle row), and coronal (bottom row) views of contoured organs for three patients (left lung [light blue], right lung [dark blue], esophagus [green], heart [pink], spinal cord [red]), each from a different institution.

cardiac toxicity and mortality is an important concern for dose escalation in lung cancer radio-therapy [66–68]. Further, higher dose to the lungs is known to be associated with higher risk of radiation-induced pneumonitis [69], and higher dose to the esophagus with higher risk of radiation-induced esophagitis [70]. Manual contours from three patients, one from each institution, are shown in Figure 1.1. Volume statistics for these structures in all 60 patients are shown in Table 1.1. This benchmark dataset is publicly available through TCIA [71]. Further description of the image data acquisition and the use of the benchmark dataset can be found in Yang et al. [72].

1.5 CLINICAL IMPLEMENTATION CONCERNS

Commercial auto-segmentation systems have been available for more than ten years. However, their routine clinical use is still very limited, mainly due to the following two reasons: poor quality of results and poor workflow integration. While varied clinical studies have shown that auto-segmentation can reduce contouring time [68, 73–75], such studies often exclude cases with large abnormalities. Thus, although time may be saved in routine clinical practice for many cases, even a moderate number of "failure" cases that require more time to edit than would be required for manual contouring can lead to frustration and discontinuation of use by the clinical end-user. Additionally, some existing solutions may suffer from poor workflow integration; any additional demands on clinical users to navigate menus, to select, or to wait for auto-segmentation can be perceived as an additional time burden compared with manual contouring that outweighs any

TABLE 1.1

Volume Statistics for the Contoured Structures on CT Scans from All 60 Patients. SD, Standard Deviation

	Minimum volume, cm³	Maximum volume, cm³	Median volume, cm³	Mean volume, cm³	Standard deviation
Left lung	128.1	4091.0	1564.1	1644.8	724.3
Right lung	892.4	3565.8	1910.2	2026.1	598.8
Esophagus	29.8	100.9	47.1	49.1	13.2
Heart	457.0	1352.2	741.0	772.5	196.4
Spinal cord	31.8	109.6	72.1	72.8	17.9

perceived benefit. Furthermore, most manual contouring tools are better suited to initial manual contouring than they are to editing of auto-generated structures. This, in turn, can lead to a poor perception of auto-segmentation, even if it is the editing tools that are the limiting factor rather than the auto-segmentation itself.

Another limiting factor for the clinical use of auto-segmentation has been the lack of a formal commissioning guideline. One important component of commissioning an auto-segmentation tool is the benchmark dataset. Anatomical sites, ground truth contours, extreme pathological cases, uncommon clinical setups, or degraded image quality could all affect the auto-segmentation accuracy. Public availability of benchmark datasets is often limited, and curation of benchmark datasets often needs extended time and follows a standard guideline. Another important component of commissioning is to objectively evaluate the performance of an auto-segmentation tool. Quantitative measures such as Dice, mean surface distance, or Hausdorff distance are often used as a gold standard. However, it has been observed that there is only weak correlation between such quantitative assessment and the time needed to edit contours to a clinically acceptable standard [76], and this quantitative assessment is greatly affected by inter-observer agreement [44, 77]. There is a need for a standard guideline in using quantitative metrics correctly for objective evaluation. Part III of this book will focus on addressing the clinical implementation concerns, including a commissioning guideline, data curation challenges, and objective evaluation of auto-segmentation.

REFERENCES

1. Ezzell, GA, et al., Guidance document on delivery, treatment planning, and clinical implementation of IMRT: report of the IMRT subcommittee of the AAPM radiation therapy committee. *Medical Physics*, 2003. **30**(8): pp. 2089–2115.
2. Mackie, TR, et al., Image guidance for precise conformal radiotherapy. *International Journal of Radiation Oncology * Biology * Physics*, 2003. **56**(1): pp. 89–105.
3. Stapleford, LJ, et al., Evaluation of automatic atlas-based lymph node segmentation for head-and-neck cancer. *International Journal of Radiation Oncology * Biology * Physics*. 2010. **77**(3): pp. 959–966.
4. Yang, J, et al., A statistical modeling approach for evaluating auto-segmentation methods for image-guided radiotherapy. *Computer Medicine Imaging Graph*, 2012. **36**(6): pp. 492–500.
5. Cardenas, CE, et al., Advances in auto-segmentation. *Seminars in Radiation Oncology*, 2019. **29**(3): pp. 185–197.
6. Weszka, JS, Survey of threshold selection techniques. *Computer Graphics and Image Processing*, 1978. **7**(2): pp. 259–265.
7. Moussallem, M, et al., New strategy for automatic tumor segmentation by adaptive thresholding on PET/CT images. *Journal of Applied Clinical Medical Physics*, 2012. **13**(5): pp. 236–251.
8. Stawiaski, J, E Decenciere, and F Bidault, Spatio-temporal segmentation for radiotherapy planning. *Progress in Industrial Mathematics at ECMI 2008*, 2010. **15**: pp. 223–228.

9. Boykov, Y Y and MP Jolly, Interactive graph cuts for optimal boundary & region segmentation of objects in N-D images. *Eighth IEEE International Conference on Computer Vision, Vol I, Proceedings*, 2001: pp. 105–112.
10. Mangan, AP and RT Whitaker, Partitioning 3D surface meshes using watershed segmentation. *IEEE Transactions on Visualization and Computer Graphics*, 1999. **5**(4): pp. 308–321.
11. Kass, M, A Witkin, and D Terzopoulos, Snakes: active contour models. *International Journal of Computer Vision*, 1987. **1**(4): pp. 321–331.
12. El Naqa, I, et al., Concurrent multimodality image segmentation by active contours for radiotherapy treatment planning. *Medical Physics*, 2007. **34**(12): pp. 4738–4749.
13. Zhang, Y, M Brady, and S Smith, Segmentation of brain MR images through a hidden Markov random field model and the expectation maximization algorithm. *IEEE Transactions on Medical Imaging*, 2001. **20**(1): pp. 45–57.
14. Yang, J, et al., A multimodality segmentation framework for automatic target delineation in head and neck radiotherapy. *Medical Physics*, 2015. **42**(9): pp. 5310–5320.
15. Sharp, G, et al., Vision 20/20: perspectives on automated image segmentation for radiotherapy. *Medical Physics*, 2014. **41**(5): p. 050902.
16. Rohlfing, T, et al., Quo vadis, atlas-based segmentation? in *Handbook of Biomedical Image Analysis*, J Suri, D Wilson, and S Laxminarayan, Editors. 2005, Springer, New York. pp. 435–486.
17. Thirion, J-P, Image matching as a diffusion process: an analogy with Maxwell's demons. *Medical Image Analysis*, 1998. **2**(3): pp. 243–260.
18. Rueckert, D, et al., Non-rigid registration using free-form deformations: application to breast MR images. *IEEE Transactions on Medical Imaging*, 1999. **18**(8): pp. 712–721.
19. Qazi, AA, et al., Auto-segmentation of normal and target structures in head and neck CT images: a feature-driven model-based approach. *Medical Physics*, 2011. **38**(11): pp. 6160–6170.
20. Han, X, et al., Atlas-based auto-segmentation of head and neck CT images. In *Medical Image Computing and Computer-Assisted Intervention – MICCAI 2008*. 2008. pp. 434–441.
21. Klein, S, et al., Automatic segmentation of the prostate in 3D MR images by atlas matching using localized mutual information. *Medical Physics*, 2008. **35**(4): pp. 1407–1417.
22. Wang, H, et al., Implementation and validation of a three-dimensional deformable registration algorithm for targeted prostate cancer radiotherapy. *International Journal of Radiation Oncology*Biology*Physics*, 2005. **61**(3): pp. 725–735.
23. Commowick, O and G Malandain, Efficient selection of the most similar image in a database for critical structures segmentation. *Medical Image Computing and Computer-Assisted Intervention – MICCAI 2007, Pt 2, Proceedings*, 2007. 4792: pp. 203–210.
24. Rohlfing, T, et al., Evaluation of atlas selection strategies for atlas-based image segmentation with application to confocal microscopy images of bee brains. *NeuroImage*, 2004. **21**(4): pp. 1428–1442.
25. Blezek, DJ and JV Miller, Atlas stratification. *Medical Image Analysis*, 2007. **11**(5): pp. 443–457.
26. Aljabar, P, et al., Multi-atlas based segmentation of brain images: atlas selection and its effect on accuracy. *NeuroImage*, 2009. **46**(3): pp. 726–738.
27. Wu, MJ, et al., Optimum template selection for atlas-based segmentation. *NeuroImage*, 2007. **34**(4): pp. 1612–1618.
28. Jia, HJ, et al., ABSORB: atlas building by self-organized registration and bundling. *NeuroImage*, 2010. **51**(3): pp. 1057–1070.
29. Yang, J, et al., Atlas ranking and selection for automatic segmentation of the esophagus from CT scans. *Physics in Medicine Biology*, 2017. **62**(23): pp. 9140–9158.
30. Commowick, O, SK Warfield, and G Malandain, Using Frankenstein's creature paradigm to build a patient specific atlas. *Medical Image Computing and Computer-Assisted Intervention – MICCAI 2009, Pt II, Proceedings*, 2009. 5762: pp. 993–1000.
31. Yang, J, et al., Automatic segmentation of parotids from CT scans using multiple atlases, in *Medical Image Analysis for the Clinic: A Grand Challenge*, B van Ginneken, et al., Editors. 2010, CreateSpace Independent Publishing Platform. pp. 323–330. www.amazon.com/Medical-Image-Analysis-Clinic-Challenge/dp/1453759395
32. Iglesias, JE and MR Sabuncu, Multi-atlas segmentation of biomedical images: a survey. *Medical Image Analysis*, 2015. **24**(1): pp. 205–219.
33. Chen, A, et al., Evaluation of multiple-atlas-based strategies for segmentation of the thyroid gland in head and neck. *Physics in Medicine and Biology*, 2012. **57**(1): pp. 93–111.
34. Yang, J, et al., Automatic contouring of brachial plexus using a multi-atlas approach for lung cancer radiation therapy. *Practical Radiation Oncology*, 2013. **3**(4): pp. e139–e147.

35. Sjoberg, C, et al., Clinical evaluation of multi-atlas based segmentation of lymph node regions in head and neck and prostate cancer patients. *Radiation Oncology*, 2013. **8**: pp. 229

36. Kirisli, HA, et al., Evaluation of a multi-atlas based method for segmentation of cardiac CTA data: a large-scale, multicenter, and multivendor study. *Medical Physics*, 2010. **37**(12): pp. 6279–6291.

37. Isgum, I, et al., Multi-atlas-based segmentation with local decision fusion – application to cardiac and aortic segmentation in CT scans. *IEEE Transactions on Medical Imaging*, 2009. **28**(7): pp. 1000–1010.

38. Sabuncu, MR, et al., A generative model for image segmentation based on label fusion. *IEEE Transactions on Medical Imaging*, 2010. **29**(10): pp. 1714–1729.

39. Warfield, SK, KH Zou, and WM Wells, Simultaneous truth and performance level estimation (STAPLE): an algorithm for the validation of image segmentation. *IEEE Transactions on Medical Imaging*, 2004. **23**(7): pp. 903–921.

40. Langerak, TR, et al., Label fusion in atlas-based segmentation using a selective and iterative method for performance level estimation (SIMPLE). *IEEE Transactions on Medical Imaging*, 2010. **29**(12): pp. 2000–2008.

41. Ramus, L and G Malandain, Multi-atlas based segmentation: application to the head and neck region for radiotherapy planning, in *Medical Image Analysis for the Clinic: A Grand Challenge*, B van Ginneken, et al., Editors. 2010, CreateSpace Independent Publishing Plaform. pp. 281–288.

42. Pekar, V, et al., Head and neck auto-segmentation challenge: segmentation of the parotid glands, in *Medical Image Analysis for the Clinic: A Grand Challenge*, B van Ginneken, et al., Editors. 2010, CreateSpace Independent Publishing Platform. pp. 273–280.

43. Raudaschl, PF, et al., Evaluation of segmentation methods on head and neck CT: auto-segmentation challenge 2015. *Medical Physics*, 2017. **44**(5): pp. 2020–2036.

44. Yang, J, et al., Autosegmentation for thoracic radiation treatment planning: a grand challenge at AAPM 2017. *Medical Physics*, 2018. **45**(10): pp. 4568–4581.

45. McCarroll, R, et al., Retrospective validation and clinical implementation of automated contouring of organs at risk in the head and neck: a step toward automated radiation treatment planning for low- and middle-income countries. *Journal of Global Oncology*, 2018. **4**: pp. 1–11.

46. Zhou, R, et al., Cardiac atlas development and validation for automatic segmentation of cardiac sub-structures. *Radiotherapy and Oncology*, 2017. **122**(1): pp. 66–71.

47. Heimann, T and HP Meinzer, Statistical shape models for 3D medical image segmentation: a review. *Medical Image Analysis*, 2009. **13**(4): pp. 543–563.

48. Pekar, V, T McNutt, and M Kaus, Automated model-based organ delineation for radiotherapy planning in prostatic region. *International Journal of Radiation Oncology*Biology*Physics*, 2004. **60**(3): pp. 973–980.

49. Freedman, D, et al., Model-based segmentation of medical imagery by matching distributions. *IEEE Transactions on Medical Imaging*, 2005. **24**(3): pp. 281–292.

50. Feng, QJ, et al., Segmenting CT prostate images using population and patient-specific statistics for radiotherapy. *2009 IEEE International Symposium on Biomedical Imaging: From Nano to Macro*, Vols 1 and 2, 2009: pp. 282–285.

51. Geremia, E, et al., Spatial decision forests for MS lesion segmentation in multi-channel magnetic resonance images. *NeuroImage*, 2011. **57**(2): pp. 378–390.

52. Criminisi, A, J Shotton, and E Konukoglu, Decision forests: a unified framework for classification, regression, density estimation, manifold learning and semi-supervised learning. *Foundations and Trends® in Computer Graphics and Vision*, 2011. **7**(2–3): pp. 147–227.

53. Li, W, et al., Learning image context for segmentation of the prostate in CT-guided radiotherapy. *Physics in Medicine and Biology*, 2012. **57**(5): pp. 1283–1308.

54. Zhang, X, et al., Interactive liver tumor segmentation from CT scans using support vector classification with watershed. *2011 Annual International Conference of the IEEE Engineering in Medicine and Biology Society (EMBC)*, 2011: pp. 6005–6008.

55. Jin, C, et al., 3D fast automatic segmentation of kidney based on modified AAM and random forest. *IEEE Transactions on Medical Imaging*, 2016. **35**(6): pp. 1395–1407.

56. Liu, JM, et al., Mediastinal lymph node detection and station mapping on chest CT using spatial priors and random forest. *Medical Physics*, 2016. **43**(7): pp. 4362–4374.

57. Krizhevsky, A, I Sutskever, and GE Hinton, ImageNet classification with deep convolutional neural networks. *Advances in Neural Information Processing Systems*, 2012. **25**: pp. 1097–1105.

58. Long, J, E Shelhamer, and T Darrell, Fully convolutional networks for semantic segmentation. *2015 IEEE Conference on Computer Vision and Pattern Recognition (CVPR)*, 2015: pp. 3431–3440.

59. Ronneberger, O, P Fischer, and T Brox, U-Net: convolutional networks for biomedical image segmentation. *Medical Image Computing and Computer-Assisted Intervention, Pt III*, 2015. **9351**: pp. 234–241.
60. Milletari, F, N Navab, and SA Ahmadi, V-Net: fully convolutional neural networks for volumetric medical image segmentation. *Proceedings of 2016 Fourth International Conference on 3D Vision (3DV)*, 2016: pp. 565–571.
61. Feng, X, et al., Deep convolutional neural network for segmentation of thoracic organs-at-risk using cropped 3D images. *Medical Physics*, 2019. **46**(5): pp. 2169–2180.
62. Cardenas, CE, et al., Head and neck cancer patient images for determining auto-segmentation accuracy in T2-weighted magnetic resonance imaging through expert manual segmentations. *Medical Physics*, 2020. **47**(5): pp. 2317–2322.
63. Clark, K, et al., The Cancer Imaging Archive (TCIA): maintaining and operating a public information repository. *Journal of Digital Imaging*, 2013. **26**(6): pp. 1045–1057.
64. Kong, F, et al. Atlases for organs at risk (OARs) in thoracic radiation therapy. 12/09/2019 [cited 2019 12/09/2019]; Available from: www.rtog.org/LinkClick.aspx?fileticket=qlz0qMZXfQs%3d&tabid=361.
65. Kong, FM, et al., Consideration of dose limits for organs at risk of thoracic radiotherapy: atlas for lung, proximal bronchial tree, esophagus, spinal cord, ribs, and brachial plexus. *International Journal of Radiation Oncology Biology Physics*, 2011. **81**(5): pp. 1442–1457.
66. Hardy, D, et al., Cardiac toxicity in association with chemotherapy and radiation therapy in a large cohort of older patients with non-small-cell lung cancer. *Annals of Oncology*, 2010. **21**(9): pp. 1825–1833.
67. Bradley, JD, et al., Standard-dose versus high-dose conformal radiotherapy with concurrent and consolidation carboplatin plus paclitaxel with or without cetuximab for patients with stage IIIA or IIIB non-small-cell lung cancer (RTOG 0617): a randomised, two-by-two factorial phase 3 study. *Lancet Oncology*, 2015. **16**(2): pp. 187–199.
68. Luo, Y, et al., Automatic segmentation of cardiac substructures from noncontrast CT images: accurate enough for dosimetric analysis? *Acta Oncology*, 2019. **58**(1): pp. 81–87.
69. Jin, HK, et al., Dose-volume thresholds and smoking status for the risk of treatment-related pneumonitis in inoperable non-small cell lung cancer treated with definitive radiotherapy. *Radiotherapy and Oncology*, 2009. **91**(3): pp. 427–432.
70. Niedzielski, JS, et al., Objectively quantifying radiation esophagitis with novel computed tomography–based metrics. *International Journal of Radiation Oncology*Biology*Physics*, 2016. **94**(2): pp. 385–393.
71. Yang, J, et al., *Data from Lung CT Segmentation Challenge*. 2017, The Cancer Imaging Archive, http://doi.org/10.7937/K9/TCIA.2017.3r3fvz08. https://wiki.cancerimagingarchive.net/display/Public/Lung+CT+Segmentation+Challenge+2017#242845390e69ea3a95bd45b5b9ac731fb837aa14
72. Yang, J, et al., CT images with expert manual contours of thoracic cancer for benchmarking auto-segmentation accuracy. *Medical Physics*, 2020. **47**(7): pp. 3250–3255.
73. La Macchia, M, et al., Systematic evaluation of three different commercial software solutions for automatic segmentation for adaptive therapy in head-and-neck, prostate and pleural cancer. *Radiation Oncology*, 2012. **7**(1): p. 160.
74. Daisne, JF and A Blumhofer, Atlas-based automatic segmentation of head and neck organs at risk and nodal target volumes: a clinical validation. *Radiation Oncology*, 2013. **8**(1): p. 154.
75. Eldesoky, AR, et al., Internal and external validation of an ESTRO delineation guideline – dependent automated segmentation tool for loco-regional radiation therapy of early breast cancer. *Radiotherapy and Oncology*, 2016. **121**(3): pp. 424–430.
76. Gooding, MJ, et al., Comparative evaluation of autocontouring in clinical practice: a practical method using the Turing test. *Medical Physics*, 2018. **45**(11): pp. 5105–5115.
77. Langmack, KA, et al., The utility of atlas-assisted segmentation in the male pelvis is dependent on the interobserver agreement of the structures segmented. *The British Journal of Radiology*, 2014. **87**(1043): pp. 20140299.

Part I

Multi-Atlas for Auto-Segmentation

2 Introduction to Multi-Atlas Auto-Segmentation

Gregory C. Sharp

CONTENTS

2.1   INTRODUCTION

Atlases have been used to perform segmentation since the mid-1990s, with multi-atlas segmentation methods appearing ten years later. Most of the basic concepts were already understood in 2005 [1], and Iglesias and Sabuncu detail the history and development of this technique in their excellent survey [2]. An atlas is defined as an image with a segmentation, while a collection of atlases is referred to as an atlas database. A modern atlas-based segmentation implementation will typically follow the form of Figure 2.1. During atlas database creation, the atlases are pre-processed to improve algorithm speed and accuracy. When a query image arrives, one or more atlases are chosen from the atlas database. Each of these atlases is deformably registered to the query image. The algorithm may then choose to perform additional atlas selection at this time, by rejecting a subset of atlases after registration. The set of segmentations are then mapped onto the query image and combined into a single segmentation through a process known as label fusion. Finally, modern algorithms will generally choose to perform post-processing on the label fusion output to achieve a final result.

Each algorithm step is subject to myriad design choices, with complex relationships between them. This chapter will briefly survey each major component of the multi-atlas segmentation pipeline.

2.2   DATABASE CONSTRUCTION

Despite its importance, relatively little is known about the art of atlas database construction. There are many major questions an architect must answer, such as database size, image quality, segmentation quality, image pre-alignment, and image resampling. It is generally believed that larger databases are better, and databases in size of up to several hundreds of cases are feasible [3, 4]. However, there are arguments that increasing the database beyond one hundred or so atlases might come with little benefit [5]. It is also possible to reduce database size while maintaining atlas diversity through clustering. Because this operation can be performed before creating ground truth segmentation, this can lead to a higher quality database for the same manual effort [6].

There is little research on the effect of image quality on atlas-base segmentation, but it is widely believed that atlas image quality should match the quality of the query images. There are many aspects to image quality: slice thickness, longitudinal and transaxial field of view, presence of

13

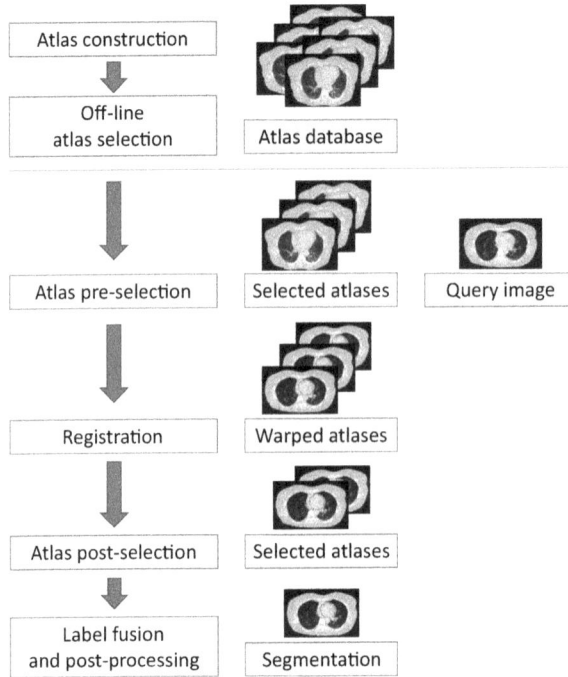

FIGURE 2.1 Most multi-atlas segmentation methods proceed in sequence through database creation, offline atlas selection, atlas pre-selection, registration, atlas post-selection, label fusion, and post-processing.

contrast, immobilization, truncation, metal artifacts, and more. There is a similar lack of research on the effect of segmentation quality within the database. However, there are community guidelines for maximizing the quality of the segmentation: adherence to professional standards, consensus results from multiple observers, and quality assurance. Furthermore, it is possible to perform statistical analysis of segmentation results to detect outliers [7]. Unfortunately, the database designer must also consider the thorny questions of missing segmentations, segmentations that abut or intersect anatomic abnormalities, and segmentations that lie near the image boundary. Opinions are mixed as to whether these should be included as a means of database enrichment or should be excluded to prevent them from being selected inappropriately. Chapter 14 discusses additional concerns for data curation, albeit with a focus on deep learning methods

Finally, while it is acknowledged that the atlas images should be pre-processed for application speed and accuracy, we know of no authoritative guidance on how this should be performed. It seems prudent to perform at least some form of rigid alignment of all atlas images to facilitate atlas selection, and if deformation metrics are also used for atlas selection, deformable registration should also be performed. It also seems prudent to crop the atlases to a region that captures the structures to be segmented. Of the large set of remaining pre-processing options, such as resampling, padding, masking, and intensity modification, these are left to each designer's choices. Chapter 5 discusses some aspects of database pre-processing within a modern multi-atlas segmentation system.

2.3 ATLAS SELECTION

Atlas selection, the process of choosing a subset of atlas images from the database, is a well-studied, though not completely understood problem. There is evidence to support that atlas selection provides superior accuracy compared to choosing a random subset of atlas images [3], and it might even be superior to choosing all atlases [8]. When deciding which atlas selection method to use, it is important to consider also that atlas selection may be performed offline. For example, the number of atlases to consider can be reduced using atlas clustering [9]. Online atlas selection may be performed after linear registration, after deformable registration, or after both. In the former case, intensity-based metrics, such as the sum of squared difference or normalized mutual information are most common. In the latter case, deformation metrics may also be used [10]. There are few concrete guidelines as to the optimal number of atlases to choose, so this parameter is usually determined empirically. Chapter 3 contains a thorough introduction to this topic.

2.4   QUERY IMAGE REGISTRATION

Almost all multi-atlas segmentation methods use deformable registration to match each selected atlas with the query image. It is usually assumed that these registrations must be high quality, or at least high quality on average. However, one or two low quality registrations may be tolerated if they are dominated by high quality registrations. Because this is the most time-consuming step of the process, the algorithm designer must optimize this step for performance.

Algorithm designers use a wide variety of registration methods, the details of which can be found in high quality literature surveys [11, 12]. There are many reasons for this: lack of consensus in benchmarking methodology, limited benchmark datasets, and algorithm variety and complexity. For example, an algorithm might be benchmarked by artificially warping an image with the goal to recover the artificial warp. However, this methodology is not universally accepted because it does not reflect real-world problem domains. As another example, it seems most modern deformable registration software implements a variant of either the demons algorithm or the B-spline algorithm. The demons algorithm is a highly flexible algorithm and has numerous variants, such as standard demons [13], accelerated demons [14], symmetric demons [15], symmetric log-domain diffeomorphic demons [16], spectral log demons [17], and so on. Given that each variant has numerous tuning parameters, the barrier to experimental benchmarking is high. Nevertheless, Chapter 4 provides insight into several algorithm parameter choices for multi-atlas segmentation.

2.5   LABEL FUSION

After multiple atlases have been matched with the query image, label fusion is used to combine these multiple segmentations into a single segmentation. Rohlfing et al. describe the earliest fusion methods: (1) choosing the best scoring atlas, and (2) segmenting according to majority voting [18]. The first of these, choosing the best scoring atlas, uses a registration metric such as intensity difference, and then segments the query image according to that atlas. As such, it can also be considered an atlas selection process. In the second of these, majority voting, each atlas votes for its preferred labeling for each voxel in the image. Then, after all atlases have voted, the label with the most votes wins. Majority voting was found to outperform other methods.

An extension to majority voting is weighted majority voting, whereby the weights of each atlas are weighted by a similarity score. Generally, this similarity score is assigned on a per-voxel basis. Even a simple method such as weighting by intensity similarity usually outperforms majority voting [19]. Other metrics that can be considered include weighting by the local deformation Jacobian [20] or metrics that summarize similarity over a local neighborhood [21]. One popular method is to combine intensity similarity with a distance metric [22]. To construct the distance metric, a distance map is created from the structure, with higher weights given to voxels more interior and lower weights given to voxels more exterior. The highest possible weights are given to voxels with similar pixel intensities and are closest to the structure center.

A third major category of fusion algorithms consists of statistical methods based on simultaneous truth and performance level estimation (STAPLE) algorithm [23]. While the original use of STAPLE was to create consensus segmentation from multiple human observers, one can easily see why it would be an attractive approach for multi-atlas segmentation. The core concept is to weigh the observations using a model of their performance and perform expectation maximization to find a probabilistic estimate of the segmentation. One extension of this technique is the selective and iterative method for performance level estimation (SIMPLE), which iteratively performs performance estimation and atlas selection [24]. This approach is only one of many that performs post-registration selection integrated within the label fusion process.

2.6   LABEL POST-PROCESSING

It is generally wise to perform post-processing on a multi-atlas segmentation. Morphological operations such as dilation and erosion are commonly used to fill holes and remove islands. The degree to which this is necessary depends on algorithmic choices, especially depending on the choice of label fusion method. Alternatively, more sophisticated post-processing can be performed, using the output of the label fusion process as a starting point. For example, the label fusion probability map can be used to initialize a graph-cut algorithm [25] or a statistical shape model [26]. With sufficient training data, more extensive post-processing can be performed using classical machine learning [27], or deep learning architectures [28].

2.7   SUMMARY OF THIS PART OF THE BOOK

This chapter provides a brief overview of the components of an atlas-based segmentation system, and the design choices faced in developing one. The remainder of this part of the book explores some of these aspects in more depth and provides tangible examples of the 2017 AAPM Thoracic Auto-segmentation Challenge data. Chapter 3 provides a comprehensive review of the atlas selection process, while Chapter 4 presents the tradeoffs in image registration parameter selection. Chapter 5 demonstrates the design of a modern multi-atlas implementation and its application to the head and neck and to the 2017 AAPM Thoracic Auto-segmentation Challenge datasets. Beyond atlas-segmentation, the third part of the book considers more general aspects related to auto-segmentation that are applicable to atlas-based method. Data curation is considered in Chapter 14 and the assessment of auto-contouring is reviewed in Chapter 15.

REFERENCES

1. Rohlfing T, Brandt R, Menzel R, Russakoff DB, Maurer CR. Quo Vadis, Atlas-based segmentation? In: Suri JS, Wilson DL, Laxminarayan S, editors. *Handbook of Biomedical Image Analysis: Volume III: Registration Models* [Internet]. Boston, MA: Springer US; 2005 [cited 2020 Dec 16]. pp. 435–86. (Topics in Biomedical Engineering International Book Series). Available from: https://doi.org/10.1007/0-306-48608-3_11
2. Iglesias JE, Sabuncu MR. Multi-atlas segmentation of biomedical images: A survey. *Medical Image Analysis*. 2015 Aug 1;24(1):205–19. https://doi.org/10.1016/j.media.2015.06.012
3. Aljabar P, Heckemann RA, Hammers A, Hajnal JV, Rueckert D. Multi-atlas based segmentation of brain images: Atlas selection and its effect on accuracy. *NeuroImage*. 2009 Jul 1;46(3):726–38. https://doi.org/10.1016/j.neuroimage.2009.02.018
4. Kirişli HA, Schaap M, Klein S, Papadopoulou SL, Bonardi M, Chen CH, et al. Evaluation of a multi-atlas based method for segmentation of cardiac CTA data: A large-scale, multicenter, and multivendor study. *Medical Physics*. 2010;37(12):6279–91. https://doi.org/10.1118/1.3512795
5. Awate SP, Whitaker RT. Multiatlas segmentation as nonparametric regression. *IEEE Transactions on Medical Imaging*. 2014 Sep;33(9):1803–17. https://doi.org/10.1109/TMI.2014.2321281
6. Kennedy A, Dowling J, Greer PB, Holloway L, Jameson MG, Roach D, et al. Similarity clustering-based atlas selection for pelvic CT image segmentation. *Medical Physics*. 2019;46(5):2243–50. https://doi.org/10.1002/mp.13494
7. Robinson R, Valindria VV, Bai W, Oktay O, Kainz B, Suzuki H, et al. Automated quality control in image segmentation: Application to the UK Biobank cardiovascular magnetic resonance imaging study. *Journal of Cardiovascular Magnetic Resonance*. 2019 Mar 14;21(1):18. https://doi.org/10.1186/s12968-019-0523-x

8. Sanroma G, Wu G, Gao Y, Shen D. Learning-based atlas selection for multiple-atlas segmentation. In: 2014 IEEE Conference on Computer Vision and Pattern Recognition. 2014. pp. 3111–7. https://doi.org/10.1109/CVPR.2014.398

9. Langerak TR, Berendsen FF, Heide UAV der, Kotte ANTJ, Pluim JPW. Multiatlas-based segmentation with preregistration atlas selection. *Medical Physics*. 2013;40(9):091701. https://doi.org/10.1118/1.4816654

10. Commowick O, Malandain G. Efficient selection of the most similar image in a database for critical structures segmentation. In: Ayache N, Ourselin S, Maeder A, editors. *Medical Image Computing and Computer-Assisted Intervention – MICCAI 2007*. Berlin, Heidelberg: Springer; 2007. pp. 203–10. (Lecture Notes in Computer Science). https://doi.org/10.1007/978-3-540-75759-7_25

11. Maintz JBA, Viergever MA. A survey of medical image registration. *Medical Image Analysis*. 1998 Mar 1;2(1):1–36. https://doi.org/10.1016/S1361-8415(01)80026-8

12. Sotiras A, Davatzikos C, Paragios N. Deformable medical image registration: A survey. *IEEE Transactions on Medical Imaging*. 2013 Jul;32(7):1153–90. https://doi.org/10.1109/TMI.2013.2265603

13. Thirion J-P. Image matching as a diffusion process: an analogy with Maxwell's demons. *Medical Image Analysis*. 1998 Sep 1;2(3):243–60. https://doi.org/10.1016/S1361-8415(98)80022-4

14. Wang H, Dong L, O'Daniel J, Mohan R, Garden AS, Ang KK, et al. Validation of an accelerated 'demons' algorithm for deformable image registration in radiation therapy. *Physics in Medicine and Biology*. 2005 Jun;50(12):2887–2905. https://doi.org/10.1088/0031-9155/50/12/011

15. Rogelj P, Kovačič S. Symmetric image registration. *Medical Image Analysis*. 2006 Jun 1;10(3):484–93. https://doi.org/10.1016/j.media.2005.03.003

16. Vercauteren T, Pennec X, Perchant A, Ayache N. Symmetric Log-Domain Diffeomorphic registration: A demons-based approach. In: Metaxas D, Axel L, Fichtinger G, Székely G, editors. *Medical Image Computing and Computer-Assisted Intervention – MICCAI 2008*. Berlin, Heidelberg: Springer; 2008. pp. 754–61. (Lecture Notes in Computer Science). https://doi.org/10.1007/978-3-540-85988-8_90

17. Lombaert H, Grady L, Pennec X, Ayache N, Cheriet F. Spectral log-demons: Diffeomorphic image registration with very large deformations. *International Journal of Computer Vision*. 2014 May 1;107(3):254–71. https://doi.org/10.1007/s11263-013-0681-5

18. Rohlfing T, Brandt R, Menzel R, Maurer CR. Evaluation of atlas selection strategies for atlas-based image segmentation with application to confocal microscopy images of bee brains. *NeuroImage*. 2004 Apr 1;21(4):1428–42. https://doi.org/10.1016/j.neuroimage.2003.11.010

19. Isgum I, Staring M, Rutten A, Prokop M, Viergever MA, Ginneken B van. Multi-atlas-based segmentation with local decision fusion—application to cardiac and aortic segmentation in CT scans. *IEEE Transactions on Medical Imaging*. 2009 Jul;28(7):1000–10. https://doi.org/10.1109/tmi.2008.2011480

20. Ramus L, Commowick O, Malandain G. Construction of patient specific atlases from locally most similar anatomical pieces. In: Jiang T, Navab N, Pluim JPW, Viergever MA, editors. *Medical Image Computing and Computer-Assisted Intervention – MICCAI 2010*. Berlin, Heidelberg: Springer; 2010. pp. 155–62. (Lecture Notes in Computer Science). https://doi.org/10.1007/978-3-642-15711-0_20

21. Artaechevarria X, Munoz-Barrutia A, Ortiz-de-Solorzano C. Combination strategies in multi-atlas image segmentation: Application to brain MR data. *IEEE Transactions on Medical Imaging*. 2009 Aug;28(8):1266–77. https://doi.org/10.1109/TMI.2009.2014372

22. Sabuncu MR, Yeo BT, VanLeemput K, Fischl B, Golland P. A generative model for image segmentation based on label fusion. *IEEE Transactions on Medical Imaging*. 2010 Jun 17;29(10):1714–29.

23. Warfield SK, Zou KH, Wells WM. Simultaneous truth and performance level estimation (STAPLE): An algorithm for the validation of image segmentation. *IEEE Transactions on Medical Imaging*. 2004 Jul;23(7):903–21. https://dx.doi.org/10.1109/TMI.2004.828354

24. Langerak TR, Heide UA van der, Kotte ANTJ, Viergever MA, Vulpen M van, Pluim JPW. Label fusion in atlas-based segmentation using a selective and iterative method for performance level estimation (SIMPLE). *IEEE Transactions on Medical Imaging*. 2010 Dec;29(12):2000–8. https://dx.doi.org/10.1109/TMI.2010.2057442

25. Lee J-G, Gumus S, Moon CH, Kwoh CK, Bae KT. Fully automated segmentation of cartilage from the MR images of knee using a multi-atlas and local structural analysis method. *Medical Physics*. 2014;41(9):092303. https://doi.org/10.1118/1.4893533

26. Fritscher KD, Peroni M, Zaffino P, Spadea MF, Schubert R, Sharp G. Automatic segmentation of head and neck CT images for radiotherapy treatment planning using multiple atlases, statistical appearance models, and geodesic active contours. *Medical Physics*. 2014;41(5):051910. https://doi.org/10.1118/1.4871623

27. Yushkevich PA, Wang H, Pluta J, Das SR, Craige C, Avants BB, et al. Nearly automatic segmentation of hippocampal subfields in in vivo focal T2-weighted MRI. *NeuroImage*. 2010 Dec 1;53(4):1208–24. https://doi.org/10.1016/j.neuroimage.2010.06.040

28. Xie L, Wang J, Dong M, Wolk DA, Yushkevich PA. Improving multi-atlas segmentation by convolutional neural network based patch error estimation. In: Shen D, Liu T, Peters TM, Staib LH, Essert C, Zhou S, et al., editors. *Medical Image Computing and Computer Assisted Intervention – MICCAI 2019*. Cham: Springer International Publishing; 2019. pp. 347–55. (Lecture Notes in Computer Science). https://doi.org/10.1007/978-3-030-32248-9_39

3 Evaluation of Atlas Selection
How Close Are We to Optimal Selection?

Mark J. Gooding

CONTENTS

Image-based atlas selection has been used for more than two decades as an approach to improve atlas-based auto-contouring. Most published investigations have shown some level of improvement in the performance of auto-contouring over the random selection of atlases when using image-based selection. In this chapter, the published research into atlas-selection is reviewed, asking the question, "How close are we to optimal atlas selection?" An experiment is presented that assesses the most common approach to atlas selection – ranking similarity based on normalized mutual information (NMI) between the atlas and the test case – using the 2017 AAPM Thoracic Auto-segmentation Challenge data. In phrasing the evaluation with respect to optimality, it is seen that while there is some improvement in auto-contouring performance, atlas selection is far from optimal.

3.1 MOTIVATION FOR ATLAS SELECTION

The primary assumption of atlas-based contouring is that the anatomy of the person represented in the atlas is the same as that represented by the patient case to be contoured. While the majority of people are broadly anatomically similar, there is substantial diversity in size between subjects. Furthermore, there will be differences in position between patients when being imaged, leading to differences in image appearance. Therefore, deformable image registration is used to account for differences in size and positioning between subjects. However, deformable image registration is not perfect, and it is generally accepted that the less deformation that is required, the better the correspondence between images will be. Thus, it has been proposed that selecting an atlas that is more similar in appearance to the patient, thereby minimizing the deformation required, will result in a lower registration error and better atlas-segmentation performance than using an atlas that differs substantially from the patient in position or anatomy.

    Exploring this in more depth, it is understood that deformable image registration is being used to account for differences in size and positioning between subjects. Yet, deformable image registration itself has no knowledge of anatomy and the goal of image registration is to map one image to another such that the appearance is the same. If taking this objective to the extreme, the optimum image similarity can be achieved by intensity sorting; the intensities of voxels in each image are put into ordered lists. Corresponding voxels would then be determined by the position in the ordered list, known as the "Completely Useless Registration Algorithm" [1]. The moving image can thus be deformed to match the fixed image, yet the deformation field and deformed image resulting from such a mapping are anatomically nonsense. Thus, deformable registration must be constrained by a regularizer to prevent non-meaningful correspondence. Neighboring voxels are expected to move in a similar way. However, this same regularizer can prevent fine scale alignment. So, while it is true that at a large scale the majority of patients (of the same sex) are anatomically similar, e.g. all patients will have a heart, and for most patients the heart lies to the left, at finer scales there are anatomical variations that may not adequately be addressed by deformable registration.

By way of example, take the degree to which the heart varies in position from patient to patient towards the left, as illustrated in Figure 3.1 for cases Train-S3-011 and Test-S3-102 from the thoracic challenge. Firstly, to anatomically match the heart from the training case, Train-S3-011, which could be used as an atlas to the test case, Test-S3-102, the heart must be moved further to the left. However, in the training case there is lung tissue between the heart and the ribcage that is not present in that location in the test case. Whether anatomically correct or not, this lung tissue must be pushed cranially to be out of the way. However, the organs below these regions of lung must not

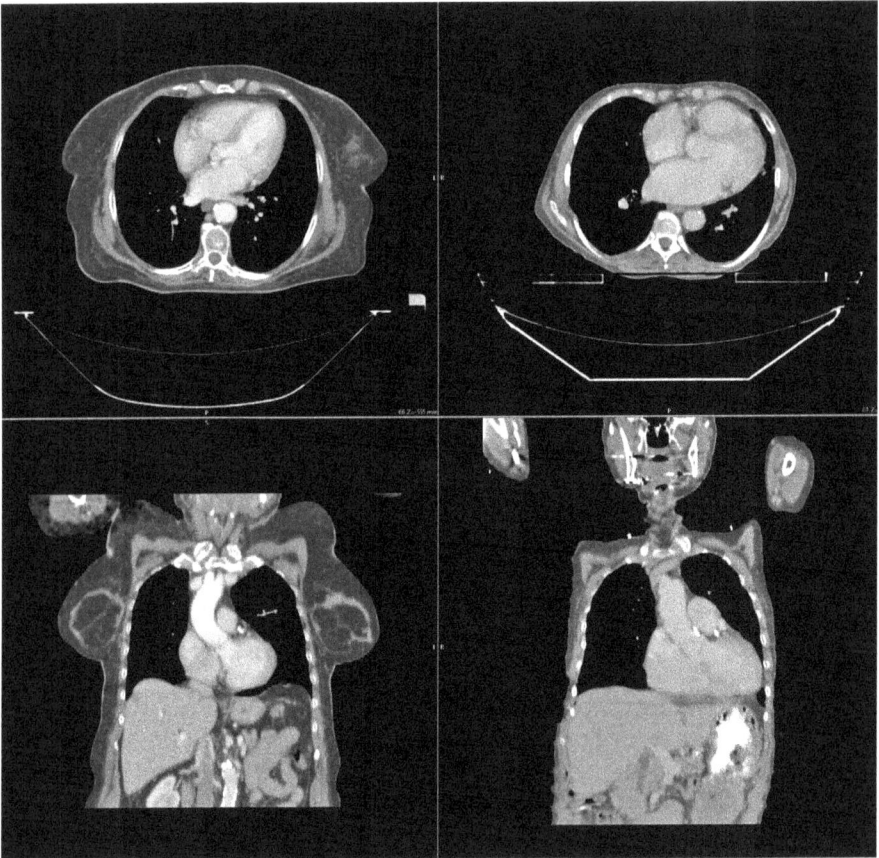

FIGURE 3.1 Cases Train-S3-011 (left) and Test-S3-102 (right) from the LSTSC. While the patients are anatomically the same, the position of the anatomy varies from subject to subject. Note particularly the difference in location, and orientation of the heart.

move cranially, nor the ribs towards the right. This anatomical difference between the patients cannot be overcome by a constrained registration. This is demonstrated in Figure 3.2 where the contours from Train-S3-011 have been propagated to Test-S3-102 following deformable registration. The left lung of the test case is reasonably well segmented on the coronal image shown using this particular image registration. Yet, to do so the regularizer of the deformable registration has dragged the caudal heart contour cranially, leading to a poor segmentation. In the axial view, the regularizer has prevented the heart moving left, leading to the lung contour going through the heart. While sliding boundary approaches have been proposed to allow for some relative organ motion, these still assume a one-to-one anatomical correspondence that may not be appropriate for inter-subject registration.

Since deformable image registration struggles to overcome significant anatomical differences, it can be expected that if a more anatomically similar atlas is used then the registration, and consequently the contouring, would be better. Taking an extreme example again, if the images/anatomy are identical then there is no registration to do and the contouring should be perfect. In practice, it is observed that re-contouring using a previous time point of the same patient in a different position results in better auto-contouring than using an atlas based on a different person. This is illustrated in Figure 3.3, where re-contouring and atlas-based contouring are shown. In this figure the anterior aspect of the heart and the posterior boundary of the right lung show degradation in contour quality for atlas contouring (bottom) but are well contoured when using the same patient (top). Therefore,

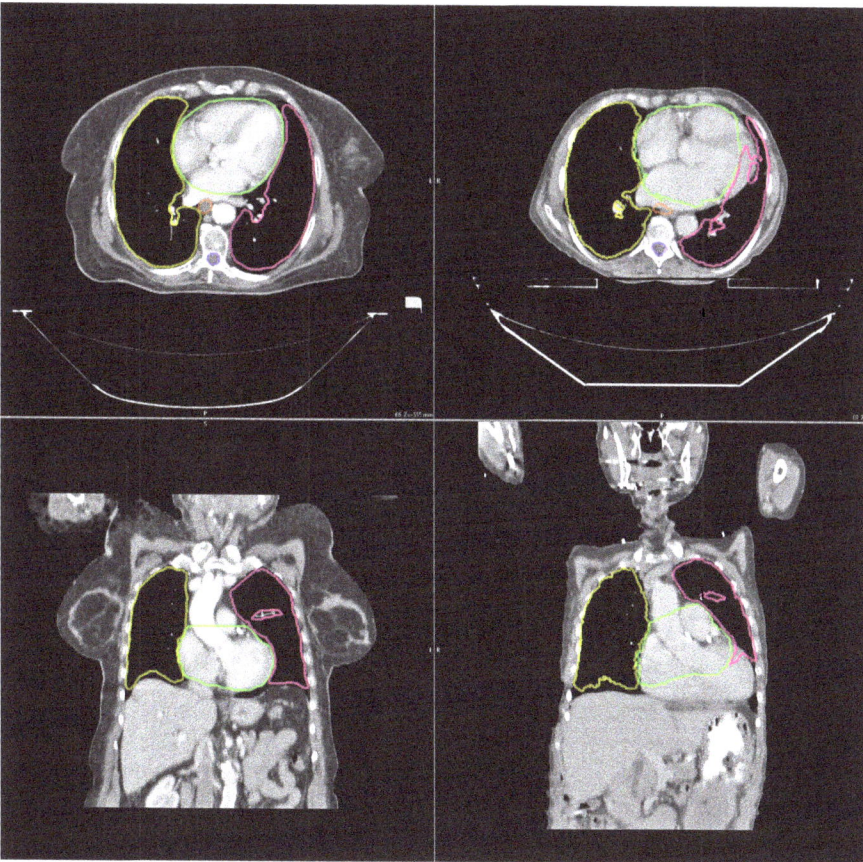

FIGURE 3.2 Failure of atlas contouring resulting from anatomical variations. LSTSC-Test-S3-102 (right) has been contoured following deformable registration with LSTSC-Train-S3-011 (right). Note the incorrect contouring of the heart.

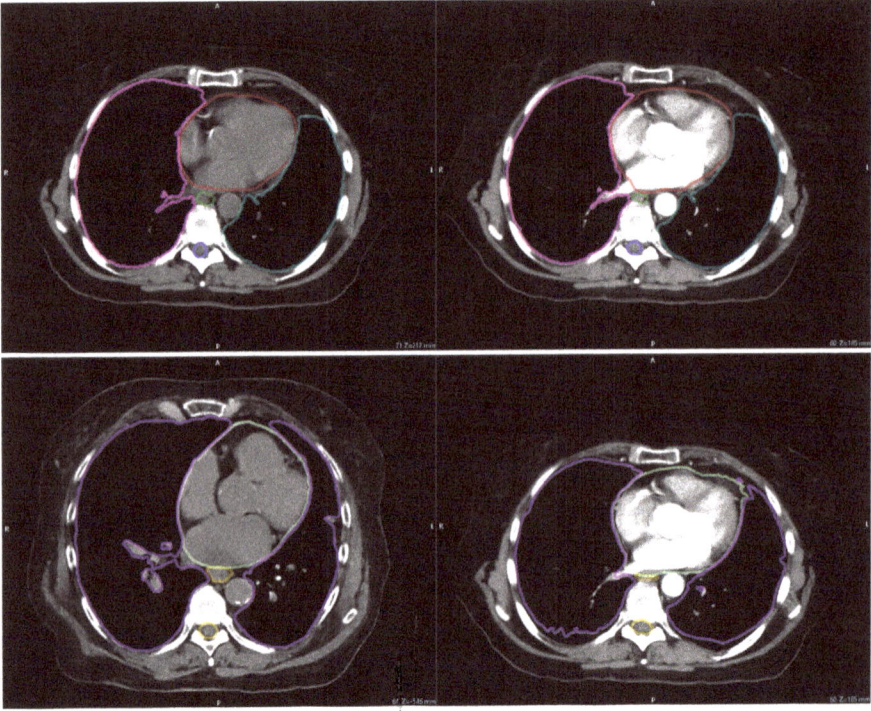

FIGURE 3.3 Comparison of re-contouring with the same patient (top) and atlas contouring with a different subject (bottom). In each case the contours on the left have been mapped to the image on the right using deformable image registration. Note, the performance of atlas contouring at the posterior of the lung and left anterior of the heart is worse compared to re-contouring.

atlas selection has been proposed as a method to improve atlas-based auto-contouring, whereby an atlas (or atlases) is selected from a larger pool of available atlases with the intent that the selected atlas(es) is more anatomically similar to the patient.

Following this line of thought, if a more similar atlas will result in better contouring performance, then the larger the size of the database of atlases, the higher the expected performance for atlas contouring will be. Thus, atlas contouring may be sufficient to "solve" auto-contouring in radiotherapy if a large enough pool of atlases is created. This line of thinking was explored by Schipaanboord et al. [2], who used extreme value theory to predict the performance of atlas-based contouring in radiotherapy assuming a database size of 5000 atlases in the presence of perfect atlas selection. They concluded that such a database could yield performance equivalent to clinical variation.

3.2 METHODS OF ATLAS SELECTION

In this section the major directions in the evolution of atlas selection methods are recapped. This is not intended as a comprehensive review, but to introduce the directions taken and the evidence of efficacy presented. Table 3.1 presents a more complete, but not necessarily exhaustive, list of studies proposing or evaluating atlas selection methods. This list is not restricted to radiotherapy since much of the work in this area has been in the context of neurology. Nevertheless, the methods proposed are generally not specific to the domain of clinical application.

Atlas selection appears to have been proposed first by Rohlfing et al. [3], who considered the impact of atlas choice in segmentation of bee brain images on a database of 20 images. Preceding work had focused either on choosing an arbitrary image as an atlas or averaging multiple atlases to

TABLE 3.1

Studies Proposing or Evaluating Atlas Selection

Reference	Year	Method	Organs	#
Rohlfing et al. [3]	2004	Global NMI, DVF after affine, DIR	Bee brain	20
Wu et al. [6]	2007	Local NMI after DIR	Brain	22
Commowick et al. [5]	2007	Local DVF wrt template after DIR	Head and neck	45
Klein et al. [29]	2008	Global NMI after DIR	Prostate	88
Aljabar et al. [9]	2009	Local NMI wrt template after affine	Brain	275
Sabuncu et al. [30]	2009	GMM of templates using ML + Generalized EM	Brain	50
Išgum et al. [31]	2009	Sequential forward selection on Jaccard index	Heart and aorta	29
Gorthi et al. [19]	2010	Global MSE after affine	Parotids	18
Langerak et al. [10]	2010	Selection based on an estimated DSC	Prostate	100
Lötjönen et al. [21]	2010	Local NMI wrt template(s) after affine, DIR	Brain	78
Ramus et al. [32]	2010	Local DVF wrt template after DIR	Head and neck	105
van Rikxoort et al. [33]	2010	Global and local MAD after affine	Heart/caudate nuc.	29/39
Wolz et al. [13]	2010	MD of local NMI wrt template after DIR	Hippocampus	826
Yang et al. [34]	2010	CC features from local PCA features	Parotids	18
Cao et al. [12]	2011	MD of local intensities after DIR	Prostate	40
Dowling et al. [35]	2011	NMI after DIR	Prostate	50
Akinyemi et al. [7]	2012	Offline selection and local CC	Heart/kidney	21/24
Hoang-Duc et al. [14]	2013	MD of global DVF wrt template	Hippocampus	826
Langerack et al. [8]	2013	Clustering and selection based on DSC	Prostate	200
Wolz et al. [36]	2013	Intensity MSE based pre-selection	Liver, spleen, pancreas, kidney	150
Asman et al. [37]	2014	PCA appearance after group-wise rigid	Spinal cord structures	67
Raudaschl et al. [22]	2014	Correlation of NMI between all atlases after rigid	Parotids	18
Sanroma et al. [15]	2014	DSC regression-based HOG after affine wrt template	Brain	183
Wang et al. [38]	2014	Graph-based clustering after DIR and selection	Brain	35
Asman et al. [39]	2015	AdaBoost on PCA appearance after affine wrt template	Brain	3464
Langerack et al. [40]	2015	Local selection based on estimated DSC	Prostate	125
Xu et al. [41]	2015	EM using GMM learned tissue prior	12 abdominal organs	100
Yan et al. [42]	2015	MD on Intensities + label after DIR	Prostate	60
Zhao et al. [23]	2016	Metric distance learning on intensities in ROI	Corpus callosum	200
Karasawa et al. [43]	2017	Jaccard index or CC on vessel structures after DIR	Pancreas	150
Yang et al. [44]	2017	Intensity CC after rigid + KL divergence after DIR	Esophagus	66
Antonelli et al. [11]	2019	Local intensity CC after DIR + DSC overlap	Prostate/left ventricle	190/45

generate an average atlas representative of the middle of the population. In that work, they posed the question as to what the best approach is to choose an atlas. In addition to considering an arbitrary fixed atlas, they introduced the idea of selecting a similar atlas based on image similarity (in this instance NMI) after registration (both affine or deformable) or on the basis of the degree of deformation required to match the atlas to the image. First, they established image similarity after deformable image registration as the best measure of similarity by evaluation of the Similarity Index (SI in Figure 3.4) – a measure now better known as the Dice Similarity Coefficient (DSC) – following atlas registration using a leave-one-out validation. Subsequently, as shown in Figure 3.4,

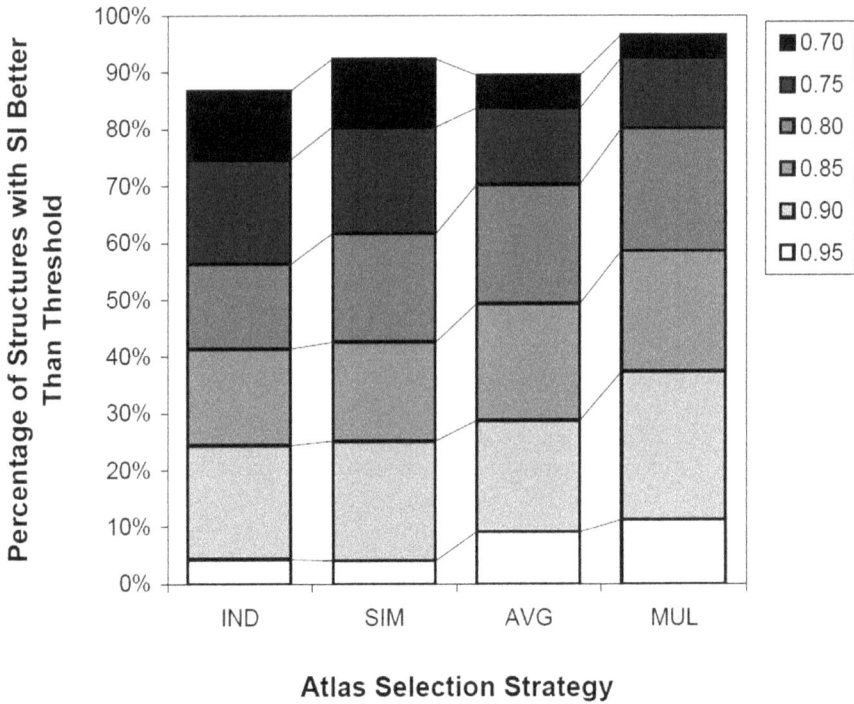

FIGURE 3.4   Atlas selection strategies investigated by Rohlfing et al. Similar images (SIM) outperformed an arbitrary individual atlas (IND) or an average atlas (AVG).

it was further demonstrated that this selection method (SIM) outperformed using a fixed single atlas (IND) or an average atlas (AVG).

Further to this, the study investigated the range of possible performance, looking both at the best and worst achievable selection using either a fixed atlas or atlas selection. A leave-one-out strategy was adopted for this whereby each image in turn was used as the fixed atlas to segment the remaining raw images and the best and worst results found assuming a fixed single atlas for all test cases. At the same time, the best and worse results possible were evaluated assuming that the atlas could be selected for each test image. This was illustrated against the observed performance, as shown in Figure 3.5. It was found that the observed performance is close to the best achievable (Best SIM), i.e. using the atlas that results in the best SI rather than selecting on the basis of the image similarity (SIM), and that image similarity-based selection outperformed the best possible fixed atlas. Finally, it is observed from Figure 3.4 that the fusion of results from multiple atlases (MUL) using the approach, presented in another study by Rolfhing et al. [4], results in better performance than any of the selection methods presented. The multiple atlas fusion approach took all 19 remaining atlases and fused their resulting segmentations in the space of the test image.

The approach to atlas selection taken by Rohlfing et al. of examining NMI between the atlas and the test case following deformable image registration requires that registration is performed on the test case for every atlas. While this is computationally tractable for low numbers (like the 20 examples used in their experiment), it may quickly become unhelpful for use in clinical practice as the length of time required to select the atlas scales linearly with the number of atlases.

Commowick et al. [5] recognized this challenge for the selection of atlases in the context of radiotherapy and sought to address it. For the purpose of selection, they defined the most similar atlas as the ones that required the least deformation to the patient image. For this to be calculated, each atlas had to be registered to the patient. This is illustrated in the right-hand side of Figure 3.6. This transformation is denoted as $T_{A_k \to P}$. Commowick et al. proposed to generate a fixed average

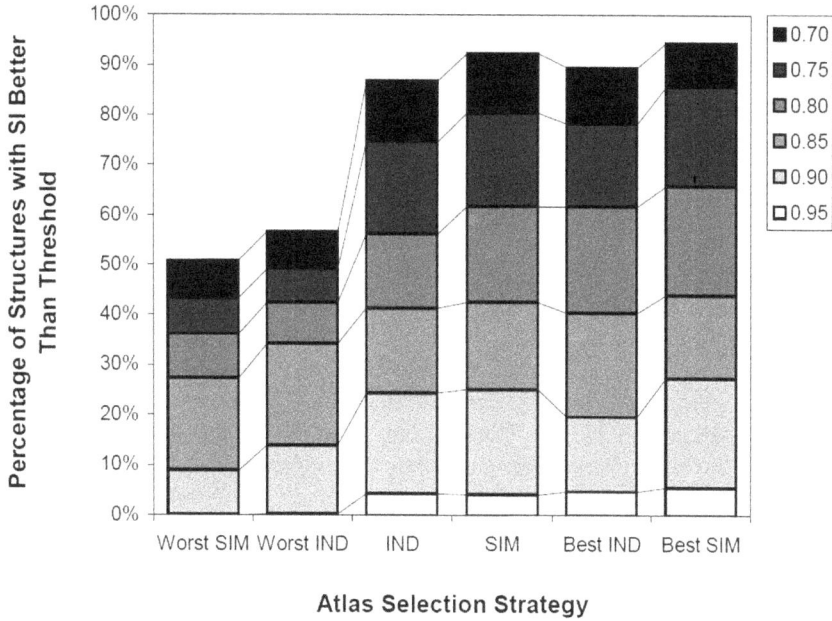

FIGURE 3.5 The range of performance with fixed and variable atlas selection strategies as reported by Rohlfing et al.

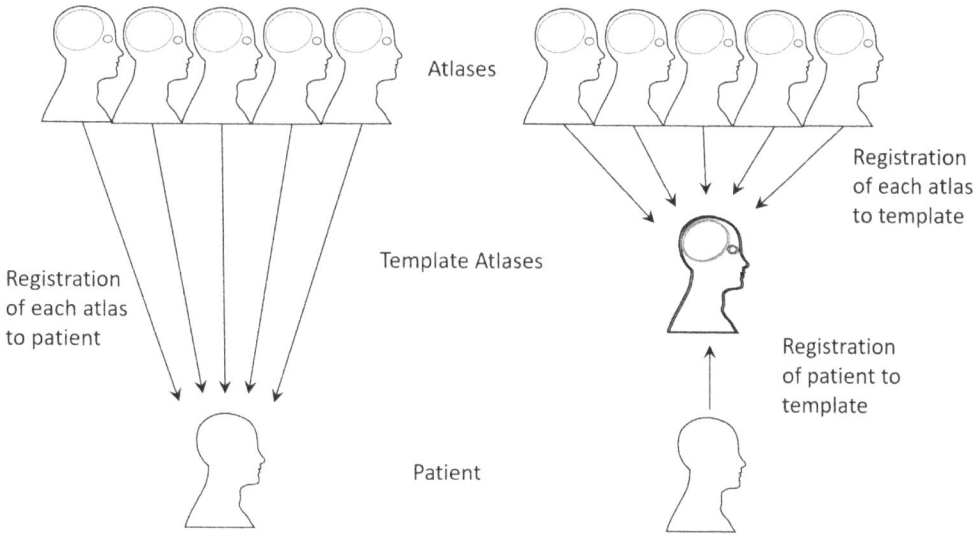

FIGURE 3.6 Evaluation of similarity for each atlas requires registration to the patient (left). This can be made computationally more efficient by registration to a template (right). The registration of atlases to the template can be performed once offline.

template image from the atlases. In doing so, all atlases would be registered to this template. This operation only needs to be performed once at the point of template creation. This transformation to the template is denoted as $T_{A_k \to M}$. Subsequently, each patient case can be registered to the template, $T_{P \to M}$. This is shown on the left-hand side of Figure 3.6. The key assumption is then made that:

$$T_{A_k \to P} \approx T_{A_k \to M} \circ T_{P \to M}^{-1}$$

This approach of using an intermediate template allows the transformation for each atlas to be computed once in an offline phase. Thus, only a single registration is required at the time of application to a patient.

Since Commowick et al. were using the deformation as the measure of similarity, this similarity can be calculated directly following composition. If image similarity is to be used, then the images can be deformed using the composite deformations and calculated in patient space. Alternatively, the patient image and atlases can be deformed to the template space and the similarity calculated in this space, avoiding the need to perform composition and allowing the atlas deformation to be performed offline, further reducing the computational complexity.

Only affine registration is evaluated in Commowick et al. [5], where the assumption that the composition of the transformation of the atlas to the template with the inverse of the transformation of the patient to the template is reasonable. Unfortunately, while the assumption seems reasonable for deformable registration, particularly diffeomorphic ones, no evidence is presented nor is it demonstrated that it would result in the same selection for deformable registration.

In reviewing the literature, care must be taken to distinguish templates used for selection, as in Commowick et al. [5], and the terminology used in neurology where the term template is used to refer to an average of a cluster of atlases used for segmenting the patient image, as in Wu et al. [6]. In the latter example, the selection of the template image is important as it is desirable to minimize the deformation $T_{P \rightarrow M}$ to ensure a good mapping from the average atlas to the patient. Thus, selection of the template is equivalent to selection of an atlas in this context, rather than the template being used for computation efficiency purposes.

Perhaps closest to the neurology concept of a template is the idea of offline atlas selection, whereby atlases are clustered to find a subset that is more descriptive of the expected population. This offline selection can be performed prior to use, requiring no knowledge of the patient. Most commonly, offline methods employ contouring performance indices such as the DSC or the Jaccard metric to evaluate the atlas contouring performance in the training phase [7, 8]. Since offline selection does not consider the patient, it is limited in its potential performance, therefore it is not considered further in this chapter.

Many of the investigations into atlas selection have considered a fairly low number of atlases. The first larger scale studies were conducted by Aljabar et al. [9] who used 275 cases in a neurology context. In this study, they researched the impact of selection on performance and compared image-based selection (NMI following affine registration) to selection based on the subject's meta data (age, sex, etc.). First, they demonstrated that image-based selection resulted in improved performance compared to random selection, as shown in Figure 3.7. For this, the performance of fusing 20 selected atlases was compared to the results of fusion for 20 random atlases. A leave-one-out approach was adopted such that 275 results of selecting 20 atlases from the remaining database of 274 atlases were considered.

However, just because the performance improves does not demonstrate that the image-based measure is a very good surrogate of performance, rather that it is better than random. Aljabar et al. express the desired quality thus: "Given two atlases, A and B, it is desirable that similarity selection determines the better of A and B as a potential segmentation atlas for Q. This means that if sim(Q,A)>sim(Q,B) then atlas-based segmentation using atlas A should generate a more accurate segmentation of Q". However, they caution that "Such a desirable property may, however, be confounded, for example by contrast differences or varying quality of the atlases or scan protocol".

Thus, they conducted a second experiment whereby the performance of atlases was assessed when ranked according to the image-similarity. For each subject, the atlases were ranked according to similarity and the performance was measured. Figure 3.8 shows the average performance over all 275 subjects for each rank plotted in order. These figures show that the performance appears to correlate with the rank of the atlas according to its similarity measure, thus the image-similarity is a good surrogate of atlas contouring performance. Finally, they compared selection based on the age of the subject to image similarity-based selection. For age-based selection, the 20 atlases with the

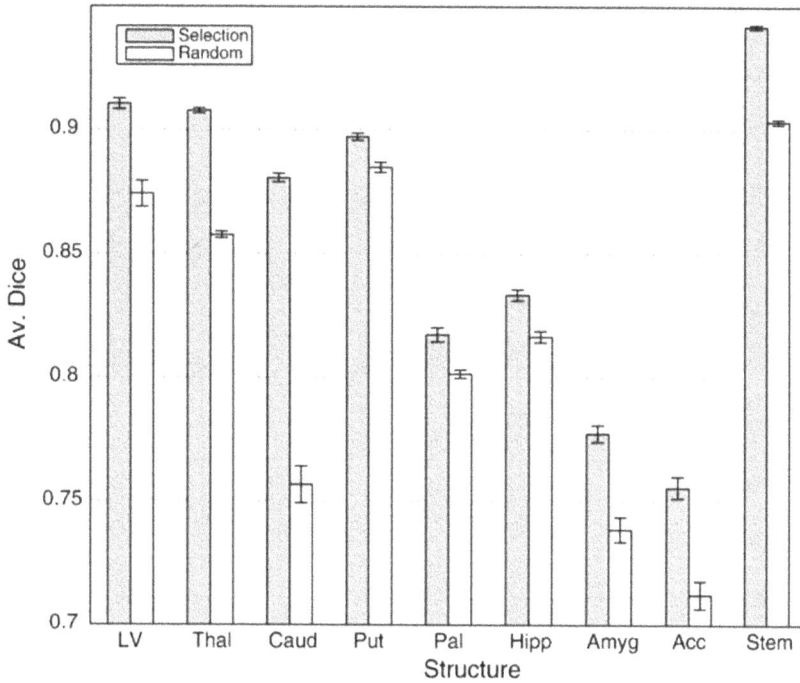

FIGURE 3.7   Image-based selection using NMI following affine registration outperformed random selection when evaluated following fusion of 20 atlases.

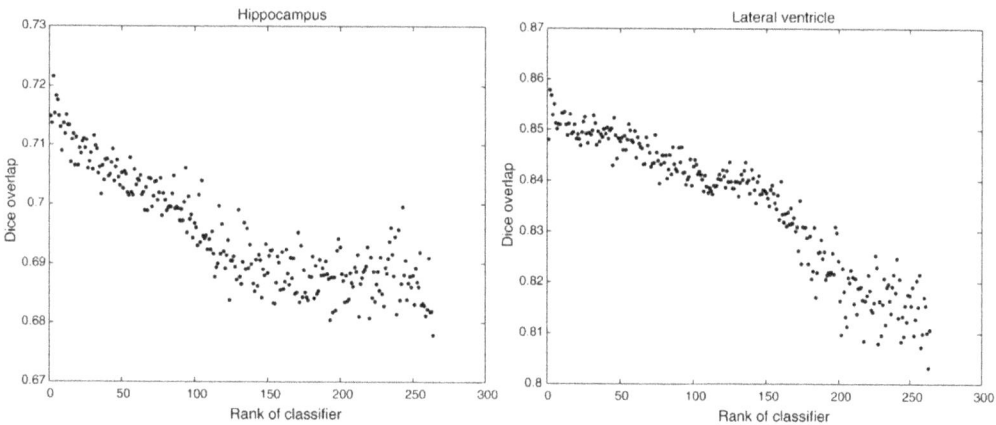

FIGURE 3.8   Average performance of atlases by rank according to the similarity measure.

most similar subject age were selected and the resulting contours fused. Figure 3.9 shows the Dice differential between these methods of selection, with the subject plotted according to age. Image-based selection appears to marginally outperform age-based selection for younger subjects, while the converse appears to be true for older subjects.

Whereas image-based selection attempts to estimate the most similar atlases in order to actively select good atlases to generate a segmentation, Langerak et al. took an alternative viewpoint of using the estimated performance to exclude poorly performing atlases [10]. In their approach all atlases are used to generate an estimated segmentation using label fusion. The performance of each atlas against this estimated contour can then be measured. Atlases which fail to meet a performance

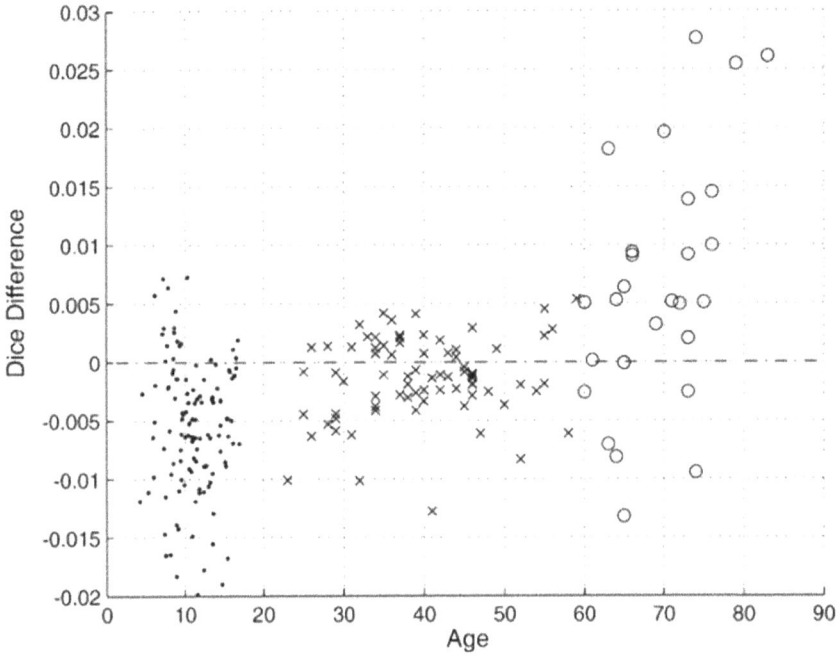

FIGURE 3.9 Comparison of results between image-based selection and age-based selection, shown against age. A positive Dice difference indicates better performance for age-based selection. The subjects were divided into the three broad age groups and their results plotted with different markers.

threshold are then excluded, and the consensus contour is re-estimated. This process is iterated until no further atlases are excluded. The authors note that the downside of this approach to select-ing atlases is that all atlases must be registered to the patient image for the contour to be esti-mated, making it computationally expensive. More recently, Antonelli et al. [11] have combined this approach with an initial image-based selection to reduce the number of atlases followed by a genetic algorithm to optimize the choice of atlases.

The use of manifolds for selection, introduced in Cao et al. [12] for selection but used previously for propagation of labels across a database in Wolz et al. [13], in some respects has overlap with the use of a template for selection. Manifolds are higher dimensional spaces into which data is mapped to give it meaning. For example, measuring distances between cities using straight lines on a 2D map may have some meaning under the assumption that the earth is locally flat. However, such mea-surement would not be correct if measuring between two distant points such as London and New York. Meaningful distances, in the sense that it is a path to be traveled on the earth's surface, exist in a spherical manifold. Manifolds may assume that the space in which atlases lie is non-Euclidean, and therefore the similarity between atlases must be mapped using the manifold. However, mani-folds have been built using the same underlying similarity measures, such as NMI [13] or the degree of deformation [14]. Manifold construction techniques, such as local linear embeddings, as used in Duc et al. [14], make the assumption that very similar atlases are similar both in Euclidean and manifold space. Therefore, assuming an atlas set suitably representative of the population such that the patient lies on the manifold near an atlas, it may be expected that the initial set of atlases selected for a patient projected into manifold space would be the same as those selected using the similarity measure directly. However, as additional atlases are selected further away from the new patient in similarity or manifold space, the differences in the selection space become more significant. This is seen in the finding of Cao et al. [12], where the performance is similar for a low number of selected atlases (n < 15) whether doing direct selection or manifold-based selection.

Perhaps the most recent innovation in atlas selection is to assume that it is not known what constitutes a similar atlas, but instead to attempt to learn which features are important. Sanroma et al. [15] used an offline training phase to learn a mapping between a high number of features to the performance measure. In their study, 10,000 features generated using the Histogram of Oriented Gradients (HOG) were initially calculated. A function was trained to predict Dice similarity between image pairs using a subset of 1000 features found to be useful on the atlas database. Given a new patient, the features were calculated and pairwise comparison was performed with each atlas to estimate their performance for segmenting the patient. Atlases were then selected based on their estimated performance. However, performance was still found to be below optimal atlas selection.

3.3 EVALUATION OF IMAGE-BASED ATLAS SELECTION

While there has been a wealth of research looking at approaches to atlas selection, few have addressed the underlying assumption as to whether the image similarity measured is a good surrogate for contouring performance. Although it has been shown that the contouring performance is improved by using a more similar image, e.g. Aljabar et al. [9], other studies have called into question whether such image-based measures work well. In motivating learned image-based selection, Sanroma et al. [15] state, "However, the problem of atlas selection still remains unexplored. Traditionally, image similarity is used to select a set of atlases. Unfortunately, this heuristic criterion is not necessarily related to the final segmentation performance". Furthermore, Ramus and Malandain [16] found greater correlation of image-based selection methods with random selection than with ground truth measures of contouring performance. Thus, the question remains as to whether image-based selection is a good method of atlas selection. This section mirrors the investigation of Schipaanboord et al. [17] in assessing how good image-based atlas selection is compared to the optimal, but uses the data from the Thoracic Contouring challenge [18] as an example.

3.3.1 IMPLEMENTATION

While the study by Schipaanboord et al. is valuable research using a large dataset, the use of proprietary code in their experiments and the restricted access to the clinical dataset means that exact recreation of the results by others is impossible. To allow full reproducibility of this study and the figures presented in this chapter, the Python code implementation has been made available on GitHub at: https://github.com/Auto-segmentation-in-Radiation-Oncology/Chapter-3.

3.3.2 BRUTE-FORCE SEARCH

First, the concept of an *Oracle* in order to be able to compare to optimal selection is introduced. The *Oracle* has perfect foreknowledge of the result of auto-contouring and is able to select atlases based on the resulting contouring performance. At the other end of the selection spectrum, there is *Random* selection whereby atlases are selected without any consideration of performance or similarity. Where image-based selection lies between these two extremes needs to be assessed.

In this experiment, the online test cases (patient IDs in the form LSTSC-Test-SX-2YY) are taken as patient cases. The training data (patient IDs in the form LSTSC-Train-SX-0YY) and the offline test cases (patient IDs in the form LSTSC-Test-SX-1YY) are used as atlases. In the challenge the offline test cases were not available to use as atlases as the contours were not provided. However, these are included as atlases to increase the size of the atlas pool now that these contours are available.

A brute-force search approach is implemented where every atlas is used to contour every patient image. As noted previously, the use of templates and manifolds may optimize this search either in terms of efficiency or projecting the similarity more appropriately far from the test patient. However, the underlying assumption remains the same. A brute-force search removes any impact that the choice of template or manifold may have, ensuring that the best possible selection using the similarity measure is assessed. Image similarity measures (root mean square error of intensities [19] and NMI [3]) are computed after both rigid and deformable registration over the whole image. Furthermore, these image measures are computed for each organ within the deformed atlas contour following deformable registration, as a measure of local image similarity. These measures can be used for atlas selection based on image similarity.

The contouring performance using each atlas is calculated against the "ground truth" contour for each patient case using the DSC [20] implemented using a voxel mask. DSC has been used in this instance as it is easy to compute, and necessary for comparison to previous publications. Additional, and perhaps more clinically relevant, measures are discussed in Chapter 15. The final DSC measure is deemed known to the *Oracle* for atlas selection.

3.3.3 ATLAS SELECTION PERFORMANCE ASSESSMENT

First using the results of selection, it is possible to produce a plot similar to Figure 3.8 following the work of Aljabar et al. [9]. Figure 3.10 shows the average DSC for all test cases plotted for atlases ranked according to NMI for the esophagus. Similar figures can be produced for all organs in the challenge case. Such figures suggest that there is broad correlation between the similarity measure (in this instance, NMI) and the performance (as measured by DSC). However, as with Aljabar et al. [9], this figure shows average performance for the test cases. While this indicates that there is some benefit to be gained on average, it reveals nothing about the impact of selection on an individual case.

Figure 3.11 shows the contouring performance for all individual cases when plotted against the atlas rank according to NMI for the esophagus. Each color in the figure represents a different test case. Showing the data in this way reveals a much weaker correlation, suggesting that the performance following selection may vary substantially. Thus, while on average the performance may be improved – for any particular case there is no guarantee of improved performance.

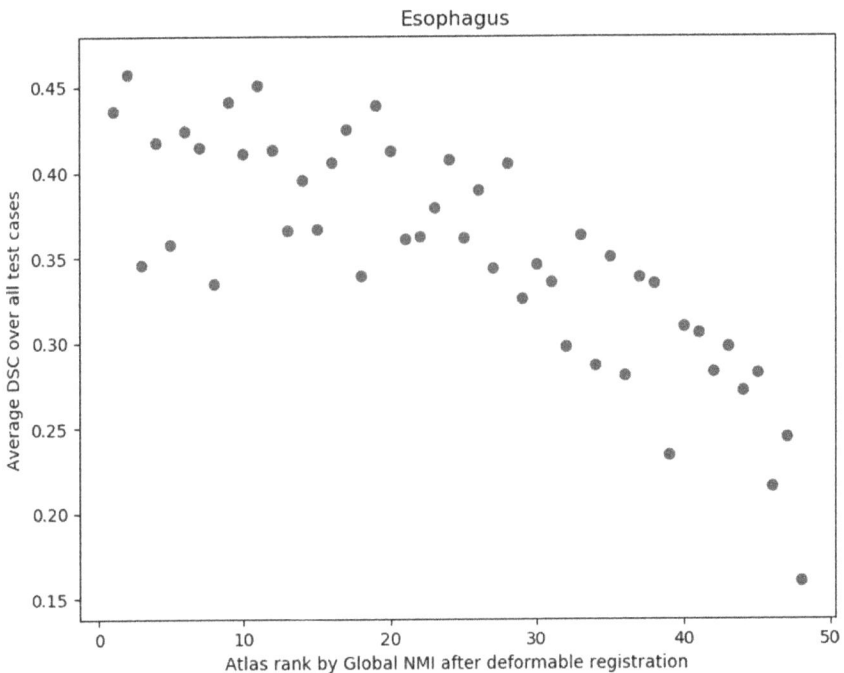

FIGURE 3.10 Average contouring performance on the esophagus over all test cases by atlases ranked according to NMI.

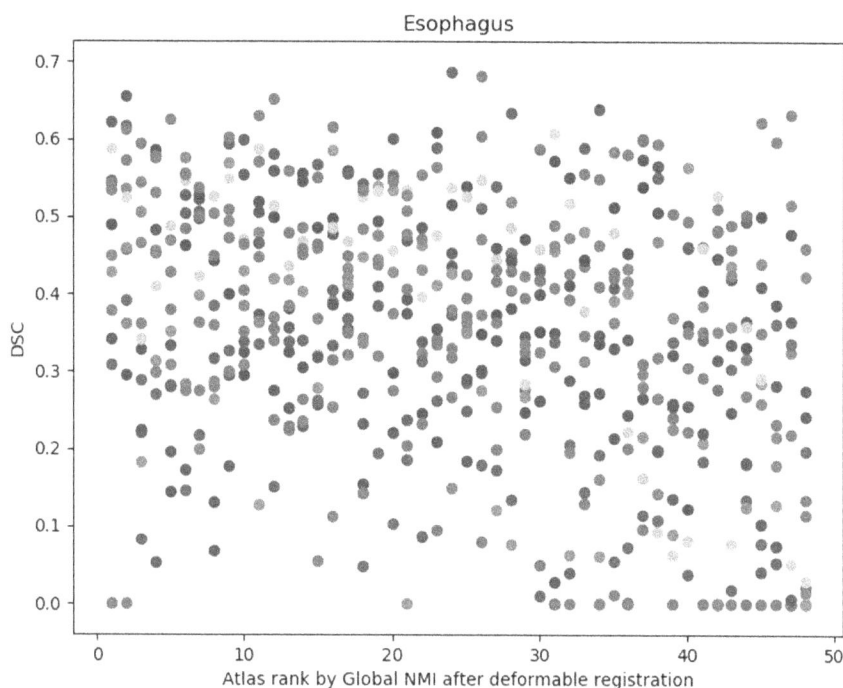

FIGURE 3.11   Contouring performance of atlases ranked by NMI. Each color represents a different test case.

Figure 3.7 also showed improvement (on average) against random atlas selection. A similar plot is shown in Figure 3.12 using the thoracic data for all six image-based selection measures implemented. The average performance of the first ten selected atlases is shown, rather than the result after contour fusion. The bars indicate the mean performance over all test cases, while the whiskers indicate the minimum and maximum observed performance over the 12 test cases. Ten atlases were selected, rather than 20 as in Figure 3.7, a result of the relatively low number of available atlases (n = 48) compared to Aljabar et al. [9] (n = 274). Figure 3.12 also includes the *Oracle*, in addition to random selection, for comparison with the best achievable performance. As with [9], it is observed that image-based selection methods perform on average better than random selection. However, it is noted that there is variation between subjects leading to large whiskers. Nevertheless, of some organs, image-based selection appears to perform close to the best achievable, as observed in Rohlfing et al. [3] and shown in Figure 3.5.

So far there appears to be conflicting information. Figure 3.10 and Figure 3.11 appear to suggest selection that is better than random, but highly variable and unlikely to result in substantial performance gains, while Figure 3.12 suggests that performance close to the best achievable can be expected for some organs. To understand this, it is necessary to look at the rank of the selected atlases. Figure 3.13 shows the average rank of the ten selected atlases rather than their contouring performance. For the *Oracle* the mean rank is 5.5 for all test cases (atlases ranked 1 to 10 would be selected). For *Random* selection, a mean rank of around 24.5 would be expected, however this varies between cases and organs leading to the whiskers shown. The mean rank of atlases selected using the image-based measures ranges from around 12 to 22, with the whiskers extending this range for individual patients from around 6 to 36. In many cases the whiskers extend to a higher average rank than random showing that for some test cases selection has performed worse than might be expected with random selection. Therefore, this figure strengthens what is observed in Figure 3.10 and Figure 3.11, that image-based selection is on average better than random, but highly variable and not robust.

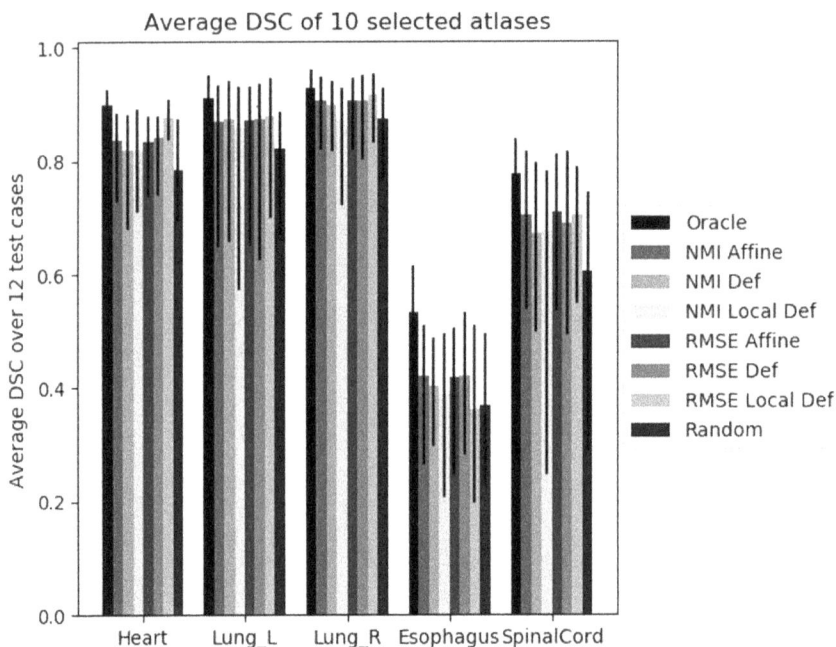

FIGURE 3.12 Average performance following selection of the ten best atlases using various selection methods. The whiskers indicate minimum and maximum performance observed within the 12 test cases. The image similarity measures (NMI, RMSE) used for selection were calculated over the whole image following affine registration (Affine) and following deformable registration (Def). The measures used for selection were also calculated within the deformed atlas contour only following deformable registration (Local Def) to give a local measure of similarity.

Next, performance with respect to atlas rank is considered. Figure 3.14 shows performance for each organ, averaged over the 12 test cases according to the *Oracle*'s ranking. The mean rank position for the ten selected atlases according to each selection method is indicated on the figure. For the lungs and heart, where DSC is normally expected to be high, as they are larger organs, the performance curve is quite shallow except for approximately the last 20% of poorly performing atlases. Therefore, the performance improvement from atlas selection is expected to be small. Conversely, a performance close to the best achievable is still expected. However, for the esophagus and spinal cord, the impact of the choice of atlas is much greater. Here the image-based selection has a larger impact on performance, but the gulf between the observed performance and the best achievable remains large. Thus, this figure links the observations back to Figure 3.12, where it was observed that following selection some organs perform close to the best achievable, despite less than perfect atlas selection. Figure 3.14 also highlights the need to achieve perfect atlas selection, if the promise of near-perfect contouring suggested in Schipaanboord et al. [2] is to be realized. It can be observed the contouring performance has a small but marked improvement for the best first or second ranked atlases compared to even the third or fourth ranked ones. When searching for an atlas similar to the patient, there will only be a very tiny percentage of the atlas population that will constitute a very good match – thus placing high importance on an exceptional selection method to achieve exceptional contouring performance.

3.3.4 DISCUSSION AND IMPLICATIONS FOR ATLAS SELECTION

The study conducted above has a few notable limitations – particularly with respect to learnt similarly measures. However, before these limitations are considered, what the study does show, rather than what it does not, should also be considered.

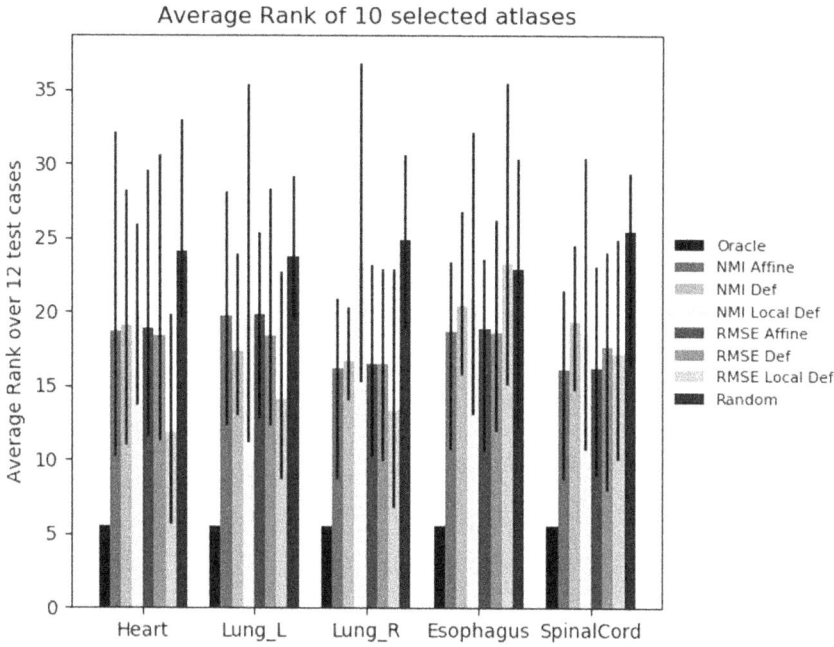

FIGURE 3.13 Average rank of ten selected atlases using various selection methods. The bar indicates the mean performance over the 12 test cases, while whiskers show the minimum and maximum performance. The image similarity measures (NMI, RMSE) used for selection were calculated over the whole image following affine registration (Affine) and following deformable registration (Def). The measures used for selection were also calculated within the deformed atlas contour only following deformable registration (Local Def) to give a local measure of similarity.

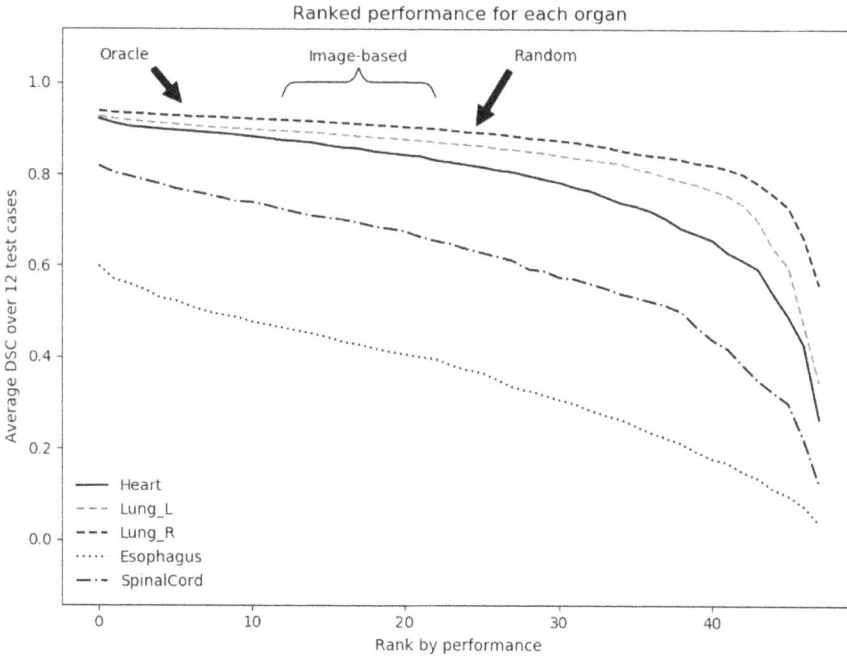

FIGURE 3.14 Performance of the *Oracle*. Performance on all organs of atlases ranked by performance averaged over all test cases.

TABLE 3.2

Clinical Impact of Atlas Contouring and Atlas Selection

Reference	Anatomy	Number of atlases	Time saving (mins)	Time saving (%)	Selection method
Teguh et al. [26]	Head and neck	10	114	63	Online selection – mutual information following rigid registration
Stapleford et al. [45]	Head and neck	1	6	35	None
Young et al. [27]	Endometrial	15	9	26	Online selection – commercial software, using mutual information
Gambacorta et al. [46]	Rectal	4	13	34	None
Hwee et al. [47]	Prostate	75	3.4	19	Online selection – commercial software, no details given
Lin et al. [48]	Prostate	98	12	45	Stratification by bladder size, followed by online selection – commercial software, no details given
Granberg et al. [49]	Prostate	15	10	26	None
Langmack et al. [50]	Prostate	11	10.7	40	None

Recapping, atlas selection was first motived by Rohlfing et al. [3]. In Figure 3.4, it was seen that NMI-based selection outperformed the use of a single fixed atlas or an average atlas. The height for the bar chart represents the percentage of structures with a DSC higher than a particular threshold. While a greater percentage of structures are above a threshold of 0.7 for image-based selection than for the average atlas, it is seen that the conclusion would change as the threshold increases. An average atlas would outperform the similarity selected one at a threshold of 0.75 or greater. At a threshold of 0.85 the performance of a fixed atlas looks equivalent to a selected one. Therefore, the conclusion could be drawn that image-based selection is capable of increasing performance by rejecting poor performance at the lower end of the performance spectrum rather than improving it at the extreme.

Rohlfing et al. also considered the best possible performance (the *Oracle*) and found that image similarity-based selection performed close to this, as shown in Figure 3.5. This is similar to the finding in this study in Figure 3.12. However, it has been seen in this study that looking at performance measures alone can be misleading in the evaluation of selection performance. The plot of performance against ranking (Figure 3.14) demonstrates the importance of understanding the performance profile for a particular structure to evaluate how close selection is to optimal.

While Rohlfing et al. considered performance against the optimum, very few subsequent studies have done so. The majority of studies choose to consider any improvement in contouring performance with respect to some reference alternative, or just report the contouring performance of the method reported. Of the studies listed in Table 3.1 subsequent to Rohlfing et al. [3], only Lötjönen et al. [21], Akinyemi et al. [7], Raudaschl et al. [22], Sanroma et al. [15], Zhao et al. [23], and Zaffino et al. [24] consider performance with respect to the optimum. Furthermore, only Sanroma et al. [15] and Zhao et al. [23] consider performance in terms of rank rather than contouring performance, finding that only about a third of the atlases selected using NMI would be considered relevant selections by rank. Thus, the focus of most studies is improving performance from the current state, without considering how much room for improvement exists.

3.3.5 LIMITATIONS

The study presented has clear limitations, only considering simple image similarity measures for selection. However, it noted that only such basic atlas selection approaches have been implemented with clinical software [25].

While it is argued that this assessment adequately addresses the potential of template and manifold type approaches, this is not demonstrated. Conversely, no study has investigated the impact of these techniques with respect to optimal selection. Therefore, this remains an area for future exploration.

Fusion of selected atlas contours is also not considered in this experiment, although it is touched on in Schipaanboord et al. [17]. Contour fusion has been repeatedly shown to result in improved contouring compared to single atlas segmentation, even in the extreme case [2], and therefore it is widely used. Yet, most atlas selection approaches ignore the subsequent fusion that is likely to take place, opting for a greedy atlas selection method, whereby the best atlases are individually chosen prior to generating the consensus. A better approach would be to choose the best set of atlases in combination [24], however this comes with a potentially prohibitive computational cost as database size increases. A recent contribution suggested exploring the large search space using a genetic algorithm to optimize the combinatorial selection [11], however, this approach has yet to be shown to result in an optimal selection. Zaffino et al. [24] trained a neural network to predict performance on groups of atlases, demonstrating that selection of a group equivalent to the single atlas oracle could be achieved – however, this was still below the performance of a group atlas oracle. Researching optimal combinatorial selection requires considerable computation, yet there could be substantial performance gains if an efficient method can be found that can be proven to be optimal, or near-optimal compared to an oracle-based selection.

Approaches of atlas selection using machine learning selection have also not been considered in this chapter, on account of the range of implementation that could be adopted and the need for an additional training set. Such methods have shown promise compared to image-based selection, but the evidence to date is that these approaches still fall short of optimal atlas selection [15].

3.4 IMPACT OF ATLAS SELECTION ON CLINICAL PRACTICE

There have been numerous studies investigating atlas-based contouring within the radiotherapy domain. Most of these studies only evaluate quantitative accuracy, yet a number have also considered the clinical impact with respect to contouring time. Table 3.2 shows studies where editing time has been investigated for atlas contouring. Although the atlas selection methods are not described in detail for commercial software, it is known that selection is being performed using mutual information as a similarity criteria by some manufacturers [26, 27].

It can be observed in Table 3.2 that for the prostate there are two studies where larger numbers of atlases have been used together with atlas selection, and two with fewer atlases and no selection. However, the reported percentage time saving is similar regardless of the number of atlases or the use of selection. While timing was not investigated, Lee et al. [28] found no improvement in performance using quantitative measures for a commercial system when increasing the size of an atlas database size from 20 to 100 atlases in steps of 20 atlases. Although that study only investigated two organs in the head and neck, these organs (mandible and thyroid) were chosen as laborious to draw but also represent organs with significant difference in contrast and appearance. Thus, the available evidence suggests that the selection methods implemented by some manufacturers have no impact on the resulting clinical workflow.

3.5 SUMMARY AND RECOMMENDATIONS FOR FUTURE RESEARCH

In this chapter the background to atlas selection has been reviewed, seeing that since the outset the underlying assumption has been that an atlas with a more similar image to the test case will have better contouring performance than a less similar atlas. It was also seen that while there was some evidence to back up this assumption, it was not fully examined. An experiment was presented to test this assumption whereby a number of image-based atlas selection methods were evaluated and compared to perfect atlas selection. This study reveals that image-based methods fall some way short of perfect selection in terms of rank, and consequently atlas selection is having negligible impact in terms of clinical efficiency.

Notwithstanding this rather negative finding, there is some cause for optimism. Works exploring learned similarity measures have also shown encouraging improvements [15] over the direct image-based measures explored here, and the search for the optimum combination of atlases to generate the best consensus has only recently been considered [11, 24].

Schipaanboord et al., using extreme value theory, suggested that contouring performance equivalent to clinical standards could be possible using a large atlas database in the presence of perfect atlas selection [2]. Thus, there is scope for significant gains to be made with further research into atlas selection. However, as demonstrated in this chapter methods proposed in the future should evaluate their performance not with respect to where they have come from (i.e. older reference implementations) but with respect to where they want to get to (i.e. perfect selection).

REFERENCES

1. T Rohlfing, "Image similarity and tissue overlaps as surrogates for image registration accuracy: widely used but unreliable," *IEEE Trans. Med. Imaging*, vol. 31, no. 2, pp. 153–163, Feb. 2012.
2. B Schipaanboord et al., "Can atlas-based auto-segmentation ever be perfect? Insights from extreme value theory," *IEEE Trans. Med. Imaging*, vol. 38, no. 1, pp. 99–106, 2019.
3. T Rohlfing, R Brandt, R Menzel, and C Maurer, "Evaluation of atlas selection strategies for atlas-based image segmentation with application to confocal microscopy images of bee brains," *Neuroimage*, vol. 21, no. 4, pp. 1428–1442, 2004.
4. T Rohlfing, DB Russakoff, and CR Maurer, "Performance-based classifier combination in atlas-based image segmentation using expectation-maximization parameter estimation," *IEEE Trans. Med. Imaging*, vol. 23, no. 8, pp. 983–994, 2004.
5. O Commowick, and G Malandain, "Efficient selection of the most similar image in a database for critical structures segmentation," *Lect. Notes Comput. Sci. (including Subser. Lect. Notes Artif. Intell. Lect. Notes Bioinformatics)*, vol. 4792 LNCS, no. PART 2, pp. 203–210, 2007.
6. M Wu, C Rosano, P Lopez-Garcia, CS Carter, and HJ Aizenstein, "Optimum template selection for atlas-based segmentation," *Neuroimage*, vol. 34, no. 4, pp. 1612–1618, 2007.
7. A Akinyemi, C Plakas, J Piper, C Roberts, and I Poole, "Optimal atlas selection using image similarities in a trained regression model to predict performance," *Proc. Int. Symp. Biomed. Imaging*, pp. 1264–1267, 2012.
8. TR Langerak, FF Berendsen, UA Van Der Heide, ANTJ Kotte, and JPW Pluim, "Multiatlas-based segmentation with preregistration atlas selection," *Med. Phys.*, vol. 40, no. 9, pp. 1–17, 2013.
9. P Aljabar, RA Heckemann, A Hammers, JV Hajnal, and D Rueckert, "Multi-atlas based segmentation of brain images: atlas selection and its effect on accuracy," *Neuroimage*, vol. 46, no. 3, pp. 726–738, 2009.
10. TR Langerak, UA Van Der Heide, ANTJ Kotte, MA Viergever, M Van Vulpen, and JPW Pluim, "Label fusion in atlas-based segmentation using a selective and iterative method for performance level estimation (SIMPLE)," *IEEE Trans. Med. Imaging*, vol. 29, no. 12, pp. 2000–2008, 2010.
11. M Antonelli et al., "GAS: a genetic atlas selection strategy in multi-atlas segmentation framework," *Med. Image Anal.*, vol. 52, pp. 97–108, 2019.
12. Y Cao, Y Yuan, X Li, and P Yan, "Putting images on a manifold for atlas-based image segmentation," *Proc. – Int. Conf. Image Process. ICIP, No. May 2016*, pp. 289–292, 2011.
13. R Wolz, P Aljabar, JV Hajnal, A Hammers, and D Rueckert, "LEAP: learning embeddings for atlas propagation," *Neuroimage*, vol. 49, no. 2, pp. 1316–1325, 2010.
14. AKHoang Duc et al., "Using manifold learning for atlas selection in multi-atlas segmentation," *PLoS One*, vol. 8, no. 8, 2013.
15. G Sanroma, G Wu, Y Gao, and D Shen, "Learning to rank atlases for multiple-atlas segmentation," *IEEE Trans. Med. Imaging*, vol. 33, no. 10, pp. 1939–1953, Oct. 2014.

16. L Ramus, and G Malandain, "Assessing selection methods in the context of multi-atlas based segmentation," in *2010 IEEE International Symposium on Biomedical Imaging: From Nano to Macro*, 2010, pp. 1321–1324.

17. B Schipaanboord et al., "An evaluation of atlas selection methods for atlas-based automatic segmentation in radiotherapy treatment planning," *IEEE Trans. Med. Imaging*, vol. 38, no. 11, pp. 2654–2664, 2019.

18. J Yang et al., "Autosegmentation for thoracic radiation treatment planning: a grand challenge at AAPM 2017," *Med. Phys.*, vol. 45, no. 10, pp. 4568–4581, Oct. 2018.

19. S Gorthi, and M Cuadra, "Multi-Atlas based segmentation of head and neck CT images using active contour framework," *MICCAI Work. 3D Segmentation Chall. Clin. Appl.*, pp. 313–321, 2010.

20. LR Dice, "Measures of the amount of ecologic association between species," *Ecology*, vol. 26, no. 3, pp. 297–302, 1945.

21. JM Lötjönen et al., "Fast and robust multi-atlas segmentation of brain magnetic resonance images," *Neuroimage*, vol. 49, no. 3, pp. 2352–2365, 2010.

22. P Raudaschl, K Fritscher, P Zaffino, GC Sharp, MF Spadea, and R Schubert, "A novel atlas-selection approach for multi-atlas based segmentation using the correlation of inter-atlas similarities," in *Proceedings of Image-Guided Adaptive Radiation Therapy Workshop*, 2014, pp. 53–60.

23. T Zhao, and D Ruan, "Learning image based surrogate relevance criterion for atlas selection in segmentation," *Phys. Med. Biol.*, vol. 61, no. 11, pp. 4223–4234, 2016.

24. P Zaffino et al., "Multi atlas based segmentation: should we prefer the best atlas group over the group of best atlases?" *Phys. Med. Biol.*, vol. 63, no. 12, 2018.

25. G Sharp et al., "Vision 20/20: perspectives on automated image segmentation for radiotherapy," *Med. Phys.*, vol. 41, no. 5, pp. 1–13, 2014.

26. DN Teguh et al., "Clinical validation of atlas-based auto-segmentation of multiple target volumes and normal tissue (swallowing/mastication) structures in the head and neck," *Int. J. Radiat. Oncol. Biol. Phys.*, vol. 81, no. 4, pp. 950–957, 2011.

27. AV Young, A Wortham, I Wernick, A Evans, and RD Ennis, "Atlas-based segmentation improves consistency and decreases time required for contouring postoperative endometrial cancer nodal volumes," *Int. J. Radiat. Oncol. Biol. Phys.*, vol. 79, no. 3, pp. 943–947, 2011.

28. H Lee et al., "Clinical evaluation of commercial atlas-based auto-segmentation in the head and neck region," *Front. Oncol.*, vol. 9, pp. 1–9, 2019.

29. S Klein, UA Van Der Heide, IM Lips, M Van Vulpen, M Staring, and JPW Pluim, "Automatic segmentation of the prostate in 3D MR images by atlas matching using localized mutual information," *Med. Phys.*, vol. 35, no. 4, pp. 1407–1417, 2008.

30. MR Sabuncu, SK Balci, ME Shenton, and P Golland, "Image-driven population analysis through mixture modeling," *IEEE Trans. Med. Imaging*, vol. 28, no. 9, pp. 1473–1487, Sep. 2009.

31. I Išgum, M Staring, A Rutten, M Prokop, MA Viergever, and B Van Ginneken, "Multi-atlas-based segmentation with local decision fusion-application to cardiac and aortic segmentation in CT scans," *IEEE Trans. Med. Imaging*, vol. 28, no. 7, pp. 1000–1010, 2009.

32. L Ramus, and G Malandain, "Multi-atlas based segmentation: application to the head and neck region for radiotherapy planning," *Med. Image Anal. Clin.*, pp. 281–288, 2010.

33. EM van Rikxoort et al., "Adaptive local multi-atlas segmentation: application to the heart and the caudate nucleus," *Med. Image Anal.*, vol. 14, no. 1, pp. 39–49, 2010.

34. J Yang, Y Zhang, L Zhang, and L Dong, "Automatic segmentation of parotids from CT scans using multiple atlases," in *Medical Image Analysis for the Clinic: A Grand Challenge*, 2010, pp. 323–330. www.amazon.com/Medical-Image-Analysis-Clinic-Challenge/dp/1453759395

35. JA Dowling et al., "Fast automatic multi-atlas segmentation of the prostate from 3D MR images," *Lect. Notes Comput. Sci. (including Subser. Lect. Notes Artif. Intell. Lect. Notes Bioinformatics)*, vol. 6963 LNCS, pp. 10–21, 2011.

36. R Wolz, C Chu, K Misawa, M Fujiwara, K Mori, and D Rueckert, "Automated abdominal multi-organ segmentation with subject-specific atlas generation," *IEEE Trans. Med. Imaging*, vol. 32, no. 9, pp. 1723–1730, 2013.

37. AJ Asman, FW Bryan, SA Smith, DS Reich, and BA Landman, "Groupwise multi-atlas segmentation of the spinal cord's internal structure," *Med. Image Anal.*, vol. 18, no. 3, pp. 460–471, 2014.

38. H Wang, JW Suh, SR Das, JB Pluta, C Craige, and PA Yushkevich, "Multi-atlas segmentation with joint label fusion," *IEEE Trans. Pattern Anal. Mach. Intell.*, vol. 35, no. 3, pp. 611–623, 2013.

39. AJ Asman, Y Huo, AJ Plassard, and BA Landman, "Multi-atlas learner fusion: an efficient segmentation approach for large-scale data," *Med. Image Anal.*, vol. 26, no. 1, pp. 82–91, Dec. 2015.

40. TR Langerak, UA Van Der Heide, ANTJ Kotte, FF Berendsen, and JPW Pluim, "Improving label fusion in multi-atlas based segmentation by locally combining atlas selection and performance estimation," *Comput. Vis. Image Underst.*, vol. 130, pp. 71–79, 2015.

41. Z Xu et al., "Efficient multi-atlas abdominal segmentation on clinically acquired CT with SIMPLE context learning," *Med. Image Anal.*, vol. 24, no. 1, pp. 18–27, 2015.

42. P Yan, Y Cao, Y Yuan, B Turkbey, and PL Choyke, "Label image constrained multiatlas selection," *IEEE Trans. Cybern.*, vol. 45, no. 6, pp. 1158–1168, 2015.

43. K Karasawa et al., "Multi-atlas pancreas segmentation: atlas selection based on vessel structure," *Med. Image Anal.*, vol. 39, pp. 18–28, 2017.

44. J Yang et al., "Atlas ranking and selection for automatic segmentation of the esophagus from CT scans," *Phys. Med. Biol.*, vol. 62, no. 23, pp. 9140–9158, 2017.

45. LJ Stapleford et al., "Evaluation of automatic atlas-based lymph node segmentation for head-and-neck cancer," *Int. J. Radiat. Oncol. Biol. Phys.*, vol. 77, no. 3, pp. 959–966, 2010.

46. MA Gambacorta et al., "Clinical validation of atlas-based auto-segmentation of pelvic volumes and normal tissue in rectal tumors using auto-segmentation computed system," *Acta Oncol. (Madr).*, vol. 52, no. 8, pp. 1676–1681, 2013.

47. J Hwee et al., "Technology assessment of automated atlas based segmentation in prostate bed contouring," *Rad. Oncol.*, vol. 6, no. 1, pp. 1–9, 2011.

48. A Lin, G Kubicek, JW Piper, AS Nelson, AP Dicker, and RK Valicenti, "Atlas-based segmentation in prostate IMRT: timesavings in the clinical workflow," *Int. J. Radiat. Oncol.*, vol. 72, no. 1, pp. S328–S329, 2008.

49. C Granberg, *Clinical Evaluation of Atlas Based Segmentation for Radiotherapy of Prostate Tumours*, M.S. Thesis, pp. 1–66, 2011.

50. KA Langmack, C Perry, C Sinstead, J Mills, and D Saunders, "The utility of atlas-assisted segmentation in the male pelvis is dependent on the interobserver agreement of the structures segmented," *Br. J. Radiol.*, vol. 87, no. 1043, 2014.

4 Deformable Registration Choices for Multi-Atlas Segmentation

Keyur Shah, James Shackleford, Nagarajan Kandasamy, and Gregory C. Sharp

CONTENTS

Registration is used for several purposes during multi-atlas segmentation. Atlases may be pre-aligned using rigid or deformable registration, and online registration may be performed during atlas selection. Furthermore, all atlas-based methods rely on deformable registration to map atlases onto the target image and the accuracy of the segmentation depends, to a large degree, on the deformable registration strategy undertaken. This chapter examines the influence of deformable registration parameter selection for two popular registration methods, the B-spline and demons algorithms. For B-spline methods, the effect of varying control point grid spacing, image subsampling rate, regularization method, and its corresponding weights are studied. The effect of Gaussian kernel width used to smooth the displacement field is explored for the demons algorithm. Experimental evaluation is performed on the Lung CT Segmentation Challenge (LCTSC) dataset.

4.1   INTRODUCTION

Deformable registration is one of the most important steps in the multi-atlas segmentation pipeline. One or more atlas images are registered to the image to be segmented, usually known as the query image, to create the segmentation. While it is widely acknowledged that deformable registration quality strongly influences the final segmentation quality, there is no consensus on the best registration techniques, such as the best objective function or transformation model. Various objective functions are used, most commonly: mean squared error (MSE) or sum of squared difference (SSD) of image intensity, mutual information (MI), normalized mutual information (NMI), and correlation coefficient (CC). Similarly, various transformation models have also been proposed, including: B-splines, thin plate splines (TPS), displacement fields, and velocity fields.

Several researchers have explored image registration approaches for atlas-based segmentation. Alven et al. presented a feature-based registration method which combines the information of the entire atlas set and efficiently finds robust correspondences and transformations between the target and all the images in the atlas set [1]. Bai et. al. compared, in their work, the performance of four different image registration algorithms: affine, B-spline, free-form deformation, and large deformation diffeomorphic metric mapping (LDDMM). Their experiments found that the LDDMM registration algorithm worked best for the mouse brain image segmentation task [2]. Datteri et al. used the adaptive bases algorithm (ABA) with an NMI objective function for registration. ABA models the deformation field as a linear combination of radial basis functions with finite support [3]. Doshi et al. used the Advanced Normalization Tools (ANTs) registration toolkit for atlas registration [4]. Heckemann et al. compare the image registration toolkit (IRTK) [5], based on B-splines and maximizing the NMI, with an algorithm they termed multi-atlas propagation with enhanced registration (MAPER), which optimizes both image intensity and tissue classification. They found that the MAPER approach provides superior results for a brain segmentation task [6]. Lötjönen et al. introduced a similarity metric based on intensity normalized images and compared it with NMI. They found a threefold reduction in the computation time with similar registration accuracy [7]. Sjöberg et al. compared a B-spline based method with a demons algorithm for registration strategy. No significant differences were reported [8]. Yeo et al. employed a generative model for the construction of a probabilistic atlas for joint registration and segmentation of images [9].

Previous works have largely considered the difference between registration methods. This chapter explores parameter choices for the B-spline and demons algorithm for atlas-based segmentation. The Plastimatch multi-atlas-based segmentation (MABS) platform is used [10]. Because the lung cancer segmentation challenge data uses CT, only the MSE similarity metric is considered. The following subsections describe the deformable registration problem, the transformation models, and the similarity metric.

4.1.1 Deformable Registration

The goal of deformable image registration is to align two or more images into the same reference frame. Given a fixed image F with voxel coordinates $\theta = (x,y,z)$ and voxel intensities $F(\theta) = f$ and moving image M with voxel coordinates $\phi = (x',y',z')$ and voxel intensities $M(\phi) = m$, the two images are said to be registered when the cost function

$$C = \sum_{\theta \in \Omega} \Psi(f,m) + S(v) \qquad (4.1)$$

is minimized with respect to a similarity metric ψ and regularization term S over the image overlap domain Ω under the coordinate mapping $T(\theta) = \theta + v$. Here, v is the dense displacement field defined for every voxel $\theta \in \Omega$, which maps from F to M.

For a unimodal registration problem such as LCTSC, the MSE is an appropriate choice. The expression for MSE is given as:

$$\Psi(f,m) = \frac{(f-m)^2}{N}, \qquad (4.2)$$

where N is the number of voxels in Ω.

Because deformable registration is an ill-posed problem, it is helpful to constrain the solution to the set of physically meaning transforms. Equation 4.1 achieves this through the use of a regularization penalty term, where the smoothness $S(v)$ is used to drive T toward smoother displacement fields. For this study, the use of curvature and third order regularizers, which are commonly used regularizers in deformable registration, are considered using the implementation within Plastimatch [11, 12]. It is considered that choosing correct regularizer weights is more important than the choice of regularizer type.

4.1.2 B-SPLINE REGISTRATION

B-spline registration is a method of deformable image registration that uses B-spline functions to define a continuous displacement field that maps the voxels in one image to those in another image. B-spline interpolation yielding the x component of the vector field for a voxel located at θ is

$$v_x(\theta) = \sum_{l=0}^{3}\sum_{m=0}^{3}\sum_{n=0}^{3}\beta_l(u_x)\beta_m(u_y)\beta_n(u_z)p_{x,l,m,n} \tag{4.3}$$

where p_x is the B-spline coefficient defining the x component of the displacement field for one of the 64 control points that influence the voxel. The β_l, β_m, and β_n terms represent the uniform cubic B-spline basis function in the x, y, and z directions, respectively and u_x, u_y, and u_z represent the normalized voxel location relative to the control point locations [13].

4.1.3 DEMONS ALGORITHM

The demons algorithm uses gradient information of a static fixed image in order to generate the demons force that deforms the moving image. Unlike the B-spline deformation model which interpolates the displacement vector field based on the control point weights, the demons algorithm generates the displacement vector field at each voxel. There are many different variants of the demons algorithm, including diffeomorphic and symmetric forms [14–16]. This investigation considers only the classic demons algorithm [17]. The vector field at location θ in the moving image is solved iteratively by updating a displacement field according to

$$v = \frac{(m-f)\nabla f}{|\nabla f|^2 + (m-f)^2} \tag{4.4}$$

where ∇f is the gradient of the fixed image at voxel θ. At each iteration, an update to the displacement field is solved according to Equation 4.3. Between iterations, the displacement field is updated by smoothing.

4.2 PLASTIMATCH MABS IMPLEMENTATION DETAILS

The Plastimatch MABS implementation of atlas-based segmentation was used for this study. Plastimatch is an open source software for image computation with focus on high-performance volumetric registration of medical images. The MABS workflow is divided into a training phase, where registration and segmentation parameters can be optimized, and the segmentation phase which uses the optimized parameters to segment new cases. Both phases perform a sequence of operations: (1) data conversion (optional), (2) pre-alignment (optional), (3) atlas selection (optional), (4) image registration, and (5) voting. A configuration file describes the parameters used for each operation [10]. The configuration settings used in this study are briefly described.

Data Conversion: Digital Imaging and Communications in Medicine (DICOM) files from the clinical scanners are converted into a compressed raw format to enable faster atlas file loading. In this study, the DICOM files were converted to the Nearly Raw Raster Data (NRRD) file format. Similarly, DICOM-RT Structure Set files containing the anatomical structures are rasterized into volumetric format using one file for each structure.

Pre-Alignment: The anatomical image within each atlas is pre-aligned to a randomly chosen reference image using rigid or affine transformation. The transformation is also applied to the structure images. In this study, the atlas images were pre-aligned to a reference image using the pre-alignment step.

Atlas Selection: A subset of atlases can be selected for a given atlas based on image similarity or displacement field metrics, such as NMI of the image intensities or mean-squared vector difference of the displacement field. Atlas selection can also be performed randomly or using a precomputed ranking. Once the atlases are ranked, a fixed number of atlases can be selected, or a variable number of atlases that meet a threshold criterion can be selected. The atlas selection step is optional; if not selected all the atlases in the database will be used. In this study, NMI was used as the metric and the five atlases with the highest NMI were selected. Chapter 3 explores atlas selection in more depth.

Image Registration: Each atlas from the selected set is registered to the query atlas using deformable image registration. Multiple registration strategies can be compared during the training phase, and the best strategy is selected using exhaustive search. Details of the image registration used in this study are found in the next section.

Voting: After the structures are warped into the reference of the query image, the final segmentations are generated through statistical algorithms. Plastimatch MABS supports Gaussian weighted (GW) and STAPLE as voting techniques. The voting parameters of either algorithm are specified in the configuration file and can be optimized during the training phase. In this study, Gaussian weighted voting was used, and voting parameters were optimized prior to investigating registration parameters.

4.3   EVALUATION METRICS

The Dice similarity coefficient and 95% Hausdorff distance were used to quantify the accuracy of the segmented structures. These metrics are described in Chapter 15. In contrast to the implementation described in Chapter 15, evaluation was performed on voxelized segmentations using Plastimatch. To overcome inconsistencies in the superior and inferior extents of manual labeled tubular structures (spinal cord and esophagus), both these segmented structures and the ground truth structures were cropped at the superior and inferior borders by 10 mm with respect to the ground truth structures.

The directed percent Hausdorff measure, for a percentile r, is the rth percentile distance over all distances from points in X to their closest point in Y.

4.4   EXPERIMENTAL STEPS

The LCTSC data [18], described in Chapter 1, was used for the experiments. The 36 training cases were used for optimizing the registration and atlas fusion parameters. The remaining 24 offline and online test cases were used to evaluate accuracy. Four different deformable image registration strategies were carried out. The first three strategies used the B-spline algorithm, and the last strategy used the demons algorithm. All strategies used multiple registration stages as shown in Table 4.3. The B-spline methods used five stages and the demons algorithm used four stages. The first two stages for all the strategies were rigid, followed by the deformable stages. Three different parameters were varied for the B-spline based model: (1) image subsampling rate (Res) was varied within a range from 2 mm to 6 mm, (2) regularization weight and type (Reg) was varied between the curvature regularizer with weights from 0–100, and the third order regularizer with weights from 1–10, and (3) B-Spline grid spacing (GS) was varied within a range from 10 mm to 100 mm. For the last strategy (demons), the width of the Gaussian kernel used to smoothen the displacement field was varied from 1 mm to 4 mm. The value ranges for each of these parameters were selected based on the authors' prior experience with image registration. In order to narrow down the range of the parameters, a few experiments were run on a subset of atlases. Table 4.1 describes the acronyms used in the chapter. The fixed and constant parameters for all registration strategies are described in Tables 4.2 and 4.3. Only one parameter is varied for a given strategy, keeping all other parameters fixed. The fixed parameter

TABLE 4.1

Acronyms Used for Each Parameter and Their Explanations

Acronym Used in Text	Explanation of Parameter
Res	Image sub-sampling rate
Reg	Regularization weight and type
GS	Grid spacing
Demons	Width of Gaussian kernel

TABLE 4.2

Parameters Varied for Each Registration Strategy

	Registration Stages				
Registration strategy	Rigid-1	Rigid-2	Deformable-1	Deformable-2	Deformable-3
Res 1	$4 \times 4 \times 3$	$4 \times 4 \times 3$	$4 \times 4 \times 3$	$2 \times 2 \times 3$	$2 \times 2 \times 3$
Res 2	$4 \times 4 \times 3$	$4 \times 4 \times 3$	$4 \times 4 \times 3$	$3 \times 3 \times 3$	$3 \times 3 \times 3$
Res 3	$4 \times 4 \times 3$	$4 \times 4 \times 3$	$4 \times 4 \times 3$	$4 \times 4 \times 3$	$4 \times 4 \times 3$
Res 4	$4 \times 4 \times 12$	$4 \times 4 \times 12$	$4 \times 4 \times 12$	$2 \times 2 \times 6$	$2 \times 2 \times 6$
Res 5	$6 \times 6 \times 6$	$6 \times 6 \times 6$	$6 \times 6 \times 6$	$3 \times 3 \times 3$	$2 \times 2 \times 3$
Res 6	$6 \times 6 \times 6$	$6 \times 6 \times 6$	$6 \times 6 \times 6$	$3 \times 3 \times 3$	$3 \times 3 \times 3$
Res 7	$6 \times 6 \times 6$	$6 \times 6 \times 6$	$6 \times 6 \times 6$	$6 \times 6 \times 6$	$6 \times 6 \times 6$
Reg 0	NA	NA	Curvature 0	Curvature 0	Curvature 0
Reg 1	NA	NA	Curvature 100	Curvature 1	Curvature 0.1
Reg 2	NA	NA	Curvature 100	Curvature 10	Curvature 0.1
Reg 3	NA	NA	Curvature 100	Curvature 10	Curvature 1
Reg 4	NA	NA	Curvature 100	Curvature 100	Curvature 100
Reg 5	NA	NA	Third order 100	Third order 10	Third order 1
Reg 6	NA	NA	Third order 1000	Third order 10	Third order 1
Reg 7	NA	NA	Third order 1000	Third order 100	Third order 1
Reg 8	NA	NA	Third order 1000	Third order 100	Third order 10
GS 1	NA	NA	$100 \times 100 \times 100$	$30 \times 30 \times 30$	$10 \times 10 \times 10$
GS 2	NA	NA	$100 \times 100 \times 100$	$30 \times 30 \times 30$	$20 \times 20 \times 20$
GS 3	NA	NA	$100 \times 100 \times 100$	$50 \times 50 \times 50$	$10 \times 10 \times 10$
GS 4	NA	NA	$100 \times 100 \times 100$	$50 \times 50 \times 50$	$20 \times 20 \times 20$
GS 5	NA	NA	$100 \times 100 \times 100$	$50 \times 50 \times 50$	$30 \times 30 \times 30$
GS 6	NA	NA	$100 \times 100 \times 100$	$50 \times 50 \times 50$	$50 \times 50 \times 50$
GS 7	NA	NA	$100 \times 100 \times 100$	$100 \times 100 \times 100$	$100 \times 100 \times 100$
Demons 1	NA	NA	2	1	NA
Demons 2	NA	NA	3	2	NA
Demons 3	NA	NA	4	3	NA
Demons 4	NA	NA	5	4	NA

settings are marked in bold in Table 4.2. For example, when the image subsampling rate is varied, the grid spacing, regularization type, and weight are held constant as seen in Table 4.3. Once the optimal parameters were determined, each query image was segmented based on the optimal parameters and the segmentations were evaluated for each anatomical structure against the ground truth.

TABLE 4.3
Constant Parameters for the Registration Strategies

				Registration Stages			
Parameter varied	Constant parameters	Rigid-1	Rigid-2	Deformable-1	Deformable-2	Deformable-3
Image subsampling (mm)	Regularizer type and weight	NA	NA	Curvature 100	Curvature 1	Curvature 0.1
	Grid spacing (mm)	NA	NA	$100 \times 100 \times 100$	$50 \times 50 \times 50$	$30 \times 30 \times 30$
Regularizer type and weight	Image subsampling (mm)	$6 \times 6 \times 6$	$6 \times 6 \times 6$	$6 \times 6 \times 6$	$3 \times 3 \times 3$	$3 \times 3 \times 3$
	Grid spacing (mm)	NA	NA	$100 \times 100 \times 100$	$50 \times 50 \times 50$	$30 \times 30 \times 30$
Grid spacing (mm)	Image subsampling (mm)	$6 \times 6 \times 6$	$6 \times 6 \times 6$	$6 \times 6 \times 6$	$3 \times 3 \times 3$	$3 \times 3 \times 3$
	Regularizer type and weight	NA	NA	Curvature 100	Curvature 1	Curvature 0.1
Width of Gaussian kernel	Image subsampling (mm)	$6 \times 6 \times 6$	$6 \times 6 \times 6$	$6 \times 6 \times 6$	$3 \times 3 \times 3$	NA

4.5   RESULTS

The voxel sampling rate was found not to affect the performance of either of the registration algorithms substantially. The B-spline method was also found to be relatively robust to variations in the control-point spacing. However, both methods were found to be most sensitive to the tuning of the regularizer. Figure 4.1 compares the average Dice Similarity Coefficient (DSC) over the five Organs-at-risk (OARs) by varying the (a) width of the Gaussian kernel and (b) curvature regularizer weight. It is observed that increasing the smoothness has an inverse impact on the performance as measured by DSC. Gaussian kernel of 1 mm width and curvature regularizer weight of 0.1 lead to the highest average DSC.

GS4 was the best overall optimization strategy. The segmentation accuracy of the lungs was significantly higher than the other structures (DSC = 0.95 ± 0.02, 95% HD = 5.29 ± 3.25). The segmentation accuracy of the esophagus and spinal cord were relatively low (spinal cord: DSC = 0.8 ± 0.08, 95% HD = 12.34 ± 13.9; esophagus: DSC = 0.59 ± 0.08, 95% HD = 9.47 ± 6.57) due to poor soft-tissue contrast. Figure 4.2 shows the average over the 24 test cases (a) DSC and (b) 95% HD achieved by the best performing strategy (GS4) for all the OARs and the average over the OARs in the form of box plots, indicating confidence interval of [5,95] % and their corresponding outliers. Figure 4.3 shows the segmentation performance for the best performing strategy. Rows 1–5 show the 5th, 25th, 50th, 75th, and 95th quartiles of the median distribution, with the 5th quartile being the worst and the 95th quartile being the best segmentation. The 5th quartile example produces a poor segmentation due to the presence of a tumor in the left lung.

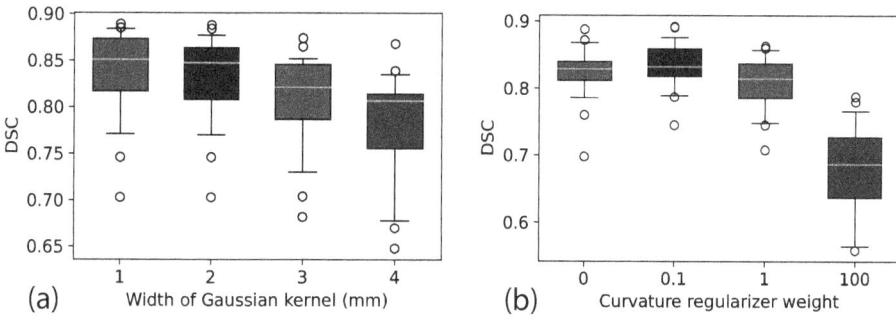

FIGURE 4.1   Average DSC over the five OARs achieved by varying (a) width of the Gaussian kernel and (b) curvature regularizer weight for the 24 test cases.

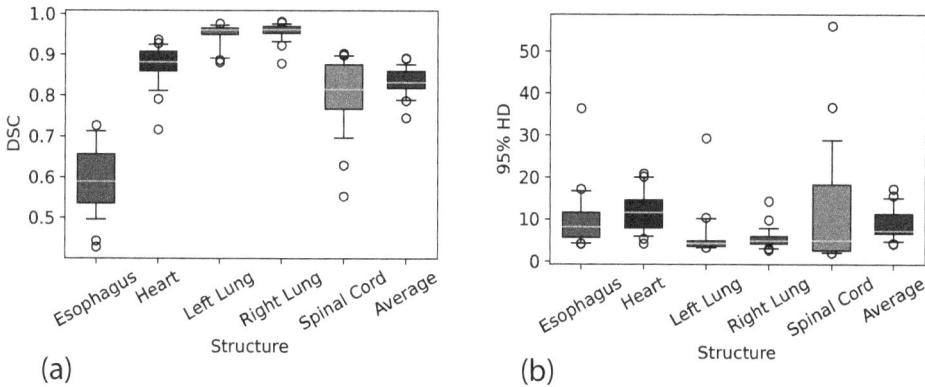

FIGURE 4.2   (a) DSC and (b) 95 % HD achieved by the best performing strategy for the five OARs and their average over the 24 test cases.

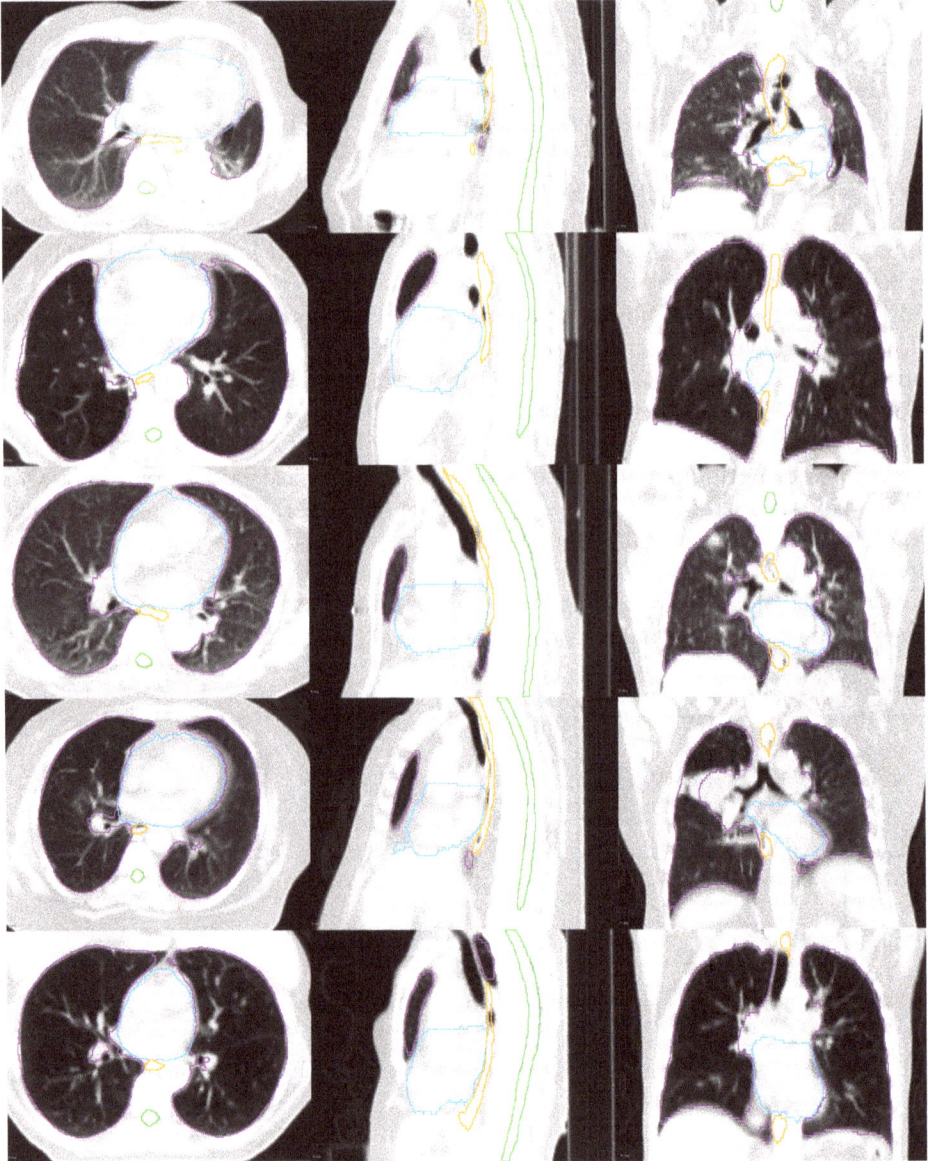

FIGURE 4.3 Segmentations generated using the best performing strategy. Rows 1–5 show the 5th, 25th, 50th, 75th, and 95th percentile images, from worst to best, from the 24 test cases.

4.6 SUMMARY

Identifying the optimal registration and voting parameters is a challenging exercise. For practical purposes, numerous algorithm details such as atlas selection and label fusion parameters must be held constant during testing. However, one must consider that interplay could exist between these algorithm settings. For example, a registration with high regularization may require a different number of atlases than a registration with low regularization.

In general, both B-spline and demons registrations performed similarly. Algorithms were not very sensitive to the voxel sampling rate, which means that faster registrations at higher sampling rates can be considered. An intermediate schedule with final grid spacing of 20 mm was preferred for B-spline grid spacing. However, final segmentation results were not found to be highly sensitive to these parameters, and average Dice similarity varied only by a few percent over fairly broad parameter setting ranges. However, both algorithms were found to be more affected by the choice of regularizer parameters, with smaller regularization penalty terms being preferred for both B-spline and demons registrations.

REFERENCES

1. Alvén, J, et al. (2016). "Überatlas: fast and robust registration for multi-atlas segmentation." **80**: 249–255.
2. Bai, J, et al. (2012). "Atlas-based automatic mouse brain image segmentation revisited: model complexity vs. image registration." **30**(6): 789–798.
3. Datteri, R, et al. (2011). "Estimation of registration accuracy applied to multi-atlas segmentation." *MICCAI Workshop on Multi-Atlas Labeling and Statistical Fusion.*
4. Doshi, J, et al. (2016). "MUSE: multi-atlas region segmentation utilizing ensembles of registration algorithms and parameters, and locally optimal atlas selection." **127**: 186–195.
5. Rueckert, D, et al. (1999). "Nonrigid registration using free-form deformations: application to breast MR images." **18**(8): 712–721.
6. Heckemann, RA, et al. (2010). "Improving intersubject image registration using tissue-class information benefits robustness and accuracy of multi-atlas based anatomical segmentation." **51**(1): 221–227.
7. Lötjönen, JM, et al. (2010). "Fast and robust multi-atlas segmentation of brain magnetic resonance images." **49**(3): 2352–2365.
8. Sjöberg, C, et al. (2013). "Multi-atlas based segmentation using probabilistic label fusion with adaptive weighting of image similarity measures." **110**(3): 308–319.
9. Yeo, BT, et al. (2008). "Effects of registration regularization and atlas sharpness on segmentation accuracy." **12**(5): 603–615.
10. Zaffino, P, et al. (2016). "Plastimatch MABS, an open source tool for automatic image segmentation." **43**(9): 5155–5160.
11. Shah, KD, et al. (2020). "A generalized framework for analytic regularization of uniform cubic B-spline displacement fields." arXiv:2010.02400
12. Shackleford, JA, et al. (2012). "Analytic regularization of uniform cubic B-spline deformation fields." *International Conference on Medical Image Computing and Computer-Assisted Intervention, Springer.*
13. Shackleford, JA, et al. (2010). "On developing B-spline registration algorithms for multi-core processors." **55**(21): 6329.
14. Joshi, S, et al. (2004). "Unbiased diffeomorphic atlas construction for computational anatomy." **23**(Supplement 1): S151–S160.
15. Avants, BB, et al. (2008). "Symmetric diffeomorphic image registration with cross-correlation: evaluating automated labeling of elderly and neurodegenerative brain." **12**(1): 26–41.
16. Vercauteren, T, et al. (2009). "Diffeomorphic demons: efficient non-parametric image registration." **45**(1): S61–S72.
17. Thirion, J-P (1998). "Image matching as a diffusion process: an analogy with Maxwell's demons." *Medical Image Analysis* **2**(3): 243–260.
18. Yang, J, et al. (2018). "Autosegmentation for thoracic radiation treatment planning: a grand challenge at AAPM 2017." **45**(10): 4568–4581.

5 Evaluation of a Multi-Atlas Segmentation System

Raymond Fang, Laurence Court, and Jinzhong Yang

CONTENTS

5.1 INTRODUCTION

Multi-atlas segmentation has been shown to minimize the effects of variability and improve segmentation accuracy, especially for low-contrast anatomy such as the brachial plexus [1–6]. Multi-atlas segmentation exploits contoured atlas data for the auto-segmentation of new patient images. Deformable image registration is a key to mapping contours from the atlases to new patients. The individual deformed contours from multiple atlases are then effectively combined to produce the best approximation of the true segmentation through various fusion approaches [7–10].

The quality of segmentation generated in multi-atlas segmentation is highly dependent on the choice of atlases. A few bad atlases may greatly adversely affect segmentation accuracy. Atlas selection can be subdivided into offline and online selection. Offline selection involves choosing a predetermined set of atlases for all target scans, while online selection involves choosing the atlases after the target image is chosen. An atlas that poorly represents the patient scan can skew and misinform the resulting segmentation [11]. Although a variety of methods has been proposed for atlas selection, determining the optimal set and number of atlases remains an unsolved problem [12].

This chapter presents a specific example to evaluate the multi-atlas segmentation system. Specifically, a two-phase online atlas selection approach is presented, to rank and select a subset of optimal atlas candidates, in terms of local anatomical similarity, for multi-atlas segmentation. In the first phase of atlas selection, the correlation coefficient of the image content in a local region between the atlas and the new image is used to evaluate their similarities and to rank the atlases. A subset of atlases based on this ranking is selected, and deformable image registration is performed to generate deformed contours and deformed images in the new image space. In the second phase of atlas selection, the Kullback–Leibler (KL) divergence is used to measure the similarity of local-intensity histograms between the new image and each deformed image; the measurements are used to rank the previously selected atlases. Deformed contours from the most to the least similar are added sequentially for overlap ratio examination. A subset of optimal atlas candidates is further identified by analyzing the variation of the overlap ratio with the number of atlases. The deformed contours from these optimal atlases are fused using a modified simultaneous truth and performance level estimation (STAPLE) algorithm [13] to produce the final segmentation.

5.2 METHODS

5.2.1 PATIENT DATA

Two sets of patient data were used for this study. The first data set included 21 patients with stage IV head and neck squamous cell carcinoma who had been treated with intensity-modulated radiation therapy (IMRT). The planning CT images of these patients had been acquired and the cervical esophagus had been contoured from the inferior edge of the cricoid cartilage down to the level of the manubrium by an experienced head and neck radiation oncologist. All CT images had an in-slice resolution of 1.0 mm and a slice spacing of 2.5 mm. For this set of data, the focus was on the segmentation of the esophagus.

The second dataset is from a public benchmark dataset used for thoracic normal-tissue segmentations for the 2017 AAPM Thoracic Auto-segmentation Challenge at the 2017 AAPM Annual Meeting (http://autocontouringchallenge.org/), as described in detail in Chapter 1. Of this dataset, the 36 training cases were used as atlases and the 24 test cases (12 offline test cases and 12 online test cases) were used to evaluate the multi-atlas segmentation performance.

5.2.2 ONLINE ATLAS SELECTION FOR MULTI-ATLAS SEGMENTATION

An overall framework of the online atlas selection approach for esophagus segmentation is illustrated in Figure 5.1. In brief, a rigid alignment between all atlas images and the test image to define a small local region was performed; a first-phase of atlas selection was performed on the basis of the similarity measurement in the local region. In the first phase of atlas selection, 12 or fewer atlases were chosen and deformable image registration was performed between these atlases and the test image to generate deformed contours and images. A second phase of atlas selection was performed by measuring the similarity of the deformed image and the test image in a region of interest (ROI), as defined by the deformed contours. The deformed contours of selected optimal atlases were then fused together to produce the final segmentation. A detailed description of the approach follows.

5.2.2.1 First Phase of Atlas Selection

The first phase of atlas selection filters out atlases dissimilar to the test image and limits the number of atlases for subsequent deformable image registration. This prescreening step avoids wasting time performing deformable registration on dissimilar atlases and reduces the adverse impact these atlases may have on contour fusion. Rigid alignment between the two 3D images is a prerequisite for this step. Rigid alignment between each atlas and the test image was performed by minimizing the mean square errors of two 2D images generated from the 3D CT images, by projecting the 3D image in sagittal and coronal views to obtain two 2D images. The rigid alignments were performed separately on these two 2D images using a gradient descent optimizer, and then further optimized with golden section search and parabolic interpolation [14] to find the optimal translations in all three directions that gave the minimal mean square error of the two registered 2D images. On the other hand, a landmark point was predefined in each atlas to record the relative shift between two

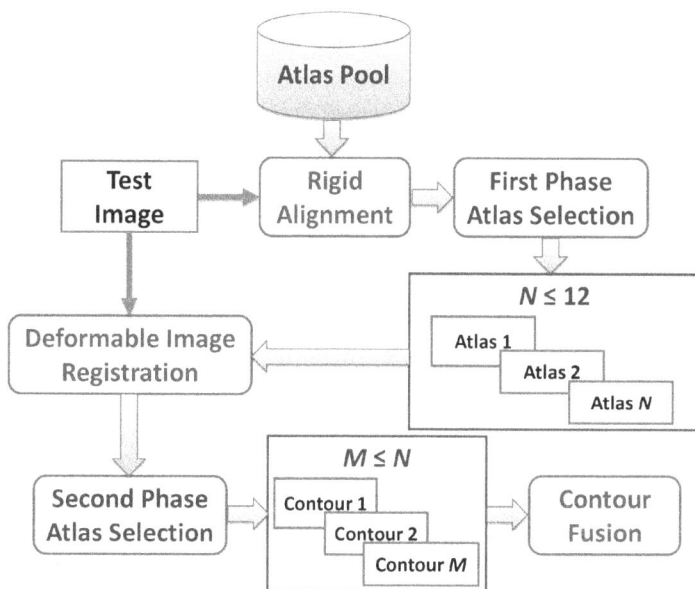

FIGURE 5.1 Overall framework of the proposed online atlas selection approach for esophagus segmentation.

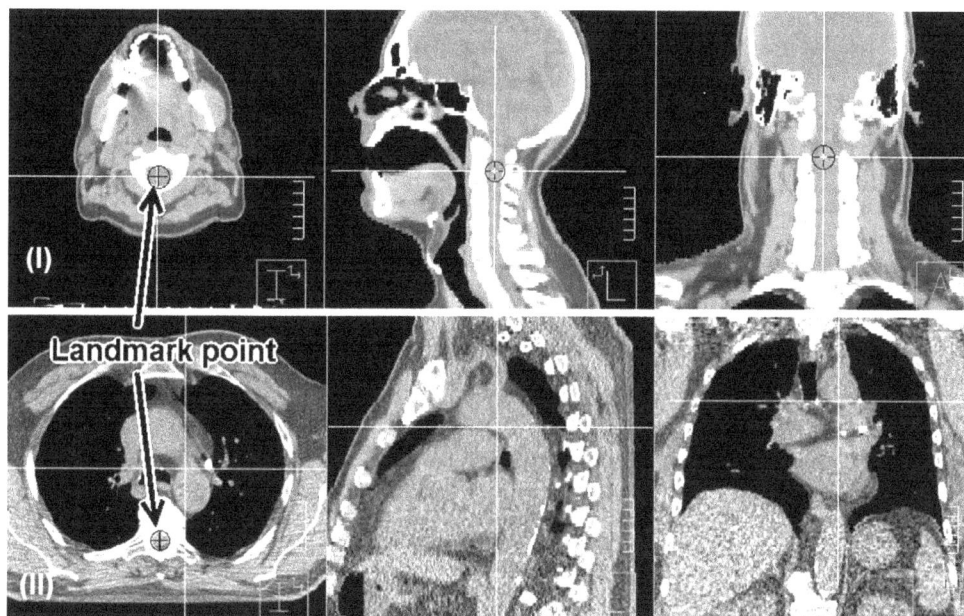

FIGURE 5.2 Illustration of landmark points in the atlases. Panel (I): for head and neck cancer patients, the landmark point is in the cord of the superior end of C2 vertebrate. Panel (II): for thoracic cancer patients, the landmark point is in the cord of the slice where the bottom of the aortic arch is located.

atlases. The landmark point in each atlas corresponds to the other. In head and neck cancer patients, the landmark point was chosen in the cord of the superior end of C2 vertebrate; in thoracic cancer patients, it was chosen in the cord of the slice where the bottom of the aortic arch is located, as shown in Figure 5.2. In the automated rigid alignment, the rigid shift between the test image and the atlas of minimal mean square errors was recorded. The rigid shift of the test image and each of the other atlases was derived from the recorded rigid shift and the relative shift between atlases.

The similarity of two images was measured in a small cubic region using the correlation coefficient of the image-intensity values in the region. This cubic region is determined as follows: after rigid alignment, all atlas contours were transformed to the test image space. The bounding box of all transformed contours was determined and formed a cubic region, which was transformed back to each atlas space to obtain the local region for each atlas. The correlation coefficient between the test image and the atlas image in the corresponding cubic regions was calculated for each atlas. Twelve atlases with the largest correlation coefficient values were selected. A previous study [15] found that the contour fusion might not benefit from more than 12 atlases; therefore, no more than 12 atlases were chosen for the next step to perform deformable registration. Atlases with a correlation coefficient less than 0.2 were removed if the total number of atlases was not less than six.

5.2.2.2   Deformable Image Registration

An intensity-based approach for the CT-to-CT deformable registration was used because Hounsfield units of CT images are calibrated to the attenuation coefficient of water so that the intensity value is consistent among images. Specifically, a dual-force demons algorithm [16] was used for the deformable registration between the test and atlas images. Before performing the deformable registration, histogram equalization was performed to match the contrast of the two images. The histogram equalization was performed locally by separating the images into small blocks. A multiresolution scheme was used to accelerate the registration and improve the robustness of the registration. The parameter settings for the deformable registration are specified in Table 5.1. Refer to Wang et al. [16] for details about this deformable registration algorithm.

The deformable registration produced a deformation vector field (DVF) pointing from the atlas image to the test image. This DVF was used to deform the atlas image, generating a deformed image in the test image space that was used for the second phase of atlas selection. The DVF was also used to deform the atlas contours to the test image space. To achieve accurate contour deformation, the 2D slice stacks of atlas contours were first converted to a closed triangular mesh [17]. The DVF was interpolated to obtain a deformation vector at each vertex of the triangular mesh and the vertex was moved along the deformation vector to the new location to complete the deformation. After the deformation, the triangular mesh was converted back to a group of slice stacks of contours by cutting through the mesh using an axial plane, slice by slice, resulting in deformed contours in the test image space. This process is similar to the work presented in Lu et al. [18].

5.2.2.3   Second Phase of Atlas Selection

The second phase of atlas selection was performed on the basis of the aforementioned deformed images and contours. First, the union of the deformed contours was expanded by 1.0 mm using morphological dilation to define an irregular local region. A local-intensity histogram was created on the basis of the voxels in the local region for the test image and deformed image, respectively, as shown in Figure 5.3. This intensity histogram was created with 64 bins. The symmetric KL

TABLE 5.1

Parameter Settings for the Dual-Force Demons Deformable Image Registration

Parameter	Value
Number of bins for histogram equalization	256
Block size for histogram equalization	20
Multi-resolution levels	6
Number of iterations	200
Upper bound of step size	1.25
Gaussian variance for regularization	1.5

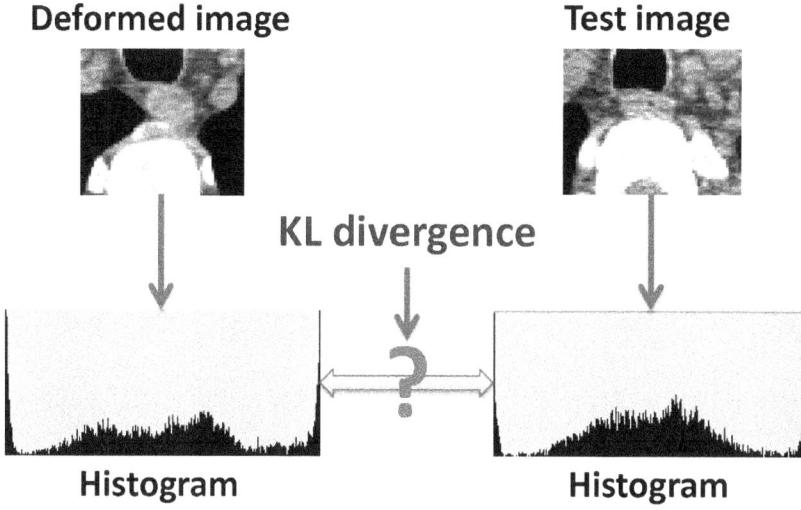

FIGURE 5.3 Illustration of using the Kullback-Leibler (KL) divergence to evaluate the similarity of the deformed image and the test image to rank the atlas in the second phase of atlas selection.

divergence was used to evaluate the similarity between the two histograms. Letting $P(i)$ denote the intensity histogram of the deformed image, where $i = 1,2,...,64$ indicates the histogram bins, and letting $Q(i)$ denote the intensity histogram of the test image, the symmetric KL divergence can be calculated as:

$$D_{KL}(P,Q) = \frac{1}{2}\left[\sum_i P(i)\ln\left(\frac{P(i)}{Q(i)}\right) + \sum_i Q(i)\ln\left(\frac{Q(i)}{P(i)}\right)\right].$$ (5.1)

The KL divergence was measured for each atlas selected in phase one and the measured KL divergence values were used to rank the atlases from the most to the least similar.

After ranking the atlases, the overlap ratio of the deformed contours was examined by sequentially overlapping atlases from the most to the least similar. The Jaccard coefficient was used to compute the overlap ratio. Letting A_i denote the volume of the ith deformed contour for $i = 1,2,...,N$, with N being the number of atlases selected in phase one, the Jaccard coefficient for a number of n overlapped deformed contours is defined as:

$$\text{Jaccard}(n) = \frac{\bigcup_{i=1}^{n} A_i}{\bigcap_{i=1}^{n} A_i}, n = 2,3,...,N.$$ (5.2)

The values of the overlap ratio were plotted versus the number of atlases, as shown in Figure 5.4. The overlap ratio curve was analyzed to further determine a subset of optimal atlases. The fundamental assumption of atlas selection in this step is that the deformed contours from different atlases should be consistent for the later contour fusion. If a less similar atlas produces a deformed contour that is largely off from the contours produced by more similar atlases, this deformed contour should not be used for contour fusion. On the basis of this principle, the following atlas selection rules were developed:

1) If Jaccard (n_0) ≥ 0.5, select at least a number of n_0 top-ranked atlases.
2) If Jaccard (n) < 0.2 for any $n > n_1$, discard all atlases ranked beyond n_1.

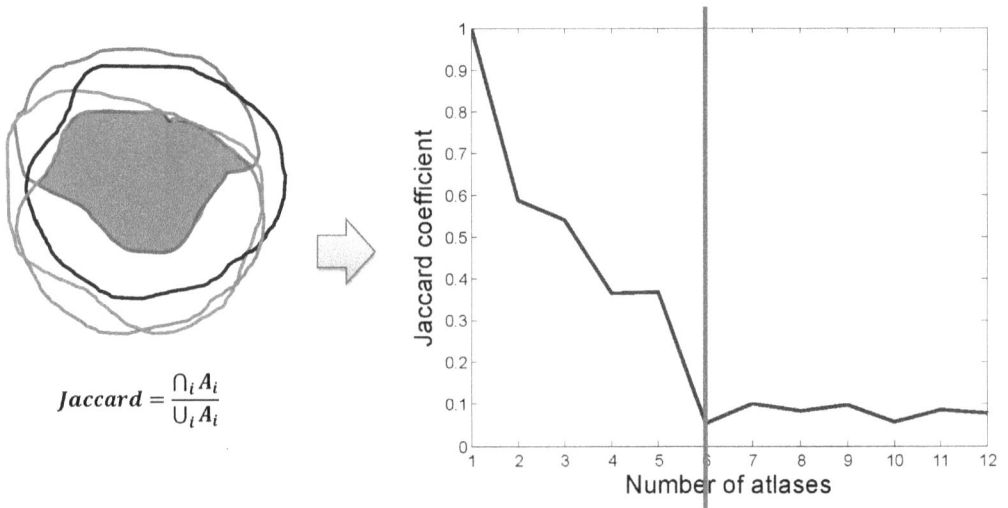

$$Jaccard = \frac{\cap_i A_i}{\cup_i A_i}$$

FIGURE 5.4 The Jaccard coefficient was used as the overlap ratio, which is defined as the intersection divided by the union of all contours. The overlap ratio was plotted as a function of the number of atlases. This example shows that using more than five atlases may not benefit multi-atlas segmentation because the overlap ratio reaches a small value.

3) Detect continuous steep slopes of the overlap ratio curve. If (a) [Jaccard ($n_2 - 1$) − Jaccard (n_2)] × 10 > tan(30°), (b) [Jaccard (n_2) − Jaccard ($n_2 + 1$)] × 5 > tan(30°), and (c) Jaccard ($n_2 + 1$) < 0.3, discard all atlases ranked beyond n_2.
4) Keep at least five atlases for contour fusion. This rule overrides the above rules.

Using these rules, a subset of optimal atlases was further identified from the atlases selected in phase one. In the example shown in Figure 5.4, the analysis selected five optimal atlases for contour fusion. The small overlap ratio values beyond the rank of five implies that those atlases are not consistent with the most similar atlas. This inconsistency can possibly deteriorate the contour fusion process.

5.2.2.4   Contour Fusion

The deformed contours of the selected optimal atlases were fused to produce the final segmentation. The contour fusion process took advantage of the STAPLE algorithm [7]. The STAPLE algorithm is based on the maximum likelihood estimates of the true positive and false negative of individual segmentations. It estimated the best agreement among individual segmentations and produced a fusion contour that was the expected truth of the segmentation. The fusion process allows the minimization of variations among segmentations obtained from different atlases and random errors in deformable image registration. However, the STAPLE algorithm does not consider the intensity information of the tissue being segmented. Its performance depends solely on the individual segmentation performances. To address this issue, a tissue appearance model (TAM) was created and integrated into the STAPLE fusion process. Specifically, the STAPLE algorithm requires providing the prior probability of true segmentation T_i for each voxel i when initializing the algorithm. In the implementation, the prior probability was initialized as:

$$f\left(T_i = 1\right) = \left(1 - w\right)R\left(i\right) + wS\left(i\right),\tag{5.3}$$

where $T_i = 1$ means voxel i belonging to the structure, $R(i)$ is the probability of voxel i belonging to the structure determined by individual segmentations, and $S(i)$ is the probability of voxel i

belonging to the structure determined by a TAM. The parameter w is the weight to balance these two terms. The $R(i)$ is defined as:

$$R(i) = \frac{\sum_j D_{ij}}{\sum_j 1},$$

(5.4)

where D_{ij} denotes the jth segmentation at voxel i that does ($D_{ij} = 1$) or does not belong to the structure ($D_{ij} = 0$). The TAM is a Gaussian model estimated from the test image as:

$$S(i) = \frac{1}{Z} \exp\left[-\frac{\left(I(i) - \mu_p\right)^2}{\sigma_p^2} \right],$$

(5.5)

where $I(i)$ is the intensity value of voxel i, Z is a normalization factor set as: $Z = \sum_i \exp\left[-\frac{\left(I(i) - \mu_p\right)^2}{\sigma_p^2} \right]$,

and the mean μ_p and variance σ_p^2 are estimated from voxels in the union region of individual segmentations. In the implementation, the weight was set at $w = 0.8$ for thoracic cancer patients to properly exclude the wrongly segmented lung tissue or high-contrast aorta tissue. For head and neck cancer patients, w was set at 0.5. A threshold of 0.5 was set to produce the final binary segmentation from the STAPLE fuzzy segmentation. If the number of atlases for contour fusion was less than six, the fusion approach was switched to majority voting with the TAM. Essentially, the binary thresholding was applied directly to the outcome of Equation (5.3).

5.2.3 EVALUATION METRICS

Auto-segmented contours were compared with manual contours using the Dice similarity coefficient (DSC), mean surface distance (MSD), Hausdorff distance (HD), and 95% HD, as the quantitative metrics. The DSC characterizes the agreement between two volumes encompassed by two contours. The MSD quantifies the average distance between the surfaces of two contours, while the HD measures the maximum distance between two contours. To cope with the sensitivity of outliers, 95% HD is often used instead of HD. Detailed definitions of these quantitative metrics can be found in Chapter 15. As mentioned in Chapter 15, difference in implementing the evaluation metrics may produce different results. To facilitate the inter-comparison with results from other chapters when their approaches were applied to the data from the 2017 AAPM Thoracic Auto-segmentation Challenge data [19], the evaluation tool provided in Chapter 15 was used to generate the evaluation metrics. Otherwise, the evaluation tool was implemented in-house.

5.2.4 ESOPHAGUS SEGMENTATION FOR HEAD AND NECK CANCER PATIENTS

The proposed multi-atlas segmentation with online atlas selection (MAS-AS) approach was used to delineate the esophagus contours in 21 head and neck cancer patients. A leave-one-out approach was used for validation. Each of the 21 patients was used as the test patient, one at a time in turn, while the remaining 20 patients were put in the atlas pool to delineate the test patient. The MAS-AS approach automatically determined a subset of optimal atlases from the 20 atlases for auto-segmentation. The auto-segmented contours were compared with the manual contours using the DSC and MSD metrics. The means and standard deviations of the values were calculated for the 21 tests. In addition, eight patients were randomly selected as a fixed set of atlases and multi-atlas segmentation (MAS) was performed for the remaining 13 patients. No atlas selection was performed, but the same rigid alignment, deformable image registration, and contour fusion described in Section 5.2.2 were used. The auto-segmentation results of these 13 patients were evaluated with DSC and MSD metrics and compared with those using the MAS-AS approach. The difference between the MAS-AS approach and the MAS approach was tested for statistical significance using a two-tailed, paired t-test, evaluated at the 0.05 significance level.

5.2.5 VALIDATION USING PUBLIC BENCHMARK DATASET

The MAS-AS was validated with the benchmark dataset used at the 2017 AAPM Thoracic Auto-segmentation Challenge. The 36 training cases were used as atlases and the 24 test cases were used for algorithm validation. In the phase 1 atlas selection, 12 optimal atlases were selected from the 36 atlases for further selection in phase 2. The entire process, including both phases of atlas selection and multi-atlas segmentation, was performed automatically. The auto-segmented contours were compared with the manual contours using DSC, MSD, and 95% HD using the evaluation tool implemented in Chapter 15 for inter-comparison with results from other chapters.

5.3 RESULTS

5.3.1 ATLAS RANKING AND SELECTION

One leave-one-out test for the head and neck cancer patients was used to illustrate the efficiency of atlas ranking and selection. The test patient had a large air pocket inside the esophagus, which made it appear different from most atlases. The first phase of atlas selection identified 12 atlases, and their contours were deformed to the test patient. These deformed contours demonstrate various agreement with the manual contour, as shown in Figure 5.5. In the second phase of atlas selection, the symmetric KL divergence of local-intensity histograms ranked atlases 12, 8, and 10 as the first, second, and third best atlases, which was consistent with the observation, as shown in Figure 5.5. The overlap ratio analysis for the 12 atlases is shown in Figure 5.4. In this analysis, a subset of five optimal atlases was further identified for contour fusion on the basis of the atlas selection rules. By comparing the auto-segmented contour generated from the five optimal atlases with the manual contour, a DSC of 0.67 and a MSD of 2.3 mm was found; the segmentation generated from the 12 atlases, without performing a second phase of atlas selection, had values of 0.31 mm and 4.0 mm, respectively. A subjective comparison is illustrated in Figure 5.6. It shows much better agreement between the auto-segmented contour and the manual contour for the approach using atlas selection than for the approach without atlas selection.

5.3.2 ESOPHAGUS SEGMENTATION FOR HEAD AND NECK CANCER PATIENTS

Table 5.2 shows the results of the leave-one-out tests of the proposed MAS-AS approach compared with the MAS approach, as applied to 21 head and neck cancer patients. The MAS-AS approach resulted in a mean DSC of 0.70 ± 0.07 and a mean MSD of 1.9 mm $\pm$ 0.7 mm. For the MAS approach, patients 14–21 were used as a fixed set of atlases to delineate the esophagus for the first 13 patients. In these 13 tests, the MAS approach achieved a mean DSC of 0.55 ± 0.15 and a mean MSD of 2.5 mm $\pm$ 0.9 mm. The results for the first 13 patients were tested to determine whether a statistically significant difference existed between the MAS-AS and MAS approaches. Significantly different DSCs and MSDs were found for these two approaches (p-value of 0.005 for DSC and 0.022 for MSD). The mean DSC and mean MSD for the MAS-AS approach applied to the first 13 patients were 0.69 ± 0.08 and 2.0 mm $\pm$ 0.7 mm, respectively. These results were better than those of MAS approach. The DSCs and MSDs for patients 2, 6, 9, 10, and 11 showed significant improvement with the MAS-AS approach. However, it was observed that the MAS-AS approach was slightly inferior to the MAS approach in some patients in terms of the DSC and MSD evaluation. In addition, the

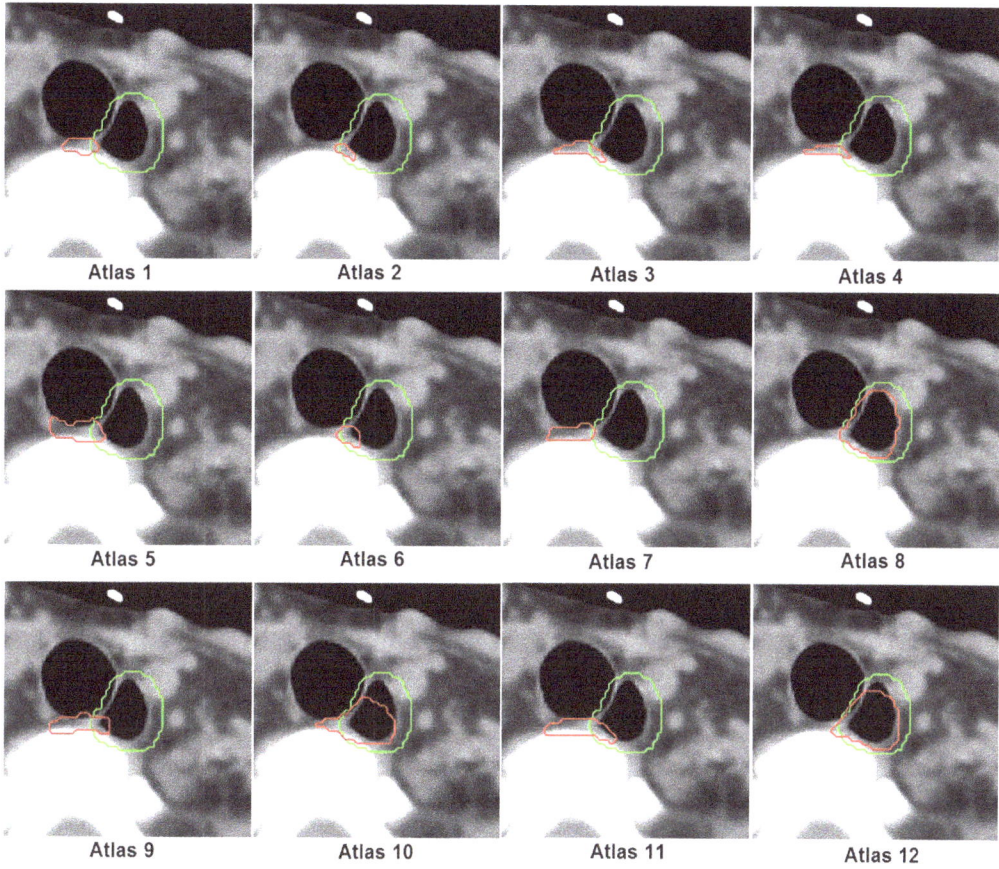

FIGURE 5.5 Esophagus contours of 12 atlases were deformed to the test image. The red and green colors indicate deformed and manual contours, respectively. In the second phase of atlas selection, atlases 12, 8, and 10 were chosen as the first, second, and third best atlases, which is consistent with the observation; they have a better agreement between the manual contour and the deform contour than other atlases do.

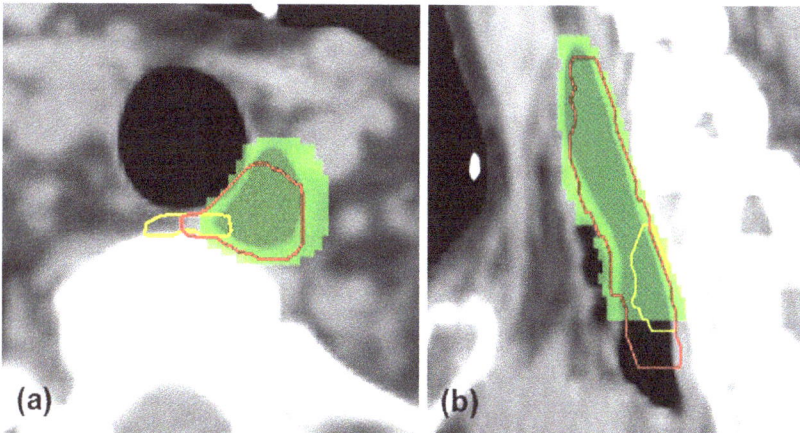

FIGURE 5.6 Comparison of auto-segmentation using five selected optimal atlases (red contour) versus that using 12 atlases without the second phase of atlas selection (yellow contour). Green colorwash indicates the manual contour. (a) Axial view. (b) Sagittal view.

TABLE 5.2

Segmentation Results of the Proposed Multi-Atlas Segmentation with the Online Atlas Selection (MAS-AS) Approach Compared with the Multi-Atlas Segmentation with a Fixed Set of Atlases (MAS)

Patient No.	Dice Similarity Coefficient			Mean Surface Distance (mm)		
	MAS-AS	MAS	Single Atlas	MAS-AS	MAS	Single Atlas
1	0.61	0.56	0.61	2.1	2.7	2.1
2	0.67	0.41	0.70	2.4	3.2	2.1
3	0.65	0.56	0.61	2.1	2.2	2.0
4	0.77	0.78	0.69	1.6	1.5	3.0
5	0.56	0.40	0.63	3.7	3.9	3.1
6	0.78	0.49	0.71	1.1	1.5	1.5
7	0.74	0.74	0.71	2.1	2.1	2.2
8	0.61	0.63	0.60	2.9	2.6	2.0
9	0.76	0.54	0.68	1.4	2.2	2.1
10	0.67	0.22	0.72	2.3	4.5	1.9
11	0.72	0.60	0.74	1.5	2.1	1.4
12	0.59	0.53	0.64	1.9	2.3	1.5
13	0.77	0.73	0.91	1.3	1.4	0.6
14	0.81	-	0.74	1.3	-	1.8
15	0.72	-	0.73	1.6	-	1.7
16	0.65	-	0.56	2.6	-	3.4
17	0.70	-	0.67	1.7	-	1.7
18	0.79	-	0.78	1.6	-	1.3
19	0.69	-	0.44	1.8	-	3.5
20	0.78	-	0.70	1.1	-	1.7
21	0.65	-	0.57	2.7	-	3.3
Median	0.70	0.56	0.69	1.8	2.2	2.0
Mean	0.70	0.55	0.67	1.9	2.5	2.1
SD	0.07	0.15	0.09	0.7	0.9	0.8

Note: Approach in 21 head and neck cancer patients. The column of "Single Atlas" reported the best result from a single atlas-based segmentation, which serves for the validation of deformable registration algorithm used in MAS-AS. The MAS approach used patients 14–21 as atlases to delineate the esophagus for the first 13 patients. SD = Standard deviation.

best overlap value from a single atlas in order to validate the deformable registration algorithm used in the proposed MAS-AS approach is reported in Table 5.2. The overlap values were measured between the deformed contour from each atlas and the best results were reported. These results showed reasonable agreements between the manual contour and the deformed contour thus justifying the use of the deformable registration algorithm.

5.3.3 VALIDATION WITH PUBLIC BENCHMARK DATASET

Table 5.3 shows the results of the MAS-AS approach validated on the 24 test cases from the 2017 AAPM Thoracic Auto-segmentation Challenge dataset, which includes contour data of the esophagus, spinal cord, left lung, right lung, and heart. For the esophagus, the DSC ranged from 0.31 to 0.83 with a median of 0.69 and an average of 0.66 ± 0.12. The MSD values ranged from 1.24 mm to 6.11 mm with a median of 2.21 mm and an average of 2.59 mm $\pm$ 1.28 mm.

TABLE 5.3

Segmentation Results of the Proposed Multi-Atlas Segmentation Method Evaluated on 24 Thoracic Cancer Patients Used in the 2017 AAPM Thoracic Auto-segmentation Challenge

Patient No.	Esophagus			Spinal Cord		
	DSC	MSD	HD95	DSC	MSD	HD95
1	0.50	3.88	15.54	0.86	0.79	2.42
2	0.60	2.65	12.68	0.91	0.58	2.02
3	0.71	1.96	6.57	0.83	0.95	2.95
4	0.31	5.94	21.15	0.88	0.85	2.37
5	0.74	1.65	6.23	0.80	0.80	2.07
6	0.76	1.28	4.34	0.84	0.89	2.67
7	0.68	2.46	11.42	0.81	1.23	3.31
8	0.79	1.35	4.35	0.80	1.00	2.98
9	0.63	2.31	6.45	0.63	0.64	1.76
10	0.74	1.79	5.41	0.82	0.88	2.65
11	0.65	2.19	9.41	0.75	0.95	2.47
12	0.60	2.98	10.42	0.80	0.74	2.42
13	0.63	3.10	21.12	0.89	0.54	1.52
14	0.76	1.49	4.44	0.85	0.81	3.19
15	0.70	2.24	10.22	0.78	0.65	2.00
16	0.74	1.77	7.07	0.87	0.85	3.18
17	0.50	6.11	28.68	0.85	0.57	2.00
18	0.71	2.10	11.85	0.86	0.72	2.26
19	0.83	1.24	3.90	0.87	0.72	1.95
20	0.66	2.89	15.66	0.85	0.90	2.94
21	0.75	1.79	5.49	0.69	0.78	2.01
22	0.74	2.15	7.58	0.79	0.74	1.98
23	0.55	3.35	14.11	0.70	0.96	2.45
24	0.60	3.44	15.09	0.71	0.82	2.33
Median	0.69	2.21	9.81	0.83	0.80	2.40
Mean	0.66	2.59	10.80	0.81	0.81	2.41
SD	0.12	1.28	6.35	0.07	0.16	0.49

(*Continued*)

TABLE 5.3 (CONTINUED)

Segmentation Results of the Proposed Multi-Atlas Segmentation Method Evaluated on 24 Thoracic Cancer Patients Used in the 2017 AAPM Thoracic Auto-segmentation Challenge

Patient No.	Left Lung			Right Lung			Heart		
	DSC	MSD	HD95	DSC	MSD	HD95	DSC	MSD	HD95
1	0.97	0.95	3.17	0.96	1.47	13.55	0.94	2.60	6.84
2	0.95	1.59	11.38	0.96	1.74	9.90	0.90	3.98	15.70
3	0.97	1.41	9.79	0.97	2.15	18.28	0.91	3.48	12.43
4	0.98	1.20	6.16	0.98	0.80	2.75	0.91	3.71	14.30
5	0.90	3.45	28.40	0.97	1.24	3.96	0.91	3.55	10.87
6	0.93	3.12	19.08	0.96	2.17	9.73	0.94	1.77	5.15
7	0.94	2.60	11.16	0.87	4.30	34.98	0.90	3.41	10.66
8	0.96	1.64	7.08	0.96	1.75	6.78	0.90	3.83	11.48
9	0.95	2.41	7.96	0.92	3.66	24.74	0.91	4.00	9.68
10	0.98	1.11	5.83	0.98	0.78	2.98	0.91	3.54	12.35
11	0.81	6.15	51.27	0.95	2.42	10.52	0.91	4.27	11.07
12	0.97	2.00	9.91	0.97	1.21	6.81	0.90	4.73	14.19
13	0.98	1.00	6.88	0.98	1.30	9.07	0.92	2.70	10.04
14	0.97	1.12	4.42	0.98	1.09	5.71	0.91	3.68	10.09
15	0.97	1.34	10.39	0.97	1.71	15.36	0.92	3.58	11.98
16	0.97	1.63	12.89	0.97	1.91	18.15	0.91	3.01	9.59
17	0.95	2.85	10.95	0.95	2.20	12.54	0.88	4.26	16.79
18	0.95	2.08	10.85	0.93	4.91	28.32	0.88	5.35	31.39
19	0.96	1.53	6.07	0.91	2.63	18.13	0.90	3.19	8.98
20	0.95	2.45	12.31	0.87	3.81	23.77	0.89	3.62	15.52
21	0.79	6.38	35.92	0.97	1.83	6.34	0.82	7.77	31.77
22	0.95	2.53	18.62	0.96	2.84	18.30	0.87	5.11	18.82
23	0.76	9.12	86.43	0.97	1.47	7.17	0.89	5.19	24.79
24	0.98	1.16	6.31	0.97	1.83	17.42	0.95	2.09	6.69
Median	0.96	1.82	10.62	0.96	1.83	11.53	0.91	3.65	11.73
Mean	0.94	2.53	16.38	0.95	2.13	13.55	0.90	3.85	13.80
SD	0.06	2.00	18.55	0.03	1.08	8.44	0.03	1.22	6.89

Note: Results are shown for the esophagus, spinal cord, left lung, right lung, and heart. DSC = Dice similarity coefficient, MSD = Mean surface distance (mm), HD95 = 95% Hausdorff distance, SD = Standard deviation.

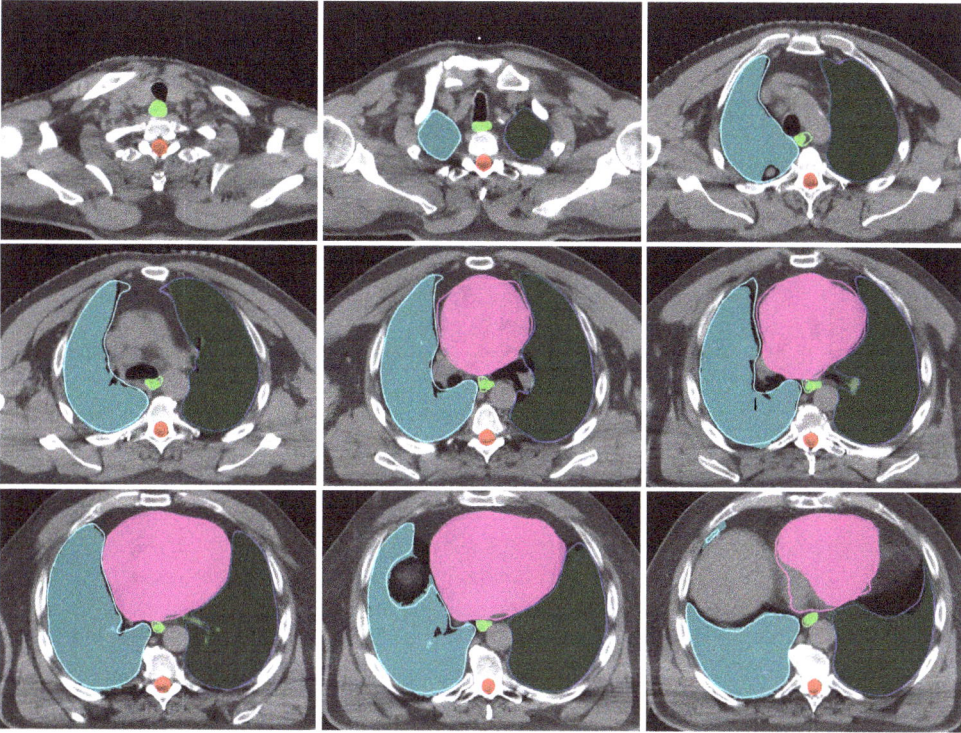

FIGURE 5.7 Illustration of auto-segmented contours (lines) compared with manual contours (colorwash) for one thoracic patient from the public benchmark dataset in several slices. The following organs are included: left lung (dark green), right lung (cyan), heart (magenta), esophagus (green), and spinal cord (red).

The 95% HD values ranged from 3.90 mm to 28.68 mm with a median of 9.81 mm and an average of 10.80 mm ± 6.35 mm.

For the spinal cord, the DSC ranged from 0.63 to 0.91 with a median of 0.83 and an average of 0.81 ± 0.07. The MSD values ranged from 0.54 mm to 1.23 mm with a median of 0.80 mm and an average of 0.81 mm ± 0.16 mm. The 95% HD values ranged from 1.52 mm to 3.31 mm with a median of 2.40 mm and an average of 2.41 mm ± 0.49 mm.

For the right lung, the DSC ranged from 0.87 to 0.98 with a median of 0.96 and an average of 0.95 ± 0.03. The MSD values ranged from 0.78 mm to 4.91 mm with a median of 1.83 mm and an average of 2.13 mm ± 1.08 mm. The 95% HD values ranged from 2.75 mm to 34.98 mm with a median of 11.53 mm and an average of 13.55 mm ± 8.44 mm. For the left lung, the DSC ranged from 0.76 to 0.98 with a median of 0.96 and an average of 0.94 ± 0.06. The MSD values ranged from 0.95 mm to 9.12 mm with a median of 1.82 mm and an average of 2.53 mm ± 2.00 mm. The 95% HD values ranged from 3.17 mm to 86.43 mm with a median of 10.62 mm and an average of 16.38 mm ± 18.55 mm. The large HD value was mainly due to the tumor exclusion. Tumors were excluded from the manual lung contours, while atlas-based segmentation, in general, is not able to handle this exclusion correctly due to different tumor locations on different patients.

For the heart, the DSC ranged from 0.82 to 0.95 with a median of 0.91 and an average of 0.90 ± 0.03. The MSD values ranged from 1.77 mm to 7.77 mm with a median of 3.65 mm and an average of 3.85 mm ± 1.22 mm. The 95% HD values ranged from 5.15 mm to 31.77 mm with a median of 11.73 mm and an average of 13.80 mm ± 6.89 mm.

Figure 5.7 illustrates the comparison of the auto-segmented contours with the manual contour for one patient in several slices. Good agreement was observed for all the structures under consideration, though with the esophagus having more variations than other structures.

5.4 DISCUSSION

In this study, an atlas ranking and selection approach to be used together with multi-atlas segmentation for the automatic delineation of organs-at-risk from CT scans was evaluated. This approach selects a subset of optimal atlas candidates from an atlas pool for multi-atlas segmentation on the basis of local anatomy, thus potentially reducing the adverse impact of inter-subject variability on multi-atlas segmentation and improving the segmentation accuracy. A STAPLE-based contour fusion approach by including the previous image was used to improve the fusion performance. This multi-atlas segmentation system was evaluated on head and neck cancer patients for esophagus segmentation and thoracic cancer patients for segmentation of heart, lungs spinal cord, and esophagus from a public benchmark dataset. The findings may have a positive impact on contouring for radiation treatment planning by saving clinicians' time and improving contouring efficiency and consistency [20–22].

In the implementation, the segmentation was run on a Windows 7-based PC with an 8-core Intel Core i7 3.4-GHz CPU and 8 GB of memory. The deformable registration was performed independently on a Windows server with an 8-core 3-GHz Intel Xeon CPU and 8 GB of memory. Multithread computing was enabled in the deformable registration algorithm, and two registration tasks were allowed to be run simultaneously on the server. Each segmentation task required around five minutes to complete in a thoracic cancer patient and around three minutes in a head and neck cancer patient, with about half the time spent on deformable registration.

The study showed that online atlas selection improved overall multi-atlas segmentation. The improvement in several poor cases was significant. For example, the MAS-AS results for patients 2, 6, 9, 10, and 11 in the head and neck cancer dataset showed significant improvement over the MAS results, as shown in Table 5.2. However, using online atlas selection did not result in improvement in every case. For example, the MAS-AS results for patients 4 and 8 were slightly inferior to the MAS results (Table 5.2). Nevertheless, online atlas selection improved the robustness of multi-atlas segmentation by preventing a worst-case scenario.

The cases showing inferior results for MAS-AS indicate that the atlas selection was not perfect. This was mostly due to the imperfect similarity metrics used to rank the atlases. For example, the cross-correlation coefficient used in the first phase of atlas selection is shift- and rotation-variant and is thus sensitive to misalignment. In addition, in rigid alignment, the rotation was not corrected. The rotation variance may play down those atlases having similar local anatomy but different scanning positions, such as different neck flexion in the scanning setup for head and neck cancer patients.

In the second phase of atlas selection, the registration error may be a major obstacle for atlas selection. The histogram-based KL divergence metric was used to reduce the impact from the imperfect deformable registration. The histogram was created from a local region for calculating the KL divergence. Choosing the proper size of this region is critical to obtaining a correct atlas ranking. A larger region may better counteract the registration error but reduce atlas selection efficacy because unrelated anatomy may be included. In the study, it was found that the local region, as defined by the union of the deformed contours, expanded by 1 mm, gave the best results in most cases, but it is acknowledged that using a different size for the local region might result in better atlas selection in different cases.

The study also found that the presence of air bubbles is one of the major obstacles in atlas-based segmentation of the esophagus. Several poor head and neck cancer cases, such as patients 5 and 10, were caused by the presence of air bubbles. One potential solution to this issue is to detect air bubbles inside the esophagus [23]. Once the air bubbles are detected, they can be replaced with similar intensity values of esophagus in the CT image; the modified CT image will be used for multi-atlas segmentation. On the other hand, it was noticed that an air bubble can significantly change the similarity between the atlas and the test image in the atlas selection process. Performing atlas selection in local segments of the esophagus may resolve the air bubble issue. In addition, a similarity comparison using the entire long winding esophagus is not always locally accurate in selecting the atlas.

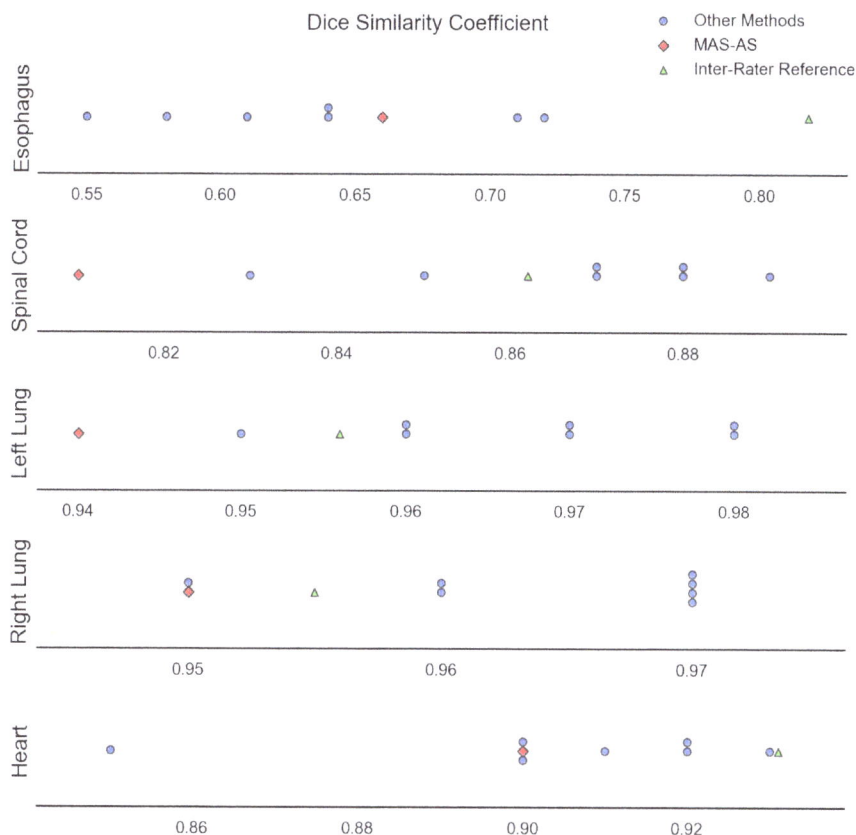

FIGURE 5.8 Comparison of the Dice similarity coefficient between different methods on the 2017 AAPM Thoracic Auto-segmentation Challenge benchmark dataset for the esophagus, spinal cord, lung, and heart. Red dot shows the results of MAS-AS on the whole dataset (24 cases), blue dots show the result of other methods on the live dataset (12 cases), and green dots show the variability of human operators on three select cases. Numbers are taken from Yang et al. [26].

In thoracic cancer patients, delineating the esophagus segment near the heart is most challenging for atlas-based segmentation. Neighboring structures such as the heart, lungs, and aorta may push or pull the esophagus, resulting in different shapes and locations from day to day and from patient to patient. In addition, the low contrast between the esophagus and the surrounding tissues makes it difficult to perform auto-segmentation. The difficulty of delineating the esophagus was further verified on the thoracic benchmark dataset. Compared to the lungs, heart, and spinal cord, the performance of MAS-AS resulted in worse DSC and MSD metrics. While automatic segmentation methods could perform close to inter-observer variability for the lungs, heart, and spinal cord, it seems that more work is needed in developing these methods to successfully segment the esophagus. One possible improvement for the MAS-AS is using nearby structures as a constraint for segmentation to assist esophagus segmentation [24, 25].

Although the atlas selection scheme presented here represents an improvement over traditional MAS schemes, comparison of MAS-AS with other available schemes on the thoracic benchmark dataset shows that MAS-AS did not perform better but similarly to the current state-of-the-art approaches. Other methods evaluating the thoracic benchmark dataset, which consisted of a combination of deep learning and other atlas-based methods, had a mean DSC ranging from 0.55 to 0.72 and a mean MSD ranging from 2.03 mm to 13.10 mm for the esophagus [26]. The DSC and MSD from the MAS-AS procedure fall within these ranges (Figures 5.8 and 5.9). Of note, the statistics

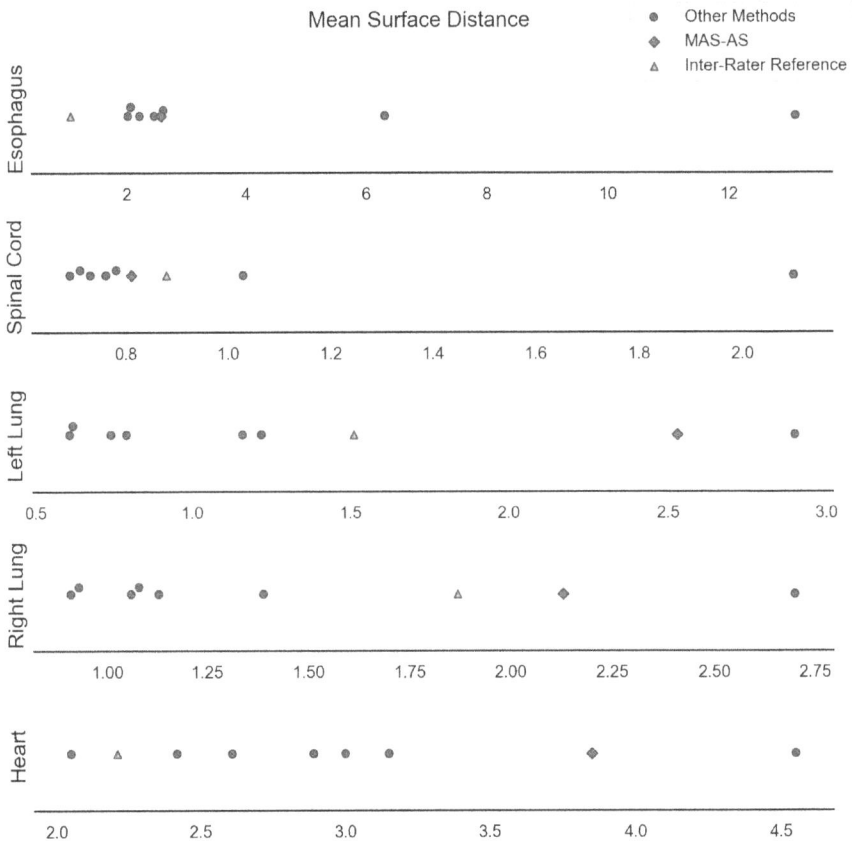

FIGURE 5.9 Comparison of the mean surface distance between different methods on the 2017 AAPM Thoracic Auto-segmentation Challenge benchmark dataset for the esophagus, spinal cord, lung, and heart. The diamond shows the results of MAS-AS on the whole dataset (24 cases), the circles show the result of other methods on the live dataset (12 cases), and the triangles show the variability of human operators on three select cases. Numbers are taken from Yang et al. [26].

from the other methods are from the online test cases, which only included 12 of the 24 cases in this analysis as offline test cases were not included. Even if the above analysis for MAS-AS is restricted to only the 12 online test cases, the conclusions remain the same. Another important comparison is with the inter-rater reference. Multiple experts contoured the three cases in the benchmark dataset and DSC and MSD were computed as the inter-rater reference. Relative to the ground truth, the mean DSC and MSD from three different raters on the three cases were 0.82 ± 0.04 and 1.07 mm $\pm$ 0.25 mm [26]. The MAS-AS approach, as well as the other methods participating in the 2017 AAPM Thoracic Auto-segmentation Challenge, performed inferiorly compared to inter-observer performance. This suggests that progress in MAS-AS and other auto-contouring methods are needed before these methods can perform equally as well as experts for esophagus delineation.

Similar findings regarding the effectiveness of MAS-AS on segmentation of the lung, heart, and spinal cord were also observed. The main difference in segmentation of these organs and the esophagus was that automatic segmentation methods compared more favorably to the inter-observer performance and occasionally generated even better metrics. The mean MSD and DSC for MAS-AS were within the range of values found from other methods with the exception that the DSC for MAS-AS was 0.01 lower than any value for the left lung and 0.02 lower than any other value for the spinal cord. However, these differences are small and the mean DSC of MAS-AS for these two organs was usually within one standard deviation of other methods [26]. Additionally, the DSC values generated by MAS-AS are comparable with those reported in literature for these organs-at-risk [27–30]. The mean inter-observer DSC was 0.96 ± 0.02 for both the left and right lung, 0.93 ± 0.02 for the heart, and 0.86 ± 0.04 for the spinal cord. In respect to MSD, the inter-observer performance was 1.51 mm ± 0.67 mm for the right lung, 1.87 mm ± 0.87 mm for the left lung, 2.21 mm ± 0.59 mm for the heart, and 0.88 mm ± 0.23 mm for the spinal cord [26]. For both the left and right lungs, the measured DSC and MSD of the MAS-AS method was within one standard deviation of the inter-observer performance. For the heart, the DSC was within two standard deviations of the inter-observer performance but the MSD was not. For the spinal cord, the DSC was within two standard deviations of inter-observer performance and the MSD was within one standard deviation. The above indicates that MAS-AS may generate contours comparable to experts but they are still inferior as an aggregate. One explanation is that while MAS-AS often generates acceptable contours, it is possible to make errors in some cases. One piece of data supporting this is that the median evaluation was superior to the mean for both DSC and MSD for every organ on the benchmark dataset.

Although the results of MAS-AS varied slightly between the three sites from which the benchmark dataset was collected, differences between the sites did not serve as the major determinant of performance. For example, DSC ranged from 0.50 to 0.83 with an average of 0.71 ± 0.10 for the esophagus at the best of the three sites in the dataset. At the worst site, DSC ranged from 0.31 to 0.76 with an average of 0.62 ± 0.15. At the third site, DSC ranged from 0.55 to 0.75 with an average of 0.66 ± 0.08. Although the performance of MAC-AS may slightly depend on site-specific scanning procedures, the above suggests that MAS-AS is robust across several different site-specific scanning protocols.

Finally, the way of drawing manual contours might offset the evaluation of segmentation results. The manual contours were drawn in axial 2D slices, which created sharp edges at the top and bottom slices of the esophagus and spinal cord when they were viewed in 3D. However, the contour deformation was processed using a 3D mesh, which necessarily processed the two ends of esophagus or spinal cord to make it smooth. This may have created a discrepancy between auto-segmented and manual contours. Disagreement was found, in most slices, at the two ends. In addition, in head and neck cancer patients, the manual esophagus contours of the inferior stopping slice may not have been consistent in different patients if the patients were positioned with different neck flexion. This potentially reduced the segmentation accuracy when these contours were used for evaluation.

5.5 CONCLUSIONS

Selecting a subset of optimal atlas candidates using local anatomy similarity improves multi-atlas segmentation. The online atlas selection approach improved the robustness of multi-atlas segmentation of thoracic organs including the esophagus, spinal cord, lung, and heart from CT scans. However, improvement in the robustness of such a method is needed to perform at the level of human experts. For example, exploring alternate image features such as the image texture and shape features that are commonly used for content-based image retrieval [31–34] for atlas ranking and selection is one possibility. Another possible improvement could be to use deep learning to measure the similarity between atlases and test images.

REFERENCES

1. Chen, A, et al., Evaluation of multiple-atlas-based strategies for segmentation of the thyroid gland in head and neck. *Physics in Medicine and Biology*, 2012. **57**(1): pp. 93–111.
2. Klein, S, et al., Automatic segmentation of the prostate in 3D MR images by atlas matching using localized mutual information. *Medical Physics*, 2008. **35**(4): pp. 1407–1417.
3. Yang, J, et al., Automatic contouring of brachial plexus using a multi-atlas approach for lung cancer radiation therapy. *Practical Radiation Oncology*, 2013. **3**(4): pp. e139–e147.

4. Sjoberg, C, et al., Clinical evaluation of multi-atlas based segmentation of lymph node regions in head and neck and prostate cancer patients. *Radiation Oncology*, 2013. **8**: p. 229.

5. Kirisli, HA, et al., Evaluation of a multi-atlas based method for segmentation of cardiac CTA data: a large-scale, multicenter, and multivendor study. *Medical Physics*, 2010. **37**(12): pp. 6279–6291.

6. Isgum, I, et al., Multi-atlas-based segmentation with local decision fusion – application to cardiac and aortic segmentation in CT scans. *IEEE Transactions on Medical Imaging*, 2009. **28**(7): pp. 1000–1010.

7. Warfield, SK, KH Zou, and WM Wells, Simultaneous truth and performance level estimation (STAPLE): an algorithm for the validation of image segmentation. *IEEE Transactions on Medical Imaging*, 2004. **23**(7): pp. 903–921.

8. Langerak, TR, et al., Label fusion in atlas-based segmentation using a selective and iterative method for performance level estimation (SIMPLE). *IEEE Transactions on Medical Imaging*, 2010. **29**(12): pp. 2000–2008.

9. Sabuncu, MR, et al., A generative model for image segmentation based on label fusion. *IEEE Transactions on Medical Imaging*, 2010. **29**(10): pp. 1714–1729.

10. Ramus, L and G Malandain, Multi-atlas based segmentation: application to the head and neck region for radiotherapy planning, in *Medical Image Analysis for the Clinic: A Grand Challenge*, B vanGinneken, et al., Editors. 2010, CreateSpace Independent Publishing Platform. pp. 281–288. www.amazon.com/Medical-Image-Analysis-Clinic-Challenge/dp/1453759395

11. Aljabar, P, et al. *Classifier Selection Strategies for Label Fusion Using Large Atlas Databases*. Berlin: Springer Berlin Heidelberg, 2007.

12. Iglesias, JE and MR Sabuncu, Multi-atlas segmentation of biomedical images: a survey. *Medical Image Analysis*, 2015. **24**(1): pp. 205–219.

13. Yang, J, et al., Atlas ranking and selection for automatic segmentation of the esophagus from CT scans. *Physics in Medicine and Biology*, 2017. **62**(23): pp. 9140–9158.

14. Forsythe, GE, MA Malcolm, and CB Moler, *Computer Methods for Mathematical Computations*. Englewood Cliffs, NJ: Prentice Hall, 1976.

15. Zhou, R, et al., Cardiac atlas development and validation for automatic segmentation of cardiac substructures. *Radiotherapy and Oncology*, 2017. **122**(1): pp. 66–71.

16. Wang, H, et al., Implementation and validation of a three-dimensional deformable registration algorithm for targeted prostate cancer radiotherapy. *International Journal of Radiation Oncology*Biology*Physics*, 2005. **61**(3): pp. 725–735.

17. Yang, J, et al., Automatic segmentation of parotids from CT scans using multiple atlases, in *Medical Image Analysis for the Clinic: A Grand Challenge*, B van Ginneken, et al., Editors. 2010, CreateSpace Independent Publishing Platform. pp. 323–330. www.amazon.com/Medical-Image-Analysis-Clinic-Challenge/dp/1453759395

18. Lu, WG, et al., Automatic re-contouring in 4D radiotherapy. *Physics in Medicine and Biology*, 2006. **51**(5): pp. 1077–1099.

19. Yang, J, et al., CT images with expert manual contours of thoracic cancer for benchmarking auto-segmentation accuracy. *Medical Physics*, 2020. **47**(7): pp. 3250–3255.

20. Yang, J, et al., A statistical modeling approach for evaluating auto-segmentation methods for image-guided radiotherapy. *Computerized and Medical Imaging Graphics*, 2012. **36**(6): pp. 492–500.

21. Chao, KSC, et al., Reduce in variation and improve efficiency of target volume delineation by a computer-assisted system using a deformable image registration approach. *International Journal of Radiation Oncology*Biology*Physics*, 2007. **68**(5): pp. 1512–1521.

22. Reed, VK, et al., Automatic segmentation of whole breast using atlas approach and deformable image registration. *International Journal of Radiation Oncology*Biology*Physics*, 2009. **73**(5): pp. 1493–1500.

23. Fieselmann, A, et al., Automatic detection of air holes inside the esophagus in CT images, in *Bildverarbeitung für die Medizin 2008*, T Tolxdorff, et al., Editors. 2008, Berlin: Springer. pp. 397–401.

24. Yang, J, LH Staib, and JS Duncan, Neighbor-constrained segmentation with level set based 3-D deformable models. *IEEE Transactions on Medical Imaging*, 2004. **23**(8): pp. 940–948.

25. Gao, Y, et al., A 3D interactive multi-object segmentation tool using local robust statistics driven active contours. *Medical Image Analysis*, 2012. **16**(6): pp. 1216–1227.

26. Yang, J, et al., Autosegmentation for thoracic radiation treatment planning: a grand challenge at AAPM 2017. *Medical Physics*, 2018. **45**(10): pp. 4568–4581.

27. Horsfield, MA, et al., Rapid semi-automatic segmentation of the spinal cord from magnetic resonance images: application in multiple sclerosis. *Neuroimage*, 2010. **50**(2): pp. 446–455.

28. Kohlberger, T, et al., Automatic multi-organ segmentation using learning-based segmentation and level set optimization. *Medical Image Computing and Computer Assisted Intervention*, 2011. **14**(Pt 3): pp. 338–345.

29. Bai, W, et al., Multi-atlas segmentation with augmented features for cardiac MR images. *Medical Image Analysis*, 2015. **19**(1): pp. 98–109.

30. Feulner, J, et al., A probabilistic model for automatic segmentation of the esophagus in 3-D CT scans. *IEEE Transactions on Medical Imaging*, 2011. **30**(6): pp. 1252–1264.

31. Newsam, SD and C Kamath, Comparing shape and texture features for pattern recognition in simulation data. *Proceedings of the. SPIE 5672, Image Processing: Algorithms and Systems IV*, 2005: pp. 106–117.

32. Howarth, P and S Rüger, Evaluation of texture features for content-based image retrieval, in *Proceedings of the International Conference on Image and Video Retrieval*, P Enser, et al., Editors. 2004, Berlin: Springer. pp. 326–334.

33. Manjunath, BS and WY Ma, Texture features for browsing and retrieval of image data. *IEEE Transactions on Pattern Analysis and Machine Intelligence*, 1996. **18**(8): pp. 837–842.

34. Yang, J, et al., Diffusion tensor image registration using tensor geometry and orientation features. *Medical Image Computing and Computer-Assisted Intervention – MICCAI, Pt II, Proceedings*, 2008. 5242(Pt 2): pp. 905–913.

Part II

Deep Learning for Auto-Segmentation

6 Introduction to Deep Learning-Based Auto-Contouring for Radiotherapy

Mark J. Gooding

CONTENTS

6.1 INTRODUCTION

The 2017 AAPM Thoracic Auto-segmentation Challenge could be regarded as a turning point for auto-segmentation in radiation oncology. This was the first challenge in the domain where deep learning-based approaches were used, but not all entries exclusively used this method. All entries to the previous similar challenge in radiotherapy, at the 2015 conference on medical imaging computing and computer assisted intervention (MICCAI), either used atlas-based or model-based segmentation approaches, while in the subsequent challenge at AAPM 2019, all entries were deep learning-based. It was also in 2017 that the first journal publications using deep learning for organ-at-risk [1] and target volume [2–4] segmentation were published, although conference publications had preceded these [5, 6]. Following this, an early clinical validation of commercial systems was reported in a journal publication as early as 2018 [7], again preceded by conference publications [8, 9].

6.2 HISTORICAL CONTEXT

Given this rapid shift in focus to using deep learning for auto-segmentation in radiotherapy, one might be led to believe that this innovation was developed within the field of radiotherapy. However, a look at references of some of the earliest papers quickly leads back to the broader fields of medical imaging and computer vision. This innovation in auto-segmentation for radiation oncology is very much standing on the shoulders of giants. Figure 6.1, reproduced from the work of Wang and Raj [10], gives an indication of the significant steps taken in the development of deep learning.

With the focus of this book being on auto-segmentation, it is not the place of this chapter, or book, to provide a detailed history of deep learning or technical introduction. For historical context,

Table 1: Major milestones that will be covered in this paper

Year	Contributer	Contribution
300 BC	Aristotle	introduced Associationism, started the history of human's attempt to understand brain.
1873	Alexander Bain	introduced Neural Groupings as the earliest models of neural network, inspired Hebbian Learning Rule.
1943	McCulloch & Pitts	introduced MCP Model, which is considered as the ancestor of Artificial Neural Model.
1949	Donald Hebb	considered as the father of neural networks, introduced Hebbian Learning Rule, which lays the foundation of modern neural network.
1958	Frank Rosenblatt	introduced the first perceptron, which highly resembles modern perceptron.
1974	Paul Werbos	introduced Backpropagation
1980	Teuvo Kohonen	introduced Self Organizing Map
	Kunihiko Fukushima	introduced Neocogitron, which inspired Convolutional Neural Network
1982	John Hopfield	introduced Hopfield Network
1985	Hilton & Sejnowski	introduced Boltzmann Machine
1986	Paul Smolensky	introduced Harmonium, which is later known as Restricted Boltzmann Machine
	Michael I. Jordan	defined and introduced Recurrent Neural Network
1990	Yann LeCun	introduced LeNet, showed the possibility of deep neural networks in practice
1997	Schuster & Paliwal	introduced Bidirectional Recurrent Neural Network
	Hochreiter & Schmidhuber	introduced LSTM, solved the problem of vanishing gradient in recurrent neural networks
2006	Geoffrey Hinton	introduced Deep Belief Networks, also introduced layer-wise pretraining technique, opened current deep learning era.
2009	Salakhutdinov & Hinton	introduced Deep Boltzmann Machines
2012	Geoffrey Hinton	introduced Dropout, an efficient way of training neural networks

FIGURE 6.1 Key stages in the development of deep learning. Table reproduced from Wang and Raj [10]. Reproduced with permission.

readers should look to overview reviews of Wang and Raj [10] or of Schmidhuber [11], while for a preliminary introduction to artificial neural networks and deep learning, there are numerous papers (e.g. [12, 13]) and books available (e.g. [14, 15]). There are also a large number of online courses (e.g. [16, 17]) and videos available (e.g. [18, 19]). Nevertheless, it is worth pursuing a brief recap, in order to motivate an analysis of why the field has migrated to deep learning from atlas-based methods and to assess what weaknesses may remain.

6.2.1 ARTIFICIAL NEURAL NETWORKS

Deep learning traces its roots back to the work done in the 1950s and 1960s on artificial neurons. The concept of a mathematical model neuron contributing to a network behavior was first explored in 1943 by McCulloch and Pitts [20]. In their theoretical paper, they mathematically modeled each neuron as an "all-or-nothing" activation, exploring conceptually what this would mean for neurology/psychology. This model of a neuron could be implemented in simple binary logic but is limited in that any response to a stimulus becomes a simple logical algorithm rather than an attempt to accurately model biological behavior. In 1958, Rosenblatt proposed the "perceptron", in which neuron behavior was modeled probabilistically instead. Importantly, this model allowed for some element of learning whereby the response to stimulus varied with experience. This perceptron model was subsequently simulated in a computer, demonstrating machine learning [21], using what could be described as a neural network.

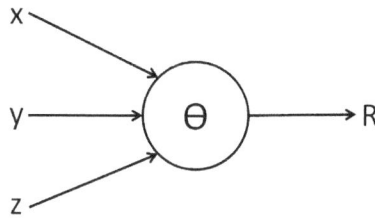

FIGURE 6.2 Model of an artificial neuron. The response R is determined based on whether the sum of the stimuli (here x, y, z) is greater than a threshold Θ.

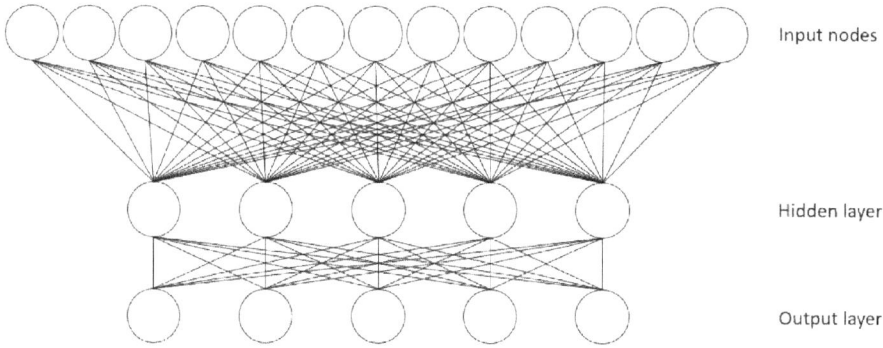

FIGURE 6.3 An example of the fully connected network architecture used in a study using neural networks to evaluate treatment plans [22]. In that study, 13 input nodes (treatment plan features) were used to predict a one-hot encoded score using five output nodes. Five hidden nodes were used, resulting in 90 weights to tune.

These artificial neurons follow a simple model, as illustrated in Figure 6.2; the response activation (R) is a binary output determined by whether the sum of all input stimuli (in this example three are shown, x, y, z; however, there could be more or less) exceed a threshold (Θ). Rosenblatt's model of each neuron has an activation response that is a binary model, 1 or 0; however, multiple neurons are combined into an "A-unit" that has a response that is the weighted sum of multiple neurons. With varying thresholds for different neurons, a weighted output system is generated.

Fast-forwarding to modern networks, the approach used in artificial neural networks does not differ much from this early model. Each neuron takes a weighted sum of its input stimuli and provides an output response based on an activation function. This activation function typically is some form of continuous threshold function, such as a sigmoid.

While there have been many notable contributions to the development of neural networks, as illustrated in Figure 6.1, two factors may be considered as critical to the success that has been achieved in auto-contouring: convolution and computation.

6.2.1.1 Convolution Neural Networks

A basic neural network design is to have every layer fully connected to the next, as illustrated in Figure 6.3. Such a network architecture works well for small numbers of inputs and a few hidden layers. However, a challenge arises with images in the number of potential inputs. A CT image, as used for radiotherapy treatment planning, is normally 512 pixels square. A planning volume may consist of 200 or so such 2D images. Thus, a 2D processing network would have 262,144 inputs to a network. Fully connecting such an input to a single hidden layer would result in 68.7 billion parameters (the input weights) to be tuned, making training intractable.

Convolutional networks recognize that "Distinctive features of an object can appear at various locations on the input image. Therefore it seems judicious to have a set of feature detectors that

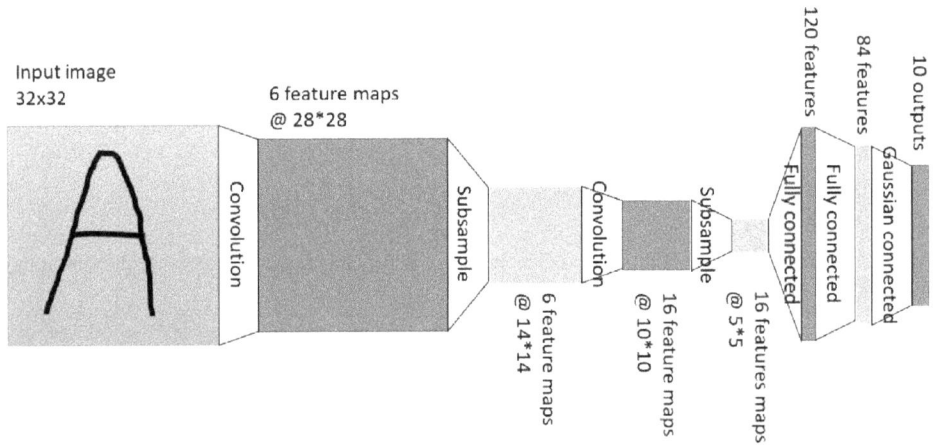

FIGURE 6.4 The network architecture introduced by LeCun et al. [24]. Multiple convolution layers were used to enable complex features to be learned, before fully connected layers were used to do character classification.

can detect a particular instance of a feature anywhere on the input plane" [23]. Therefore, the same weights are required throughout the image. This sharing of weights between connections is convolution [24]. This achieves two things; the number of parameters for each layer is vastly reduced since each layer only learns the required convolution kernels, and the network becomes less sensitive to the location of objects within the input array. Stacking of multiple convolution layers, as shown in Figure 6.4, enables complex features to be learned and provides greater learning capacity, while at the same time keeping the number of parameters more tractable. Recent networks used for organ-at-risk segmentation have only had in the order of 10–20 million weights to tune [25].

6.2.1.2 Computational Power

Neural networks, and particularly convolutional neural networks, are inherently suited to parallel processing, both through task parallelization and data parallelization. During training, the gradient updates for each of the neurons can be computed in parallel. Furthermore, the computation can be split according to the data, with each item of data providing independent updates during training. Such parallel processing lends itself to the use of graphics processing units (GPUs) for computation. As shown in Figure 6.5, the processing power of GPUs has grown enormously in the past decade. At the same time, memory capacity, and bandwidth have also increased. Meanwhile, the costs have remained relatively stable, enabling a wider participation of researchers without the need to access high-performance computers.

The IBM 704 used by Rosenblatt for the initial experiments in neural networks had a processing power of 4 kFLOPS [26], cost $2m [27] and only 123 were produced [27], whereas the NVIDIA Titan RTX produced in 2018 had a processing speed of 16.3 TFLOPS [28], and cost around $2,500 on first release. While NVIDIA does not disclose number of units produced, distribution can be assumed to be in the millions (although only a fraction will be used for deep learning) with the company's revenues for 2018 totaling $2.21bn [29]. Such readily available processing power has not only enabled more researchers to participate in this field, but also facilitated training of deeper networks with many more degrees of freedom. In turn, with greater capacity to learn, more complex tasks have been tackled such as organ-at-risk segmentation for radiotherapy.

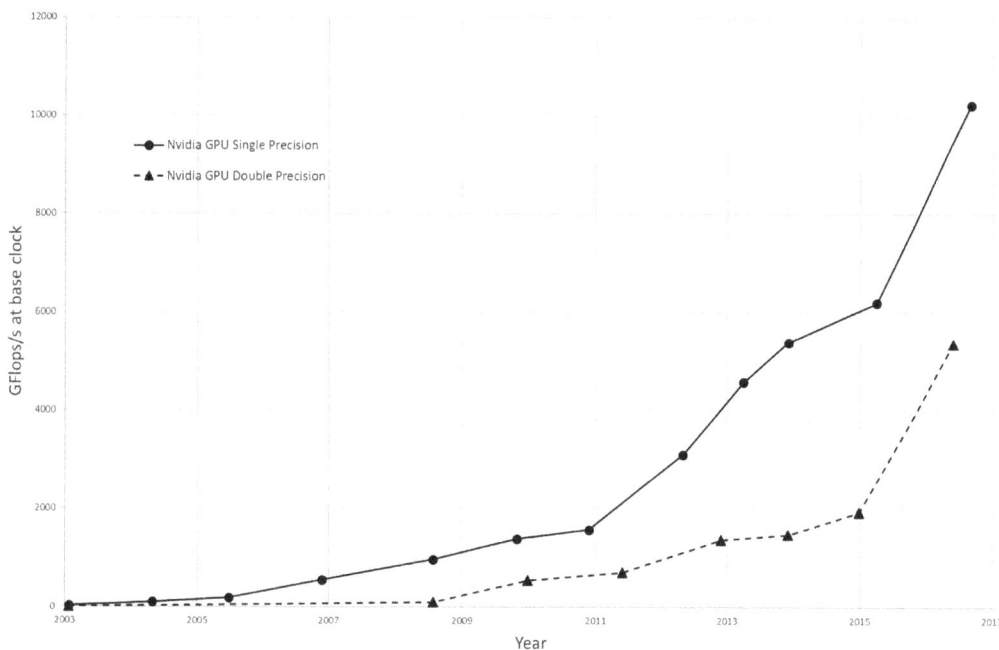

FIGURE 6.5 Improvement in GPU processing power has been rapid over the last decade, enabling large networks to be trained and used in reasonable amounts of time. Figure reproduced using approximated data from the NVIDIA Cuda C Programming Guide [30].

TABLE 6.1

High-Level Consideration of the Properties of Various Segmentation Methods

Segmentation Type	Assumptions	Use of Data	Degrees of Freedom
Model-based	Organ shape can be represented by modes of variation	Data used to train parameters. Data not required at run-time	In the order of thousands
Atlas-based	Atlas case(s) anatomical similar to patient case	Data aligned to each patient at run-time	In the order of tens to hundreds of thousands depending on the registration method
Deep learning-based	Information in the image can identify the object	Data used to train parameters. Data not required at run-time	In the order of tens of millions

6.3 WHAT MAKES DEEP LEARNING-BASED CONTOURING SO DIFFERENT TO ATLAS-BASED OR MODEL-BASED APPROACHES?

Deep learning-based, model-based, and atlas-based segmentation are very different techniques; therefore, fully understanding the differences between them that lead to difference in performance would require a substantial degree of theoretical analysis and experimental research. However, at a high level it is possible to assess their underlying assumptions, use of data, and the order of magnitude of the number of degrees of freedom to get an appreciation of what makes deep learning-based contouring so powerful. These attributes are summarized in Table 6.1.

6.3.1 UNDERLYING ASSUMPTIONS

The underlying assumptions of each segmentation approach are fundamental to what makes each approach so distinct. While one atlas-based segmentation method may differ from another or one deep-learning architecture from another, the class of segmentation approach remains the same, as do the main assumptions in which they are grounded. Model-based segmentation typically makes the assumption that while there is variation between patients in anatomy, these differences can be modeled by a limited number of modes of variation. Furthermore, the assumption is made that the segmentation at a specific anatomical location can be made locally. Together these assumptions constrain the segmentation to fit the model so as to best fit the local appearance similarity while overall constraining the global shape to be a plausible variation. That the training data fully represents the likely variations in anatomy is implicitly assumed. For atlas-based segmentation the assumption is that the atlas is anatomically similar to the patient, and that any differences in anatomy can be overcome by deformable image registration. Where multi-atlas fusion is used, the assumption is further made that the differences between contours represent random errors resulting either from inaccurate contouring of the atlas or the inability of the registration to align the atlas to the patient. Like model-based segmentation, deep learning contouring also assumes that the training data adequately reflects the variation of anatomy and image appearance observed in the patient population. Less assumption is made as to the nature of the data, with the network free to learn what is important. However, deep learning is inherently statistical; variation used infrequently in training will not strongly influence the weights learned in the model. The representative nature of the data used to train, or as input, to the segmentation method is common to all approaches. However, the methods differ in how strongly this input is expected to closely represent a new patient case.

6.3.2 USE OF DATA

As noted in the previous section, all the categories of methods considered here use previous cases to inform the segmentation of new ones. However, the methods differ in how they use this data. Both model-based and deep learning-based approaches use the data to train parameters of a segmentation system. While the deep learning-based contouring may have many more parameters to train and make weaker assumptions on how the image and contour relate, both methods encode knowledge as a segmentation model. While the previous data is used for training, only the model is required at run-time to segment a new patient case. In contrast, atlas-based segmentation uses the input cases directly, as previously, on the segmentation and no attempt is made to encode important features of the input data. This means that negligible effort is required to prepare a segmentation system before processing the patient case. However, the full data for all the atlases are required at run-time and must be available wherever the system is deployed. In one respect this could make atlas-based methods potentially more powerful; no information is lost in creating a model. Unfortunately, the limitation that the atlas data must be available restricts the use of "big data" in commercial systems. Furthermore, the strong assumption that deformable image registration is sufficient to overcome difference in appearance between the atlas and the patient is a limitation which affects how well this "full information" is used.

6.3.3 DEGREES OF FREEDOM

For any of the methods considered, their complexity makes it nearly impossible to calculate the true degrees of freedom provided. While each method may have a large number of free parameters, there are also correlations in behavior between local outputs as a result of regularizers that constrain the freedom. For example, the mesh nodes locations in an active shape model are constrained and correlated by restricting the number of eigenmodes to form more likely shapes. Therefore, each node cannot be said to have complete freedom with respect to another. Despite this, examination of the parameters, such as the number of mesh nodes, can give an indication of the number of degrees of freedom available.

Taking heart segmentation as an example, the differences in degrees of freedom between these three classes can be gauged. Zhao et al. used a modification of the active shape model for segmentation of the heart to aid radiotherapy planning [31]. While the number of eigenmodes used is not reported, it can be seen that around 900 surface nodes were used. Each node has three degrees of freedom; therefore, it could be assumed that there are approximately three thousand degrees of freedom to the model. In contrast, Rikxoort et al. used an atlas-based method for cardiac segmentation on CT. Since registration provides the ability to adapt to the patient, this can be examined to understand the method's degrees of freedom. A b-spline registration is used in the Rikxoort method. At its finest resolution, a spline control point is placed every four pixels, following subsampling of the image by a factor of two. This would result in approximately 80,000 degrees of freedom, assuming a typical image size of 150 slices of 512×512 pixels. If each atlas can be considered to provide independent information, the use of 10 atlases would raise this to 800,000 degrees of freedom. Finally, considering deep learning-based segmentation, for the methods used in the 2017 AAPM Thoracic Auto-segmentation Challenge, albeit for segmentation of five organs and not just the heart, the number of free parameters to be tuned were reported to be between 14 million and 66 million [25]. Therefore, it can be understood that the power of deep learning methods to outperform the other approaches is likely to come from its additional, by an order of magnitude, flexibility to encode knowledge.

6.4 SUMMARY OF THIS PART OF THE BOOK

This part of the book explores the various choices to be made when designing a deep learning network for organ-at-risk segmentation. Figure 6.6 illustrates the process for the training and use of a deep learning-based segmentation system. Data acquired for the purpose of training a deep learning system is usually separated into three categories: training, validation, and test. Where insufficient data is available, data augmentation techniques are commonly used to ensure that the network is robust to variations in the data and to prevent overfitting. Augmentation can be performed both during training, but also during testing. The training data are passed into the deep learning network resulting in a segmentation. This output is compared to an expected output during training using a loss function, and the loss function is used to determine the gradients to update the network parameters. The same loss function can be used with the validation data to ensure that the network is not overfitting to the training set. Alternatively, quantitative validation approaches can be used to give an indication of potential clinical performance. The independent test set can then be evaluated with the same quantitative measures to confirm the performance observed during training.

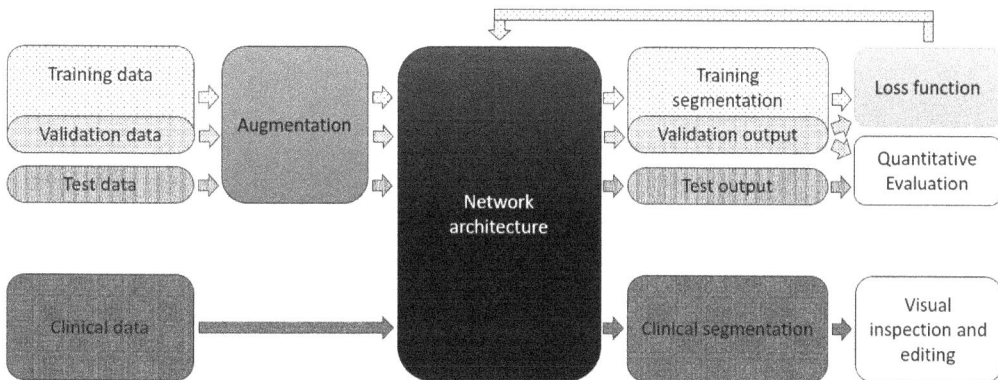

FIGURE 6.6 The training and use of a deep learning segmentation system. Many aspects of training and use are covered within this part of the book. Evaluation and curation of data will be considered in the third part.

When used in the clinic, a known segmentation is not available for evaluation of performance, rather, the contours are inspected and edited for clinical use. This subjective clinical evaluation can also provide insight into how well the network is performing.

Data curation for training, validation and testing, quantitative evaluation measures, and appropriate clinical use of deep-learning segmentation are considered within the third part of this book. In this part of the book, the more technical aspects are reviewed. First, the network architecture is considered. Chapter 7 gives a detailed overview of the various types of network designed to date. No attempt is made at experimental comparison on account of the difficulty in performing such a comparison in a meaningful way. However, the strengths and weaknesses of each approach are considered, and the results of benchmarking reported previously are considered. Recognizing that the U-net [32] is a popular choice for image segmentation; Chapter 8 evaluates the difference in performance between 2D and 3D implementations of this particular architecture. Various constraints could be considered as to what is a "fair" evaluation. In this chapter, what can be achieved for a given graphics card memory is the primary consideration. Chapter 9 considers another design aspect, whether it is better to perform segmentation a single organ at a time or using a multi-label approach. Again, the U-net architecture is evaluated, this time using images resampled to the same input resolutions. Whether multi-class or multi-label segmentation is performed has an impact on the choice of loss function used in network optimization.

Going beyond the network design, Chapter 10 reviews the choices available, considering the strengths and weaknesses with respect to the task of image segmentation for a range of loss functions. Chapter 11 then considers how data augmentation can be used to improve training, particularly where small quantities of data are available. A wider range of data augmentation strategies are discussed, and an evaluation of the impact of a simple data augmentation approach is presented. Finally, Chapter 12 touches on clinical use, considering what might cause organ segmentation to fail. An underlying assumption of deep learning segmentation is that the training data is representative of the use-case scenario. This chapter examines how far a network can be stretched beyond those boundaries.

REFERENCES

1. B Ibragimov, L Xing, Segmentation of organs-at-risks in head and neck CT images using convolutional neural networks. *Med. Phys.* 44 (2017):547–557.
2. R Trullo, C Petitjean, D Nie, D Shen, S Ruan, Joint segmentation of multiple thoracic organs in CT images with two collaborative deep architectures. In: *Deep Learn. Med. Image Anal. Multimodal Learn. Clin. Decis. Support*, 2017:21–29. https://doi.org/10.1007/978-3-319-67558-9_3.
3. K Men, J Dai, Y Li, Automatic segmentation of the clinical target volume and organs at risk in the planning CT for rectal cancer using deep dilated convolutional neural networks. *Med. Phys.* 44 (2017):6377–6389. https://doi.org/10.1002/mp.12602.
4. K Men, X Chen, Y Zhang, T Zhang, J Dai, J Yi, Y Li, Deep deconvolutional neural network for target segmentation of nasopharyngeal cancer in planning computed tomography images. *Front. Oncol.* 7 (2017):1–9. https://doi.org/10.3389/fonc.2017.00315.
5. B Ibragimov, F Pernuš, P Strojan, L Xing, Development of a novel deep learning algorithm for autosegmentation of clinical tumor volume and organs at risk in head and neck radiation therapy planning. *Int. J. Radiat. Oncol.* 96 (2016):S226. https://doi.org/10.1016/j.ijrobp.2016.06.561.
6. R Trullo, C Petitjean, S Ruan, B Dubray, D Nie, D Shen, Segmentation of organs at risk in thoracic CT images using a SharpMask architecture and conditional random fields. *Proc. – Int. Symp. Biomed. Imaging.* (2017):1003–1006. https://doi.org/10.1109/ISBI.2017.7950685.
7. T Lustberg, J van Soest, M Gooding, D Peressutti, P Aljabar, J van der Stoep, W van Elmpt, A Dekker, Clinical evaluation of atlas and deep learning based automatic contouring for lung cancer. *Radiother. Oncol.* 126 (2018):312–317. https://doi.org/10.1016/j.radonc.2017.11.012.
8. T Lustberg, J Van der Stoep, D Peressutti, P Aljabar, W Van Elmpt, J Van Soest, M Gooding, A Dekker, EP-2124: time-saving evaluation of deep learning contouring of thoracic organs at risk. *Radiother. Oncol.* 127 (2018):S1169. https://doi.org/10.1016/s0167-8140(18)32433-2.

9. D Peressutti, P Aljabar, J van Soest, T Lustberg, J van der Stoep, A Dekker, W van Elmpt, MJ Gooding, TU-FG-605-7 deep learning contouring of thoracic organs at risk. *Med. Phys.* 44(6) (2017):3159.

10. H Wang, B Raj, On the origin of deep learning. (2017):1–72. http://arxiv.org/abs/1702.07800.

11. J Schmidhuber, Deep learning in neural networks: an overview. *Neural Networks* 61 (2015):85–117. https://doi.org/10.1016/j.neunet.2014.09.003.

12. A Maier, C Syben, T Lasser, C Riess, A gentle introduction to deep learning in medical image processing. *Z. Med. Phys.* 29 (2019):86–101. https://doi.org/10.1016/j.zemedi.2018.12.003.

13. H Wang, B Raj, A survey: time travel in deep learning space: an introduction to deep learning models and how deep learning models evolved from the initial ideas. (2015):1–43. http://arxiv.org/abs/1510.04781.

14. SK Zhou, H Greenspan, D Shen, eds., *Deep Learning for Medical Image Analysis*, Academic Press, 2017.

15. AC Müller, S Guido, *Introduction to Machine Learning with Python: A Guide for Data Scientists*, O'Reilly Media, 2016.

16. A Ng, K Katanforoosh, YB Mourri, Deep learning specialization. (n.d.). www.coursera.org/specializations/deep-learning (accessed September 4, 2020).

17. A Amini, A Soleimany, Introduction to deep learning, MIT. (2020). http://introtodeeplearning.com/ (accessed September 4, 2020).

18. A Amini, Introduction to deep learning, YouTube. (2020). www.youtube.com/watch?v=njKP3FqW3Sk (accessed September 4, 2020).

19. J Johnson, Detection and segmentation, YouTube. (2017). www.youtube.com/watch?v=nDPWyw-WRIRo (accessed September 8, 2020).

20. WS McCulloch, W Pitts, A logical calculus of the ideas immanent in nervous activity. *Bull. Math. Biophys.* 5 (1943):115–133.

21. F Rosenblatt, Perceptron simulation experiments. *Proc. IRE.* 48 (1960):301–309.

22. TR Willoughby, G Starkschall, NA Janjan, II Rosen, Evaluation and scoring of radiotherapy treatment plans using an artificial neural network. *Int. J. Radiat. Oncol. Biol. Phys.* 34 (1996):923–930. https://doi.org/10.1016/0360-3016(95)02120-5.

23. Y LeCun, B Boser, JS Denker, D Henderson, RE Howard, W Hubbard, LD Jackel, Backpropagation applied to digit recognition. *Neural Comput.* 1 (1989):541–551. www.ics.uci.edu/~welling/teaching/273ASpring09/lecun-89e.pdf.

24. Y LeCun, L Bottou, Y Bengio, P Haffner, Gradient-based learning applied to document recognition, *Proc. IEEE.* 86 (1998):2278–2324. https://doi.org/10.1109/5.726791.

25. J Yang, H Veeraraghavan, SG Armato, K Farahani, JS Kirby, J Kalpathy-Kramer, W van Elmpt, A Dekker, X Han, X Feng, P Aljabar, B Oliveira, B van der Heyden, L Zamdborg, D Lam, M Gooding, GC Sharp, Autosegmentation for thoracic radiation treatment planning: a grand challenge at AAPM 2017. *Med. Phys.* 45 (2018):4568–4581. https://doi.org/10.1002/mp.13141.

26. IBM, *704 Data Processing System, IBM Arch.* (n.d.). www.ibm.com/ibm/history/exhibits/mainframe/mainframe_PP704.html (accessed October 3, 2020).

27. Computer History Museum, IBM 704 electronic data processing system. *Comput. Hist. Museum.* (n.d.). www.computerhistory.org/revolution/early-computer-companies/5/113/489 (accessed October 3, 2020).

28. NVIDIA, *The Ultimate PC GPU NVIDIA Titan RTX.* (2019). www.nvidia.com/content/dam/en-zz/Solutions/titan/documents/titan-rtx-for-creators-us-nvidia-1011126-r6-web.pdf.

29. NVIDIA, NVIDIA announces financial results for fourth quarter and fiscal 2019. (2019). https://nvidianews.nvidia.com/news/nvidia-announces-financial-results-for-fourth-quarter-and-fiscal-2019.

30. NVIDIA, Cuda C programming guide, 2015.

31. X Zhao, Y Wang, G Jozsef, Robust shape-constrained active contour for whole heart segmentation in 3-D CT images for radiotherapy planning. In: *2014 IEEE Int. Conf. Image Process.*, IEEE, 2014: 1–5. https://doi.org/10.1109/ICIP.2014.7024999.

32. O Ronneberger, P Fischer, T Brox, U-net: convolutional networks for biomedical image segmentation. *Lect. Notes Comput. Sci. (Including Subser. Lect. Notes Artif. Intell. Lect. Notes Bioinformatics)* 9351 (2015):234–241. https://doi.org/10.1007/978-3-319-24574-4_28.

7 Deep Learning Architecture Design for Multi-Organ Segmentation

Yang Lei, Yabo Fu, Tonghe Wang, Richard L.J. Qiu,
Walter J. Curran, Tian Liu, and Xiaofeng Yang

CONTENTS

7.1   INTRODUCTION

An increasing number of deep learning (DL) techniques have been proposed in the computer vision field in recent years. Inspired by their success, researchers have extended them into organ segmentation tasks in medical images [1–20]. DL-based methods adaptively explore deep features from medical images to represent the image structural information in detail. This fundamental change in methodology enables them to achieve state-of-the-art performances in medical image segmentation, especially in multi-organ segmentation. Based on convolutional neural networks as a basic component, current studies have proposed a variety of architectures of DL networks. These architectures vary in network structure, complexity, and implementation, resulting in variations in, and task-dependent, performance. Reviewing the network architectures developed recently will indicate the progress in current DL and facilitate the clinical transition of auto-segmentation methods. This review will also reveal the limitations in current network design that need to be addressed in future studies. In this chapter, popular deep learning network designs for multi-organ segmentation are summarized. Specifically, thoracic organ segmentations are used as an example to discuss network performances and challenges of organ-at-risk (OAR) auto-contouring for thoracic radiation treatment planning. This survey aims to:

- Summarize the latest architectural developments in DL-based medical image multi-organ segmentation
- Highlight contributions, identify challenges, and outline future trends
- Provide benchmark evaluations of recently published DL-based multi-organ segmentation methods

7.2   DEEP LEARNING ARCHITECTURE IN MEDICAL IMAGE MULTI-ORGAN SEGMENTATION

The task of medical image multi-organ segmentation is typically defined as assigning each voxel of the medical images to one of the several labels that represent the objects of interest. Segmentation is one of the most commonly studied DL-based applications in the medical field. Therefore, there are a wide variety of methodologies with many different network architectures.

There are many ways to categorize the DL-based multi-organ segmentation methods according to their properties such as network architecture, training process (supervised, semi-supervised, unsupervised, transfer learning), input size (patch-based, whole volume-based, 2D, 3D), and so on. In this chapter, approaches are classified into six categories based on their architecture, namely: (1) auto-encoder, (2) convolutional neural network, (3) fully convolutional network, (4) generative adversarial network (GAN), (5) regional convolutional neural network, and (6) hybrid DL-based methods. In each category, a comprehensive table is provided, listing all the surveyed works belonging to this category and summarizing their important features. Besides multi-organ segmentation methods, single-organ segmentation methods are also included since single-organ segmentation can be easily transformed to multi-organ segmentation by replacing the last layer's binary output to a multi-channel binary output. The difference between multi-organ and single organ approaches is considered in Chapter 4. Similarly, medical image object detection methods were included as they could be used to first obtain the region of interest (ROI) to aid the segmentation procedure and improve the segmentation accuracy.

Before diving into the details of each category, a detailed overview of DL-based medical image multi-organ segmentation methods with their corresponding components and features is provided in Figure 7.1. The aim of Figure 7.1 is to give the readers an overall understanding by listing important features of each category. The definition, features, and challenges of each category are also listed in Table 7.1.

Works cited in this review were collected from various databases, including Google Scholar, PubMed, Web of Science, Semantic Scholar, and so on. To collect as many works as possible,

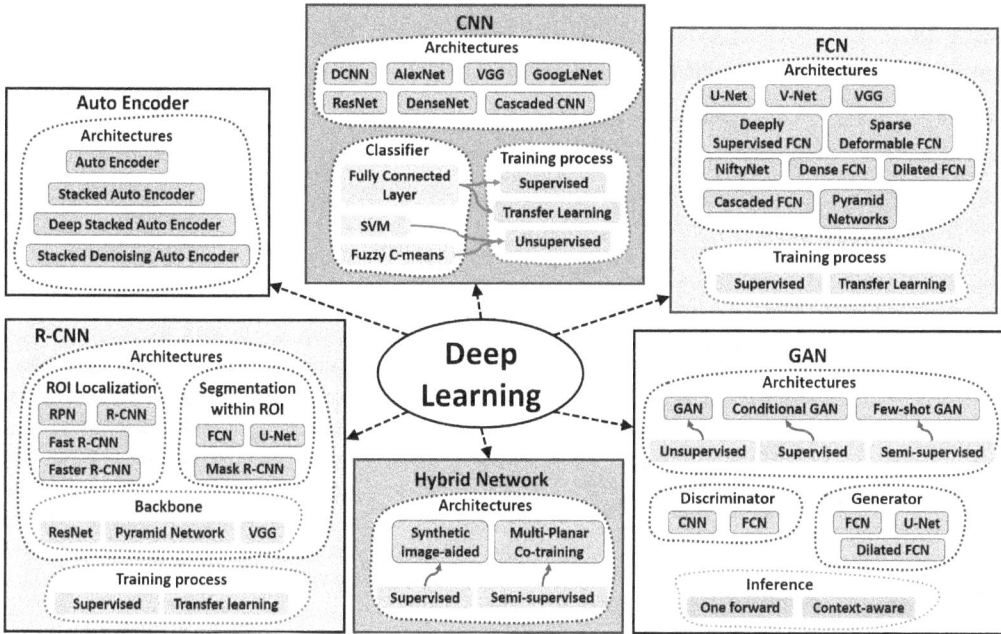

FIGURE 7.1 Overview of six categories of DL-based methods in medical image segmentation.

a variety of keywords was used, including but not limited to deep learning, multi-organ, medical segmentation, convolutional neural network, and so on. Over 180 papers that are closely related to DL-based medical image segmentation and over 40 papers that are closely related to multi-organ segmentation were reviewed. Most of these works were published between 2017 and 2019. The number of multi-organ publications is plotted against year by stacked bar charts in Figure 7.2, with the number of papers in each of the six categories shown. The dotted line in Figure 7.2 indicates increased interest in DL-based multi-organ segmentation methods over the years, highlighting the increase in the dramatic growth in publications recently.

7.2.1 Auto-Encoder Methods

7.2.1.1 Auto-Encoder and Its Variants

In the literature, the autoencoder (AE) and its variants have been extensively studied and continue to be utilized in medical image analysis [21]. AEs are often used for unsupervised [22] and semi-supervised [23] neural network learning. As shown in Figure 7.3, AEs usually consist of neural network encoder layers that transform the input into a latent or compressed representation by minimizing the reconstruction errors between input and output values of the network, and network decoder layers that restore the original input from the low-dimensional latent space. By constraining the dimension of latent representation, AEs can discover relevant patterns from the data.

To prevent an AE from learning an identity function, several improved AEs were proposed. The most widely used network model in deep unsupervised architecture is stacked AE (SAE). An SAE is constructed by organizing AEs on top of each other, also known as deep AEs. SAEs consist of multiple AEs stacked into multiple layers where the output of each layer is wired to the inputs of the successive layers [22]. To obtain good parameters, SAEs use greedy layer-wise training. The benefit of an SAE is that it represents a deeper network with more hidden layers, therefore it has greater expressive power. Furthermore, it usually captures a useful hierarchical grouping of the input [22].

Denoising autoencoders (DAEs) are another variant of the AE and are used to constitute better higher-level representation and extract useful features [24]. DAEs prevent the model from learning

TABLE 7.1

Summary of Six Categories of DL-Based Methods in Medical Image Segmentation

Category	Definition	Feature	Challenges
Auto-encoder (AE)	Single neural network encoder/ decoder layer	Low model complexity	Poor performance on target contours with large shape variability Large computation complexity when stacking multiple AEs for deeper network
Convolutional neural network (CNN)	Input/output layers and multiple hidden layers including convolutional layers, max pooling layers, batch normalization layers, dropout layers, fully connected layers, and normalization layers	Facility for deeper networks	The fully connected layer requires the classification step to be performed voxel-wise or patch-wise
Fully convolutional network (FCN)	CNN with fully connected layer replaced by convolutional layer	Enables end-to-end segmentation	Fixed receptive size Voxel-wise loss introduces boundary leakage in low contrast regions
Generative adversarial network (GAN)	Two competing networks, a generator, and a discriminator	High accuracy at low contrast region attributed to adversarial loss provided by discriminator	Less effective in simultaneous multi-organ segmentation due to imbalance of loss function among different organs
Region-CNN (R-CNN)	CNN using selective search to extract candidate regions	Enables simultaneous multi region-of-interest detection and multi-organ segmentation	Large computational burden when training with 3D image volumes
Hybrid	Two or more networks with different architectures for different functional propose	Better performance and lower demand on training data size	High model complexity

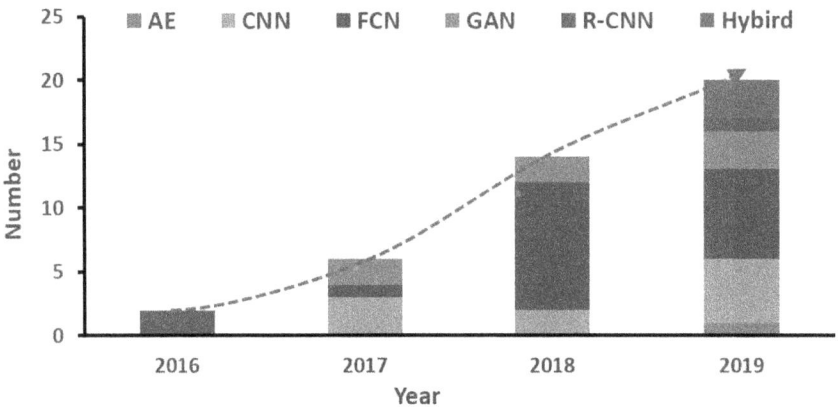

FIGURE 7.2   Overview of number of publications in DL-based multi-organ segmentation.

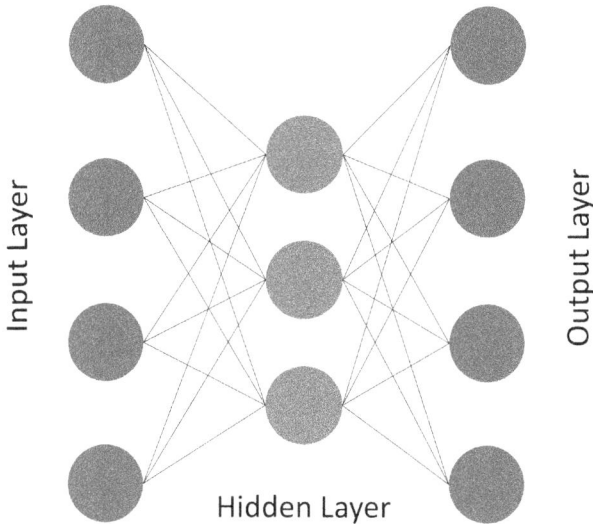

FIGURE 7.3 Diagram showing the basic architecture of auto-encoder.

a trivial solution where the model is trained to reconstruct a clean input from the corrupted version from noise or another corruption [23]. A stack denoising autoencoder (SDAE) is a deep network utilizing the power of DAEs [25].

One of the limitations of AEs is the relatively small number of hidden units resulting from their fully connected nature and graphics card memory limitations. This restricts the depth of the network and the information that can be learned from the input data. To overcome this limitation, other constraints, such as prior knowledge, can be imposed on the network, to facilitate the network learning deep information. A sparse constraint is a typically used constraint in a sparse AE. The aim of a sparse autoencoder is to make a large number of neurons have a low average output so that neurons may be inactive most of the time [26]. Sparsity can be achieved by introducing a loss function during training or manually zeroing a few of the strongest hidden unit activations.

SAE requires layer-wise pre-training since training of SAEs may be time consuming as they are built with fully connected layers. Li et al. first investigated training convolutional auto encoders (CAE) directly in an end-to-end manner without pre-training [27]. Guo et al. suggested that CAEs are beneficial to learn features for images, preserve local structures, and avoid distortion of feature space [28]. Wang et al. proposed an automated chest screening based on a hybrid model of transfer learning and CAE [29].

7.2.1.2 Overview of Works

Since abnormalities, e.g. abnormal tissue types and irregular organ shapes, are often present in medical images, it is challenging to obtain ground truth labels of multi-organs for supervised learning. However, organ segmentation in such abnormal datasets is meaningful in radiation therapy. Shin et al. applied an SAE method for organ detection in magnetic resonance imaging (MRI) [22]. Their method was used to detect the locations of the liver, heart, kidney, and spleen for MRI scans of the abdominal region containing liver or kidney metastatic tumors. Only weakly supervised training is required to learn visual and temporal hierarchical features that represent object classes from unlabeled multimodal dynamic contrast-enhanced magnetic resonance imaging (DCE-MRI) data. A probabilistic patch-based method was employed for multiple organ detection, with the features learned from the SAE model.

Accurate and automated segmentation of glioma from MRI is important for treatment planning and monitoring disease progression. Vaidhya et al. used an SDAE to solve the challenge of variable

TABLE 7.2

Overview of AE Methods

Ref.	Year	Network	Supervision	Dimension	Site	Modality
[22]	2013	SAE	Weakly supervised	3D patch	Abdomen	4D DCE-MRI
[25]	2015	SDAE	Supervised	3D patch	Brain Gliomas	MRI
[23]	2017	SDAE	Semi-supervised	2D patch	Brain lesion	MRI
[30]	2017	SAE	Transfer learning	2D slice	Liver	CT
[29]	2018	CSDAE	Transfer learning	2D slice	Thoracic	Chest X-rays
[26]	2019	SSAE	Unsupervised	2D patch	Vertebrae	CT
[31]	2019	SSAE	Unsupervised	2D patch	Vertebrae	CT
[32]	2019	Hierarchical 3D neural networks	Supervised	N/A	Head and neck	CT

shape and texture of glioma tissue in MRI for this segmentation task [25]. 3D patches were extracted from multiple MRI sequences and were then fed into the SDAE model to obtain the glioma segmentation. During training, two SDAE models were supervised in this task, one for high grade glioma (HGG) data, the other one for a combination of HGG and low-grade glioma (LGG) data. During testing, the segmentation was obtained by a combination of predictions from two trained networks via maximum a posteriori (MAP) estimation. Simultaneously, Alex et al. applied an SDAE for brain lesion detection, segmentation, and false-positive reduction [23]. An SDAE was pre-trained using a large number of unlabeled patient volumes and fine-tuned with 2D patches drawn from a limited number of patients. LGG segmentation was achieved using a transfer learning approach in which an SDAE network pre-trained with LGG data was fine-tuned using LGG data.

Ahmad et al. proposed a deep SAE (DSAE) for CT liver segmentation [30]. First, deep features were extracted from unlabeled data using the AE. Second, these features are fine-tuned to classify the liver among other abdominal organs.

In order to efficiently detect and identify normal levels during mass chest screening of lung lesions of chest X-rays (CXRs), Wang et al. proposed a convolutional SDAE (CSDAE) to determine to which three levels of the images (i.e. normal, abnormal, and uncertain cases) the CXRs belong [29].

Accurate vertebrae segmentation in the spine is essential for spine assessment, surgical planning, and clinical diagnostic treatment. Qadri et al. proposed a stacked SAE (SSAE) model for the segmentation of vertebrae from CT images [26]. High-level features were extracted via feeding 2D patches into the SSAE model in an unsupervised way. To improve the discriminability of these features, a further refinement using a supervised fashion and fine-tuning was integrated. Similarly, Wang et al. proposed to localize and identify vertebrae by combining SSAE contextual features and structured regression forest (SRF) [31]. Contextual features were extracted via SSAE in an unsupervised way and were then fed into SRF to achieve whole-spine localization (Table 7.2).

7.2.1.3   Discussion

In contrast to previous approaches to machine learning, whose performance depends on hand-craft features, an AE can learn the important contextual features of a medical image, improving their contextual discrimination ability [31].

For the segmentation of public BraTS 2013 and BraTS 2015 data [33], which are multi-modality brain MRI tumor segmentation datasets, an SDAE can provide good segmentation performance [23]. For segmenting liver on CT images, DSAEs showed high classification accuracy and can speed up the clinical task [30].

For detecting and identifying normal levels during mass chest screening of lung lesions of CXRs, the CSDAE method achieves promising results in terms of precision of 98.7% and 94.3% based on the normal and abnormal cases, respectively [29]. The results achieved by the proposed framework show superiority in classifying the disease level with high accuracy. CSDAEs can potentially save radiologists time and effort, allowing them to focus on higher-level risk CXRs.

Validated on the public MICCAI CS2014 dataset, which includes a challenging dataset of 98 spine CT scans, the SSAE method could effectively and automatically locate and identify spinal targets in CT scans, and achieve higher localization accuracy while maintaining low model complexity without making any assumptions about visual field in CT scans [26].

Although AEs have many benefits, they face some challenges and limitations in medical multi-organ segmentations. One of the limitations is related to data regularity. For example, in cases of anatomical structures like lung, heart, and liver, even if the inter-subject variability of the dataset is high, the shape variety of segmentation masks would remain low. Unlike organs which tend to have similar structure, irregular lesions and tumors with large shape variability are difficult for AEs to encode and remain challenging for the unsupervised AE methods. Furthermore, the number of layers can be limited due to the large computation complexity associated with the fully connected networks used in AE methods, compared to convolutional neural networks (CNNs) which use convolution kernels with shared learnable parameters.

7.2.2 CNN METHODS

7.2.2.1 Network Designs

CNNs derive their name from the type of hidden layers they consist of. The hidden layers of a CNN typically consist of convolutional layers, max pooling layers, batch normalization layers, dropout layers, and normalization layers. Fully connected layers are normally used at later stages, if used at all. The last layer of a CNN is typically a sigmoid or softmax layer for classification/segmentation and tanh layer for regression. Figure 7.4 demonstrates a typical CNN architecture [34].

Convolution layers are the core of CNNs and are used for feature extraction [2]. The convolution layer extracts variant feature maps depending on its learned convolution kernels. The pooling layer performs a down-sampling operation by using maximum or average of the defined neighborhood as the value to reduce the spatial size of each feature map. A rectified linear unit (ReLU) and its modifications such as leaky ReLU are among the most commonly used activation functions, which transform data by clipping any negative input values to zero while positive input values are passed as output [35]. The fully connected layer connects every neuron in the previous layer to every neuron in the next layer. Neurons in a fully connected layer are fully connected to all activations in the previous layer. They are placed before the classification output of a CNN and are used to flatten the results before a prediction is made using linear classifiers. Via several fully connected layers, the previous feature maps extracted from convolutional layers are converted to a probability-like representation to classify the medical image or medical image patch or voxels.

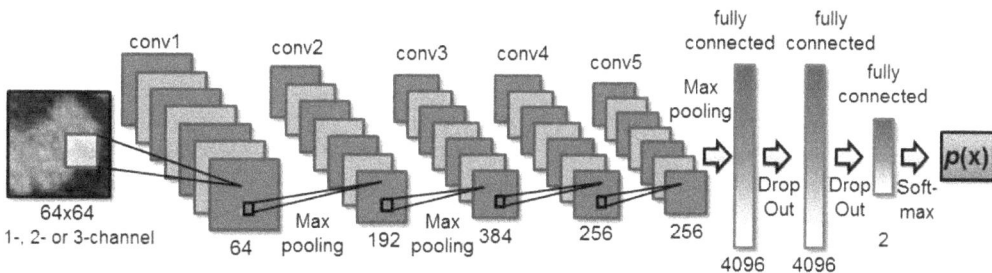

FIGURE 7.4 An exemplary diagram of CNN architecture. Reprinted by permission from Springer Nature Customer Service Centre GmbH: Springer Nature, DeepOrgan: Multi-level Deep Convolutional Networks for Automated Pancreas Segmentation by Roth et al. [34], copyright 2020.

During training of a CNN architecture, the model predicts the class scores for training images, computes the loss using the selected loss function, and finally updates the weights using the gradient descent method by back-propagation. Cross-entropy loss is one of the most widely used loss functions, and stochastic gradient descent (SGD) and Adam gradient descent optimizations are the most popular method to operate gradient descent.

LeCun et al. first proposed a CNN model, named LeNet, for hand-written digit recognition [36]. LeNet is composed of convolution layers, pooling layers, and fully connected layers. With the development of computer hardware and the increase in the amount of data available for neural network training, in 2012, Krizhevsky et al. proposed AlexNet and won the ILSVRC-2012 image classification competition [37] with a far lower error rate than the second place [38]. The improvements of AlexNet as compared to LeNet include (1) ReLU layer for nonlinearity and sparsity, (2) data augmentation to enlarge the dataset variety, (3) dropout layer to reduce learnable parameters and prevent overfitting, (4) local response normalization to normalize the nearest data and, (5) overlapping pooling. Since then, CNNs have begun to attract widespread attention, and variants of CNNs have been developed and have achieved the-state-of-art performances in various image processing tasks. Additionally, Zeiler and Fergus proposed ZFNet to improve the performance of AlexNet [39] and proved that a shallow network is able to learn edge, color, and texture features of images and a high-level network can learn abstract features of images. In addition, they demonstrated better performance can be achieved via a deeper network. The main improvement of ZFNet is a deconvolution network used to visualize the feature map.

Simonyan and Zisserman proposed the VGG to further explore the performance of the deeper network model [40]. The main innovation of VGG is a thorough evaluation of networks of increasing depth using an architecture with very small (3×3) convolution filters, which shows that a significant improvement on the prior-art configurations can be achieved by pushing the depth from 16 to 19 layers. Similarly, GoogLeNet was proposed to broaden the network structure [41]. By integrating the proposed inception module, GoogLeNet won the winner of the ImageNet Large-Scale Visual Recognition Challenge 2014 (ILSVRC14), which is an image classification and detection competition. The inception module is helpful for the CNN model to better describe the input data content while further increasing the depth and width of the network model.

Many of the previous developments of CNNs were to increase the depth and width of CNN to improve the performance. However, simply increasing the depth would lead to vanishing/exploding gradients. To ease the difficulty of training deep CNNs and solve the degradation effect caused by increasing network depth, He et al. proposed a residual network (ResNet) for image recognition [42]. ResNet, which is mainly composed of residual blocks, is demonstrated to be able to break through a 100-layer barrier and even reach 1000 layers.

Inspired by residual networks, Huang et al. later proposed a densely connected convolutional network (DenseNet) by connecting each layer to every other layer [43]. In contrast to residual blocks, which would focus on learning the structural difference between the input features and desired output, DenseNet aimed to combine both low-frequency and high-frequency feature maps from previous and current convolutional layers via dense blocks.

7.2.2.2 Overview of Works

In medical image segmentation, CNNs can be used to classify each voxel or patch in the image individually, by presenting the network with patches extracted around that voxel or patch. Roth et al. proposed a multi-level deep CNN approach for pancreas segmentation in abdominal CT image [34]. A dense local image patch label was obtained by extracting an axial-coronal-sagittal viewed patch in a sliding window manner. The proposed CNN learned to assign class probabilities for each center voxel of its patch. Finally, a stacked CNN leveraged the joint space of CT intensities and dense probability maps. The CNN architecture used consists of five convolutional layers which are followed by max pooling, three fully connected layers, two drop out layers, and a softmax operator to perform binary classification. This CNN architecture can be introduced into multi-organ segmentation frameworks by specifying more tissue types since CNN naturally supports multi-class classifications [44].

In contrast to 2D input, which would lose spatial information, Hamidian et al. proposed using a 3D patch-based CNN to detect lung pulmonary nodules for chest CT images [45] using volumes of interest extracted from the lung image database consortium (LIDC) dataset [46]. They extended a previous 2D CNN to three dimensions which would be more suitable for volumetric CT data.

For highly pathologically affected cases, segmenting and classifying the lytic and sclerotic metastatic lesions from CT images is challenging, because these lesions are ill-defined. Therefore, it is hard to extract relevant features that can well represent texture and shape information for traditional machine learning-based methods. In order to solve this problem, Chmelik et al. applied deep CNN (DCNN) to segment and classify these kinds of lesions [47]. The CNN architecture takes three perpendicular 2D patches for each voxel of 3D CT image as input and output classification of three categories (healthy, lytic, and sclerotic) for that voxel. The proposed CNN consisted of several convolutional layers which are followed by ReLU and max pooling to extract features, several fully connected layers with dropout layers to combine the feature maps to feature vectors, and a last fully connected layer to convert the feature vector to a three-element output of class scores. A high score correlates to a high probability to the corresponding class. L2 regularized cross-entropy and class error loss are used for optimization. Mini-batch gradient descent with a momentum back-propagation algorithm is used to optimize the learnable parameters of this CNN.

During radiotherapy for nasopharyngeal carcinoma (NPC) treatment, accurate segmentation of OARs in head and neck (H&N) CT images is a key step for effective planning. Due to low-contrast and surrounding adhesion tissues of the parotids, thyroids, and optic nerves, automatically segmenting these regions is challenging and will result in lower accuracy for these regions as compared to other organs. In order to solve this challenge, Zhong et al. proposed a cascaded CNN to delineate these three OARs for NPC radiotherapy by combining boosting algorithm [48]. In their study, CT images of 140 NPC patients treated with radiotherapy were collected. Manual contours of three OARs were used as the learning target. A hold-out test was used to evaluate the performance of the proposed method, i.e. the datasets were divided into a training set (100 patients), a validation set (20 patients), and a test set (20 patients). In the boosting method for combining multiple classifiers, three cascaded CNNs for segmentation were combined. The first network was trained with the traditional approach. The second one was trained on patterns (pixels) filtered by the first net, that is, the second machine recognized a mix of patterns (pixels), 50% of which were accurately identified by the first net. Finally, the third net was trained with the new patterns (pixels) screened jointly by the first and second networks. During the test, the outputs of the three nets were considered to obtain the final output. A 2D patch-based ResNet [42] was used to build the cascaded CNNs.

For multi-OARs segmentation in thoracic radiotherapy treatment, Harten et al. proposed a combination of 2D and 3D CNNs for automatic segmentation of OARs (including esophagus, heart, trachea, and aorta) on thoracic treatment planning CT scans of patients diagnosed with lung, breast, or esophageal cancer [49]. The two CNNs are summarized as follows: one 3D patch-based network that contains a deep segment of residual blocks [50] with a sigmoid layer to perform multi-class binary classification, and one 2D patch-based (2D patch extracted from axial, coronal, and sagittal planes) network containing dilated convolutions [51] with softmax layer to perform classification. A hold-out validation (40 data for training and 20 data for testing) was used to evaluate the performance of the proposed method (Table 7.3).

7.2.2.3 Discussion

For the multi-OARs segmentation of CT H&N, Dice similarity coefficient (DSC), 95th percentile of the Hausdorff distance (95% HD), and volume overlap error (VOE) were used to assess the performance of a cascaded CNN [48]. The mean DSC values were above 0.92 for parotids, above 0.92 for thyroids, and above 0.89 for optic nerves. The mean 95% HDs were approximately 3.08 mm for parotids, 2.64 mm for thyroids, and 2.03 mm for optic nerves. The mean VOE metrics were

TABLE 7.3

Overview of CNN Methods

Ref.	Year	Network	Supervision	Dimension	Site	Modality
[52]	2017	Deep deconvolutional neural network (DDNN)	Supervised	2D slice	Brain	CT
[34]	2015	Multi-level DCNN	Supervised	2D patch	Pancreas	CT
[53]	2016	Holistically nested CNN	Supervised	2D patch	Pancreas	CT
[45]	2017	3D CNN	Supervised	3D patch	Chest	CT
[54]	2017	3D DCNN	Supervised	N.A.*	Abdomen	CT
[55]	2017	CNN	Supervised	3D patch	Head and neck	CT
[56]	2017	Fuzzy-C-Means CNN	Supervised	3D patch	Lung nodule	CT
[57]	2017	DCNN	Supervised	2D Slice	Body, chest, abdomen	CT
[58]	2018	Fusion Net	Supervised	2D patch	100 ROIs	HRCT
[47]	2018	DCNN	Supervised	2D patch	Spinal lesion	CT
[59]	2018	DCNN	Supervised	2D slice	Malignant pleural mesothelioma	CT
[60]	2018	2D and 3D CNN	Supervised	2D slice, 3D volume	Artery/vein	CT
[61]	2018	3D ConvNets	Transfer learning	3D volume	Brain	MRI
[62]	2018	CNN with specific fine-tuning	Supervised or Unsupervised	2D slice, 3D volume	Brain, abdomen	Fetal MRI
[63]	2018	2D and 3D DCNN	Supervised	2D slice, 3D volume	Whole body	CT
[64]	2019	Deep fusion Network	Supervised	2D slice	Chest	CXR
[65]	2019	DCNN	Supervised	2D slice	Abdomen	CT
[66]	2019	2.5D CNN	Supervised	2.5D patch	Thorax	CT
[48]	2019	Cascaded CNN	Supervised	2D slice	Head and neck	CT
[49]	2019	2D and 3D CNN	Supervised	2D slice, 3D volume	Thorax	CT
[67]	2019	U-net neural network	Supervised	3D patch	Lung	CT

*N.A.: not available, i.e. not explicitly indicated in the publication.

approximately 14.16% for parotids, 14.94% for thyroids, and 19.07% for optic nerves. From the comparison in Zhong et al. [48], the proposed boosting-based cascaded CNN outperformed U-net [68] in segmenting the three OARs. Despite the powerful accuracy performance of the boosting structure, its pixel-based classification took more time as compared to U-net. This is because all classifiers in the boosting structure need to classify all pixels in the image.

In the study by Yang et al. [49], researchers evaluated the performance for 2D CNN, 3D CNN, and a combination of 2D and 3D CNNs individually and demonstrated the combination network produces the best results. The DSC of the esophagus, heart, trachea, and aorta were 0.84 ± 0.05, 0.94 ± 0.02, 0.91 ± 0.02, and 0.93 ± 0.01, respectively. These results demonstrate potential for automating segmentation of OARs in routine radiotherapy treatment planning.

A drawback of CNN is classification needs to be performed on every voxel or small patch. By sliding a window with a huge overlap between two neighboring patches, the CNN models perform classifications on each voxel of the whole volume. This approach is inefficient since it requires repeated forward network prediction on every voxel of the image. Fortunately, the convolution and dot product are both linear operators and thus inner products can be written as convolutions and vice versa [69]. By rewriting the fully connected layer as convolutions, the traditional CNNs can take input and images larger than its training image and produce a likelihood map, rather than an output for a single voxel. However, this may lead to an output with a far lower resolution than input due to the pooling layers used.

7.2.3 FCN METHODS

7.2.3.1 Network Designs

As discussed previously, the CNN methods input the downsized input image or patch into the convolutional layers and fully connected layers, and subsequent output-predicted label. Shelhamer et al. first proposed a CNN whose last fully connected layer is replaced by a convolutional layer. Since all layers in this CNN are convolutional layers, the new network is named as a fully convolutional network (FCN). Due to the major improvement of deconvolution kernels used to up-sample the feature map, an FCN allows the model to have a dense voxel-wise prediction from the full-size whole volume instead of a patch-wise classification as in a traditional CNN [70]. This segmentation is also called "end-to-end segmentation". By using an FCN, the segmentation of the whole image can be achieved in just one forward pass. To achieve better localization performance, high-resolution activation maps are combined with up-sampled outputs and then passed to the convolution layers to assemble more accurate output.

One of the most well-known FCN structures using the concept of deconvolution for medical image segmentation is the U-net, initially proposed by Ronneberger et al. [68]. The U-net architecture is built upon the elegant architecture of FCN, including an encoding path and a decoding path. Besides the increased depth of network with 19 layers, the U-net introduced a design of long skip connections between the layers of equal resolution in the encoding path to those in the decoding path. These connections provide essential high-resolution features to the deconvolution layers. The improvement of U-net overcomes the trade-off between organ localization and the use of context. This trade-off arises in patch-based architectures since the large size patches require more pooling layers and consequently will reduce the localization accuracy. On the other hand, small size patches can only observe a small context of input.

Inspired by the study of U-net, Milletari et al. proposed an improved network based on the U-net, called the V-net [71]. The V-net architecture is similar to the U-net. It also consists of encoding path (compression) and decoding path and the long skip connection between the encoding and decoding paths. The improvement of V-net as compared to U-net is that at each stage of encoding and decoding path, V-net involves a residual block as a short skip connection between early and later convolutional layers. This architecture ensures convergence compared with non-residual learning network, such as the U-net. Secondly, V-net replaces the max pooling operations with convolutional layers to force the network to have a smaller memory footprint during training, as no switches mapping the output of pooling layers back to their inputs are needed for back-propagation. Thirdly, in contrast to binary cross entropy loss used in the original U-net method, the proposed V-net used Dice loss. Therefore, weights to samples of different organs to establish balance between multi-organs and background voxels are not needed.

As another improvement of the U-net, Christ et al. proposed a new FCN that by cascading two U-nets to improve the accuracy of segmentation [72], called a cascade FCN. The main idea of the cascade FCN is to stack a series of FCNs in the way that each model utilizes the contextual features extracted by the prediction map of the previous model. A simple design is to combine FCNs in a cascade manner, where the first FCN segments the image to ROIs for the second FCN, where the organ segmentation is done. The advantage of using such a design is that separate sets of filters can be applied at each stage and therefore the quality of segmentation can be significantly improved.

The main idea of deep supervision in deeply supervised FCN methods [10, 13] is to provide the direct supervision of the hidden layers and propagate it to lower layers, instead of employing only one supervision at the output layer for traditionally supervised FCNs. In this manner, supervision is extended to deep layers of the network, which would enhance the discriminative ability of feature maps to differentiate multiple classes in multi-organ segmentation tasks. In addition, recently, attention gates were used in an FCN to improve performance in image classification and segmentation [73]. The attention gate could learn to suppress irrelevant features and highlight salient features useful for a specific task.

7.2.3.2 Overview of Works

Zhou et al. proposed a 2.5D FCN segmentation method to automatically segment 19 organs in CT images of the whole body [74]. In that work, a 2.5D patch, which consists of several consecutive slices in the axial plane, was used as multi-channel input for the 2D FCN. A separate FCN was designed for each 2D sectional view, resulting in three FCNs. Ultimately, the segmentation results of three directions were fused to generate the final segmentation output. The technique produced higher accuracy for big organs such as the liver (a Dice value of 0.937) but yielded lower accuracy while dealing with small organs, such as the pancreas (a Dice value of 0.553). In addition, by implementing the convolution kernels in a 3D manner, the FCN has also been used for multi-organ segmentation for 3D medical images [54].

An FCN which has been trained on whole 3D images has high class imbalance between the foreground and background, which results in inaccurate segmentation of small organs. One possible solution to alleviate this issue is applying two-step segmentation in a hierarchical manner, where the second stage uses the output of the first stage by focusing more on boundary regions. Christ et al. performed liver segmentation by cascading two FCNs, where the first FCN detects the liver location, estimating the ROI, and the second FCN extracts features from that ROI to obtain the liver lesions segmentation [72]. This system has achieved 0.823 in Dice for lesion segmentation in CT images and 0.85 in MRI images. Similarly, Wu et al. investigated the cascaded FCN to improve the performance in fetal boundary detection in ultrasound images [75]. Their results have shown better performance compared to other boundary refinement techniques for ultrasound fetal segmentation.

Transrectal ultrasound (TRUS) is a versatile and real-time imaging modality that is commonly used in image-guided prostate cancer interventions (e.g. biopsy and brachytherapy). Accurate segmentation of the prostate is key to biopsy needle placement, brachytherapy treatment planning, and motion management. However, the TRUS image quality around the prostate base and apex region is often affected by low contrast and image noise. To address these challenges, Lei et al. proposed a deep supervision V-net for accurate prostate segmentation [10]. To cope with the optimization difficulties of training the DL-based model with limited training data, a deep supervision strategy with a hybrid loss function (logistic and Dice loss) was introduced to the different stages of decoding path. To reduce possible segmentation errors at the prostate apex and base in TRUS images, a multi-directional-based contour refinement model was introduced to fuse transverse, sagittal, and coronal plane-based segmentation.

Similarly, for the task of MRI pelvic segmentation, segmentation of the prostate is challenging due to the inhomogeneous intensity distributions and variation in prostate anatomy. Wang et al. proposed a 3D FCN with deep supervision and group dilated convolution to segment the prostate on MRI [13]. In this method, the deep supervision mechanism was introduced into FCN to effectively alleviate the common exploding or vanishing gradient problems in training deep models, which forces the update process of the hidden layer filters to favor highly discriminative features. A group dilated convolution which aggregates multi-scale contextual information for dense prediction was proposed to enlarge the effective receptive field. In addition, a combined loss (including cosine and cross entropy) was used to improve the segmentation accuracy from the direction of similarity and dissimilarity. Its architecture is shown in Figure 7.5 as an example of FCN implementation.

Segmenting glands is essential in cancer diagnosis. However, accurate automated DL-based segmentation of glands is challenging because a large variability in glandular morphology across tissues and pathological subtypes exist, and a large number of accurate gland annotations from several tissue slides is required. Binder et al. investigated the idea of cross-domain (organ type) approximation that aims at reducing the need for organ-specific annotations [76]. Two proposed dense-U-nets are trained on hematoxylin- and eosin-strained colon adenocarcinoma samples focusing on the gland and stroma segmentation. Unlike U-net, dense-U-nets use an asymmetric encoder and decoder. The encoder is designed to learn the spatial hierarchies of features automatically and adaptively from low to high level patterns coded within the image. The encoder uses transition

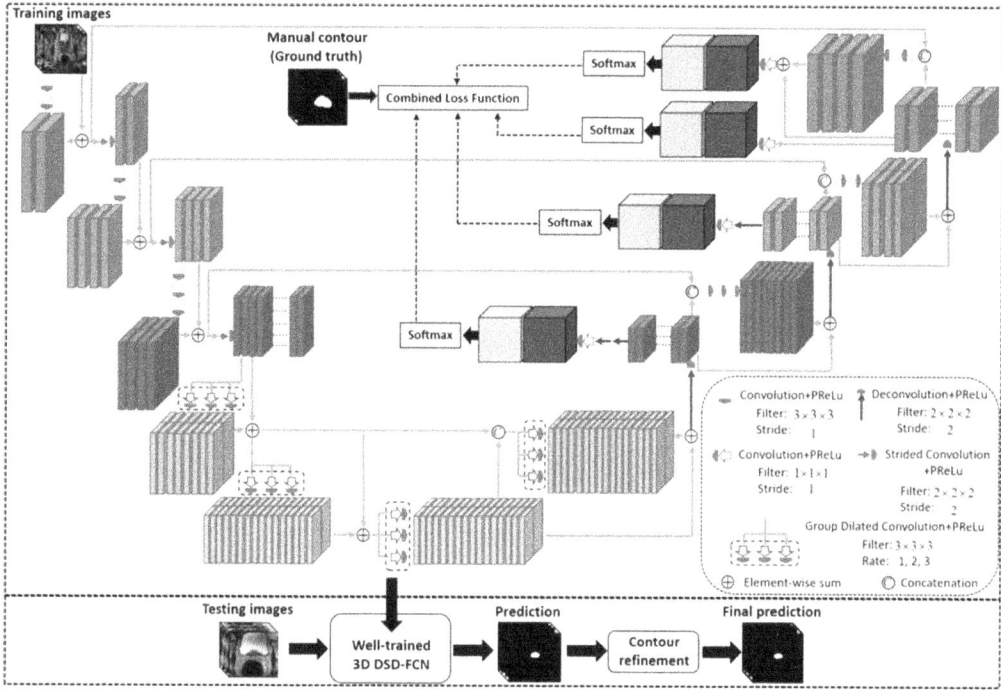

FIGURE 7.5 Schematic representation of an exemplary FCN architecture. Reprinted by permission from John Wiley and Sons: Medical Physics, "Deeply supervised 3D fully convolutional networks with group dilated convolution for automatic MRI prostate segmentation" by Wang et al. [13], copyright 2020.

layer (convolution with stride size 2) and dense convolution blocks consecutively to extract the compressed encoded feature representation. The dense-convolution blocks from DenseNet [43] are used to strengthen feature propagation, encourage feature reuse, and substantially reduce the number of parameters. The decoder is composed of deconvolution layers and convolution blocks. The skip connection between the encoder and the decoder side allows for feature reuse and information flow. The architecture has two decoders, one to predict the relevant gland locations, and a second to predict the gland contours. Thus, the decoders output a gland probability map and a contour probability map. The network is supervised to jointly optimize the prediction of gland locations and gland contours (Table 7.4).

7.2.3.3 Discussion

A problem of the FCN is that the receptive size is fixed so if the object size changes then the FCN struggles to detect it all. One solution is multi-scale networks, where input images are resized and fed to the network. Multi-scale techniques can overcome the problem of the fixed receptive size in an FCN [90]. However, sharing the parameters of the same network on a resized image is not a very effective way as the object of different scales requires different parameters to process. Another solution for images with a larger field of view than the receptive field is for the FCN to be applied as a sliding window across the entire image [45].

As compared to a single FCN architecture, the advantage of cascade FCNs is that separate sets of filters can be applied at each stage and therefore the quality of segmentation can be significantly increased. For example, Trullo et al. proposed using two collaborative FCNs to jointly segment multiple organs in thoracic CT images, one is used for organ localization and the other one is used to segment the organ within that ROI [105]. However, because of the additional network involved and two or more steps, the computation time of this kind of method would be longer than a single

TABLE 7.4

Overview of FCN Methods

Ref.	Year	Network	Supervision	Dimension	Site	Modality
[68]	2015	U-net	Supervised	2D slice	Neuronal structure	Electron microscopic
[72]	2016	Cascaded FCN	Supervised	3D volume	Liver and lesion	CT
[77]	2016	3D U-net	Supervised	3D volume	Kidney	Xenopus
[78]	2017	Dilated FCN	Supervised	2D slice	Abdomen	CT
[79]	2017	3D FCN Feature Driven regression forest	Supervised	3D patch	Pancreas	CT
[74]	2017	2D FCN	Supervised	2.5D slices	Whole body	CT
[80]	2018	Foveal fully convolutional nets	Supervised	N/A*	Whole body	CT
[81]	2018	DRINet	Supervised	2D slice	Brain, abdomen	CT
[82]	2018	3D U-net	Supervised	3D volume	Prostate	MRI
[83]	2018	Dense V-net	Supervised	3D volume	Abdomen	CT
[84]	2018	NiftyNet	Supervised	3D volume	Abdomen	CT
[85]	2018	PU-net, CU-net	Supervised	2D slice	Pelvis	CT
[86]	2018	Dilated U-net	Supervised	2D slice	Chest	CT
[87]	2018	3D U-JAPA-Net	Supervised	3D volume	Abdomen	CT
[88]	2018	U-net	Supervised	2D slice	Pelvis	CT
[89]	2018	Cascade 3D FCN	Supervised	3D patch	Abdomen	CT
[90]	2018	Multi-scale pyramid of 3D FCN	Supervised	3D patch	Abdomen	CT
[91]	2018	Shape representation model constrained FCN	Supervised	3D volume	Head and neck	CT
[92]	2018	Hierarchical dilated neural networks	Supervised	2D slice	Pelvis	CT
[8]	2018	CNN with correction network	Supervised	2D slice	Abdomen	MRI
[93]	2019	Dilated FCN	Supervised	2D slice	Lung	CT
[76]	2019	Dense-U-net	Supervised	2D slice	Head and neck	Stained colon adenocarcinoma dataset
[94]	2019	2D and 3D FCNs	Supervised	2D slice and 3D volume	Pulmonary nodule	CT
[95]	2019	Dedicated 3D FCN	Supervised	3D patch	Thorax/abdomen	DECT
[96]	2019	2D FCN (DeepLabV3+)	Transfer learning	2D slice	Pelvis	MRI
[97]	2019	2D FCN	Supervised	2D patch	Pulmonary vessels	CT

(Continued)

TABLE 7.4 (CONTINUED)
Overview of FCN Methods

Ref.	Year	Network	Supervision	Dimension	Site	Modality
[98]	2019	Dual U-net	Supervised	2D slice	Glioma nuclei	Hematoxylin and eosin (H&E)-stained histopathological image
[99]	2019	Consecutive deep encoder-decoder Network	Supervised	2D slice	Skin lesion	CT
[100]	2019	U-net	Supervised	2D slice	Lung	HRCT
[101]	2019	3D U-net	Supervised	3D volume	Chest	CT
[7]	2019	3D U-net with multi-atlas	Supervised	3D volume	Brain tumor	Dual-energy CT
[102]	2019	Triple-branch FCN	Supervised	N/A	Abdomen/torso	CT
[10]	2019	2.5D deeply supervised V-net	Supervised	2.5 patch	Prostate	Ultrasound
[13]	2019	Group dilated deeply supervised FCN	Supervised	3D volume	Prostate	MRI
[11]	2019	3D FCN	Supervised	3D volume	Arteriovenous malformations	Contract-enhanced CT
[14]	2019	3D FCN	Supervised	3D volume	Left ventricle	SPECT
[15]	2019	DeepMAD	Supervised	2.5D patch	Vessel wall	MRI
[103]	2019	3D U-net	Supervised	3D volume	Head and neck	CT
[104]	2019	OBELISK-Net (sparse deformable convolution)	Supervised	3D volume	Abdomen	CT

*N/A: not available, i.e. not explicitly indicated in the publication

FCN architecture. In addition, if the first network of cascade FCNs is used for organ localization, the performance of this method will largely rely on the accuracy of localization.

Although FCN-based prostate segmentation [10, 13] can offer good performance, limitations of this kind of method still exist. First, due to three or more stages of deep supervision and the corresponding up-sampling convolutional kernels involved, the computation complexity is higher than the U-net and V-net methods. In addition, when using 2.5D patch as input, the segmented contours of each of the three directions may not be well matched. Introducing an adaptive and non-linear contour refinement model, such as a conditional random field, would be a future work if researchers use this kind of method for multi-organ segmentation. Second, FCNs used voxel-wise loss such as cross entropy for segmentation. However, in the final segmentation map, there is no guarantee of spatial consistency. Recently, in FCN-based methods, conditional random field and graph cut methods have been used as segmentation refinement into the FCN-based workflow by incorporating spatial correlation. The limitation of these kinds of segmentation refinement is that they only consider pair-wise potential which can allow boundary leakage in low contrast regions.

7.2.4 GAN METHODS

7.2.4.1 Network Designs

GANs have gained a lot of attention in medical imaging due to their capability for data generation without explicitly modeling the probability density function. The adversarial loss brought by the discriminator provides a way of incorporating unlabeled samples into training and imposing higher order consistency. This has been proven to be useful in many cases, such as image reconstruction [106], image enhancement [107, 108], segmentation [12, 109], classification and detection [110], augmentation [111], and cross-modality synthesis [112].

A typical GAN consists of two competing networks: a generator and a discriminator [113]. The generator is trained to generate artificial data that approximate the target data distribution from a low-dimensional latent space. The discriminator is trained to distinguish the artificial data from actual data. The workflow is shown in Figure 7.6 [12]. The discriminator encourages the generator to predict realistic data by penalizing unrealistic predictions. Therefore, the discriminative loss could be considered as a dynamic network-based loss term. The two networks compete with each other in a zero-sum game. Multiple variants of GAN can be summarized into three

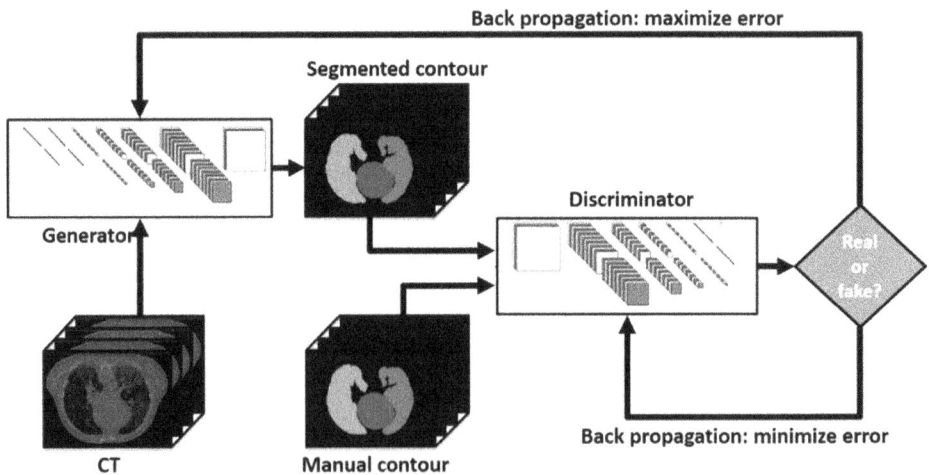

FIGURE 7.6 The process of generative adversarial network. Reprinted by permission from John Wiley and Sons: Medical Physics, "Automatic multiorgan segmentation in thorax CT images using U-net-GAN" by Dong et al. [12], copyright 2020.

categories: (1) variants of discriminator's objective, (2) variants of generator's objective, and (3) variants of architecture, which are summarized in Yi et al. [114].

7.2.4.2 Overview of Works

As discussed in the section on FCN methods, one of the challenges of medical image segmentation using traditional FCN methods is that these methods may introduce boundary leakage in low contrast regions. Using adversarial loss introduced via a discriminator can take into account high order structures that can potentially solve this problem [115]. The adversarial loss can be regarded as a learned similarity measurement between the segmented contours and the annotated ground truth (manual contours) for medical image segmentation tasks. Instead of only measuring the similarity (such as Dice loss and cross entropy loss) in the voxel domain, the additional discriminator maps the segmented and ground truth contours to a low dimensional feature space to represent the shape information and then uses logistic loss to measure the similarity of the feature vector between segmented contours and manual contours. The idea is similar to the perceptual loss. The difference is that the perceptual loss is computed from a pre-trained classification network on natural images whereas the adversarial loss is computed from a network that trained adaptively during the evolvement of the generator.

Dai et al. proposed the structure correcting adversarial network (SCAN) to segment lung fields and the heart in CXR images [109]. The SCAN approach used an FCN as the generator to generate the binary mask of segmented organs and incorporated a critic network (discriminator) to discriminate the structural regularities emerging from human physiology. During training, the critical network learns to discriminate between the ground truth organ annotations from the masks synthesized by the segmentation network. Through this adversarial process, the critical network is able to learn the higher order structures and to guide the segmentation model to achieve realistic segmentation outcomes.

In medical image multi-organ segmentation, a major limitation of traditional DL-based segmentation methods is their requirement for a large amount of paired training images with ground truth contours as learning targets. In order to solve this challenge, Mondal et al. proposed a GAN-based method for 3D multimodal brain MRI segmentation from a few-shot learning perspective [116]. The main idea of this work is to leverage the recent success of GANs to train a DL-based model with highly limited training set of labeled images, without sacrificing the performance of full supervision. The proposed adversarial network encourages the segmentation to have a similar distribution of outputs for images with and without annotations, thereby helping generalization. In addition, the few-shot learning method seeks good generalization on problems with a limited labeled dataset, typically containing just a few training samples of the target classes.

Dong et al. proposed an adversarial network to train deep neural networks for the segmentation of multiple organs on thoracic CT images [12]. The proposed design of adversarial networks, called a U-net-generative-adversarial-network (U-net-GAN), jointly trains a set of U-nets as generators and FCNs as discriminators. A U-net-GAN is a conditional GAN. Specifically, the generator, composed of a U-net produces an image segmentation map of multiple organs by an end-to-end mapping learned from the CT image and its labeled organs. The discriminator, structured as FCN, discriminates between the ground truth and segmented organs produced by the generator. The generator and discriminator compete against each other in an adversarial learning process to produce the optimal segmentation map of multiple organs (Table 7.5).

7.2.4.3 Discussion

In segmentation tasks, GAN is efficient at the prediction stage since it only needs to perform a forward pass through the generator network for segmentation. The discriminator network is only used during training. Using adversarial loss as a shape regulator can benefit more when the learning target (organ) has a regular shape, e.g. for lung and heart, but would be less useful for other irregular objects, such as vessels and catheters.

TABLE 7.5

Overview of GAN Methods

Ref.	Year	Network	Supervision	Dimension	Site	Modality
[109]	2015	SCAN	Supervised	2D slice	Chest	X-rays
[117]	2017	Multi-connected adversarial networks	Unsupervised	2D slice	Brain	Multi-modality MRI
[118]	2017	Dilated GAN	Supervised	2D slice	Brain	MRI
[119]	2017	Conditional GAN	Supervised	2D slice	Brain tumor	MRI
[120]	2017	GAN	Supervised	2D patch	Retinal vessel	Fundoscopic
[115]	2017	Adversarial image-to-image network	Supervised	3D volume	Liver	CT
[121]	2017	Adversarial FCN-CRF nets	Supervised	2D slice	Mass	Mammograms
[122]	2018	GAN	Supervised	N/A*	Brain tumor	MRI
[116]	2018	Few-shot GAN	Semi-supervised	3D patch	Brain	MRI
[123]	2018	Context-aware GAN	Supervised	2D cropped slices	Cardiac	MRI
[124]	2018	Conditional generative refinement adversarial networks	Supervised	2D slice	Brain	MRI
[125]	2018	SegAN	Supervised	2D slice	Brain	MRI
[126]	2018	MDAL	Supervised	2D slice	Left and right ventricular	Cardiac MRI
[127]	2018	TD-GAN	Unsupervised	2D slice	Whole body	X-ray
[12]	2019	U-net-GAN	Supervised	3D volume	Thorax	CT
[128]	2019	Conditional GAN	Supervised	2D slice	Nuclei	Histopathology Images
[129]	2019	Distance-aware GAN	Supervised	2D slice	Chest	CT

*N/A: not available, i.e. not explicitly indicated in the publication

In Dong et al. [12], GAN was applied to delineate the left and right lungs, spinal cord, esophagus, and heart using 35 patients' chest CTs. The averaged DSC for the above five OARs are 0.97, 0.97, 0.90, 0.75, and 0.87, respectively. The mean surface distance of the five OARs obtained with GAN method ranges between 0.4 and 1.5 mm on average among all 35 patients. The mean dose differences on the 20 SBRT lung plans using the segmented results ranged from −0.001 to 0.155 Gy for the five OARs. This demonstrates that GAN is a potentially valuable method for improving the efficiency of the lung radiation therapy treatment planning.

Patient movement, such as translation and rotation, does not change the relative position among organs. Including the transformed data could help avoid overfitting and help the segmentation algorithm learn this invariant property. However, for multi-organ segmentation, due to the size and shape differences among different organs and variation among patients, it is difficult to balance the loss function among different organs. Integrating all the segmentations into one network complicates the training process and reduces segmentation accuracy. To simplify the method, the GAN method of Dong et al. [12] grouped OARs of similar dimensions, and utilized three subnetworks for segmentation, one for lungs and heart, and the other two for esophagus and spinal cord, respectively. This approach improves segmentation accuracy at the cost of computation efficiency. However, it also introduces additional computation time for both training and prediction. This could become an issue if the method was required to segment more OARs simultaneously. Simultaneously determining the location of organs and segmenting the organs within that location for multi-organ segmentation would be a future direction for research.

7.2.5 R-CNN Methods

7.2.5.1 Network Designs

In medical image multi-organ segmentation, as discussed above, simultaneous segmenting of multiple organs is challenging, because it requires the correct detection of all organs in an image volume while also accurately segmenting the organs within that detection. It is similar to the classical computer vision of tasks of instance segmentation, which include two subtasks: one is the object detection with the goal of classifying individual objects and localizing each using a bounding box (ROI to medical image), the other one is the semantic segmentation with the goal of classifying each pixel into a fixed set of categories without differentiating object instances. Recently, the development of the region-CNN (R-CNN) family introduced a simple and flexible way to solve this challenge.

An R-CNN is a network based on ROIs [130]. To bypass the problem of selecting a large number of regions, the R-CNN utilized a selective search [131] to extract 2000 candidate regions from each image. These regions were called region proposals. By warping to the same size, these regional proposals were then fed into a CNN to extract a 4096-dimensional feature vector as output. The CNN acts as a feature extractor. The output dense layer consists of the features extracted from the image and the extracted features are fed into a support vector machine (SVM) to classify the presence of the object within that region proposal. In addition to predicting the presence of an object within the region proposals, the algorithm also predicts four values (2D version) which are offset values to increase the precision of the bounding box.

An R-CNN needs a long computation time to train the network with 2000 region proposals per 2D image slice. To address this issue, Girshick et al. proposed a faster objection algorithm called the Fast R-CNN [132]. Compared to the R-CNN, instead of selecting region proposals and feeding them into a CNN, the region proposals were obtained by first feeding the original image into an FCN to obtain the convolutional feature map to identify the regional proposals, followed by warping them into squares and reshaping them into a fixed size using an ROI pooling layer. By using a fully connected layer, the regional proposal was projected to an ROI feature vector. Finally, a softmax layer was used to predict the class of that region proposal

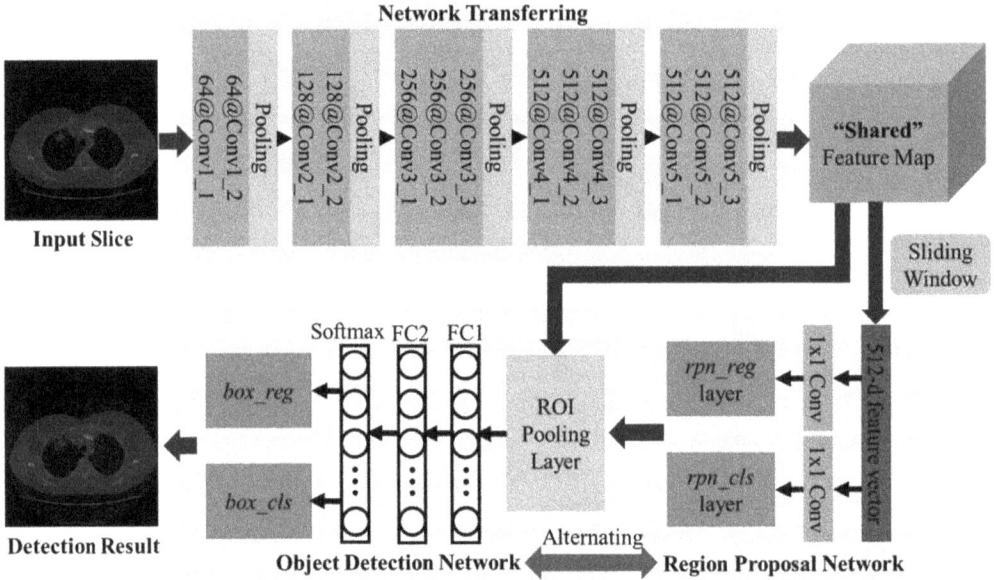

FIGURE 7.7 The architecture of Faster R-CNN. Reprinted by permission from Elsevier: Computerized Medical Imaging and Graphics, "Fast and fully-automated detection and segmentation of pulmonary nodules in thoracic CT scans using deep convolutional neural networks" by Huang et al. [134], copyright 2020.

and also the offset values for the bounding box. The reason that a Fast R-CNN is faster than an R-CNN is because it does not need to feed 2000 region proposals to the CNN for each feeding image. Instead, the convolution operation is done only once per image and a feature map is generated from it.

Both the R-CNN and the Fast R-CNN use selective search to identify the region proposals. However, selective search is time-consuming. To solve this problem, Ren et al. proposed an object detection algorithm that eliminates the selective search and lets the network learn the regional proposals, called the Faster R-CNN [133]. Similar to the Fast R-CNN, the Faster R-CNN first fed the image into an FCN to extract convolutional feature map. Instead of using selective search on the feature map to identify the region proposals, a separate network was used to predict the region proposals. The predicted region proposals were then reshaped using an ROI pooling layer which was then used to classify the image within the proposed region and predict the offset values for the bounding boxes. The architecture of Faster R-CNN is illustrated in Figure 7.7 [134].

Based on the ground works of feature extraction and regional proposals identification built by the Faster R-CNN, performing image segmentation within the detected bounding box (ROI) is easy to achieve. After an ROI pooling layer in the Faster R-CNN, He et al. integrated two more convolution layers to build the semantic segmentation within the ROI, called the Mask R-CNN [135]. Another major contribution of the Mask R-CNN is the refinement of the ROI pooling. In the previous Faster R-CNN, Fast R-CNN, and R-CNN methods the ROI warping is digitalized: the cell boundaries of the target feature map are forced to realign with the boundary of the input feature maps. Therefore, each target cell may not be of the same size. Mask R-CNN uses ROI Align which does not digitalize the boundary of the cells and make every target cell the same size. It also applies interpolation to better calculate the feature map values within the cell.

7.2.5.2 Overview of Works

In order to solve the problem of low-quality CT images, the lack of annotated data, and the complex shapes of lung nodules, Liu et al. applied a 2D Mask R-CNN for lung pulmonary nodule segmentation in a transfer learning manner [136]. The Mask R-CNN was trained on the common objects in context (COCO) data set, which is a natural image dataset, and was then fine-tuned to segment pulmonary nodules. As an improvement, Kopelowitz and Engelhard applied a 3D Mask R-CNN to handle a 3D CT image volume to detect and segment the lung nodules [137].

Xu et al. proposed a novel heart segmentation pipeline which combined the Faster R-CNN and the U-net, abbreviated as CFUN [138]. Due to the Faster R-CNN's precise localization ability and the U-net's powerful segmentation ability, CFUN needs only one-step detection and segmentation inference to get the whole heart segmentation result, obtaining good results with significantly reduced computational cost. Furthermore, CFUN adopts a new loss function based on edge information, 3D Edge-loss, as an auxiliary loss to accelerate the convergence of training and improve the segmentation results. Extensive experiments on a public dataset show that CFUN exhibits competitive segmentation performance in a sharply reduced inference time. Similarly, Bouget et al. proposed a combination of a Mask R-CNN and a U-net for the segmentation and detection of mediastinal lymph nodes and anatomical structures in CT data for lung cancer staging [139].

Li et al. proposed a lung nodule detection method based on the Faster R-CNN for thoracic MRI in a transfer learning manner [140]. A false positive (FP) reduction scheme based on anatomical characteristics was designed to reduce FPs and preserve the true nodule. Similarly, the Faster R-CNN was also used for pulmonary nodule detection on CT images [134].

Xu et al. proposed an efficient detection method for multi-organ localization in CT images using a 3D regional proposal network (RPN) [141]. Since the proposed RPN is implemented in a 3D manner, it can take advantage of the spatial context information in a CT image. AlexNet was used to build a backbone network architecture that is able to generate high-resolution feature maps to further improve the localization performance of small organs. The method was evaluated on abdomen and brain site datasets and achieved high detection precision and localization accuracy with fast inference speed (Table 7.6).

TABLE 7.6

Overview of R-CNN Methods

Ref.	Year	Network	Supervision	Dimension	Site	Modality
[136]	2018	Mask R-CNN	Transfer learning	2D slice	Lung nodule	CT
[138]	2018	Combination of faster R-CNN and U-net (CFUN)	Supervised	3D volume	Cardiac	CT
[139]	2019	Combination of U-net and mask R-CNN	Supervised	2D slice	Chest	CT
[134]	2019	Faster R-CNN	Supervised	2D slice	Thorax/pulmonary nodule	CT
[137]	2019	3D mask R-CNN	Supervised	3D volume	Lung nodule	CT
[140]	2019	3D faster R-CNN	Supervised	3D volume	Thorax/lung nodule	MRI
[142]	2019	Mask R-CNN	Supervised	N/A*	Chest	X-Ray
[141]	2019	3D RPN	Supervised	3D volume	Whole body	CT
[143]	2019	Multiscale mask R-CNN	Supervised	2D slice	Lung tumor	PET

*N/A: not available, i.e. not explicitly indicated in the publication

7.2.5.3 Discussion

In the work of Liu et al. [136], researchers used a Mask R-CNN to segment lung nodules for the first time. After a series of comparative experiments, ResNet101 and feature pyramid network (FPN) were selected as the backbone of a Mask-R-CNN. Experimental results showed that it not only identified the location of nodules, but also provided nodule contour information. It provided more detailed information for cancer treatment. The proposed method was validated on the Lung Image Database Consortium – Image Database Resource Initiative (LIDC-IDRI) data set and achieved the desired accuracy. However, due to the 2D network design, some spatial information of the CT image will be lost. 3D contexts play an important role in recognizing nodules. A 3D Mask R-CNN would perform better than a 2D version as it also captures the crani-caudal information.

Limitations still exist for the detection of lung nodules using Faster R-CNN. First, small and low contrast nodules are not successfully detected by the Faster R-CNN. This challenge may also occur for other multi-organ segmentation problems when there are small organs, such as the esophagus in lung segmentation. Second, the researchers found that some air artifacts and juxta-cardiac tissues may be falsely detected as nodules. In order to alleviate these problems, Li et al. designed a filter to improve the image quality and remove these artifacts [140]. In addition, a multi-scale strategy was introduced in the whole detection system to increase the detection rate of small and low contrast nodules.

The R-CNN family of methods could be an efficient tool for several multi-organ segmentation and detection tasks. However, technical adjustments and optimizations may be required to make the extended model to achieve comparable performance to the methods dedicated to organ segmentation. Due to the higher data dimensionality and larger number of weight parameters, training a 3D R-CNN-based model is more time-consuming than a 2D version. However, it may have significant advantages, such as higher localization accuracy and higher prediction speed. To speed up the training procedure of the proposed method, one potential solution is to apply batch normalization after each convolutional layer in the backbone network to improve the model convergence, and conduct most calculations on a graphics processing unit with parallel computing [141].

7.2.6 Hybrid Methods

7.2.6.1 Network Designs

Recently some methods used hybrid designs to solve the challenge of poor image quality, such as low contrast around the organ boundary. The hybrid design involves two or more networks for different functional proposes. For example, one network aims to enhance the image quality, and the other one to segment the OARs from the enhanced image. This has the potential to be a new trend for multi-organ segmentation.

7.2.6.2 Overview of Works

Accurate segmentation of pelvic OARs on CT images for treatment planning is challenging due to the poor soft-tissue contrast [144, 145]. MRI has been used to aid prostate delineation, but its accuracy is limited by MRI-CT registration errors [146, 147]. Lei et al. developed a deep attention-based segmentation strategy based on CT-based synthetic MRI (sMRI) images created using a cycle GAN [112]. This was done to address the segmentation of low contrast soft tissue organs (such as bladder, prostate, and rectum) without MRI acquisition [5]. This hybrid method included two main steps: first, a CycleGAN was used to estimate sMRI from CT images. Second, a deep attention FCN was trained based on sMRI and manual contours deformed from MRIs. Attention models were introduced to pay more attention to the prostate boundary. Inspired by this method, Dong et al. developed an sMRI-aided segmentation method for male pelvic CT multi-organ segmentation [4]. Similarly, Lei et al. introduced this kind of method for the multi-organ segmentation of cone-beam computed tomography (CBCT) pelvic data for CBCT-guided adaptive radiation therapy workflow [3].

TABLE 7.7

Overview of Hybrid Methods

Ref.	Year	Network	Supervision	Dimension	Site	Modality
[4, 5]	2019	Synthetic MRI-aided FCN	Supervised	2.5D patch	Pelvic	CT
[3]	2019	Synthetic MRI-aided deep attention FCN	Supervised	3D volume	Pelvic	CBCT
[148]	2019	Deep multi-planar co-training (DMPCT)	Co-training	3D volume	Abdomen	CT

In multi-organ segmentation of abdominal CT scans, as discussed earlier, supervised DL-based algorithms require lots of voxel-wise annotations, which are usually difficult, expensive, and slow to obtain. However, massive unlabeled 3D CT volumes are usually easily accessible. Zhou et al. proposed deep multi-planar co-training (DMPCT) in a semi-supervised learning manner to solve this problem [148]. The DMPCT network architecture includes three steps: (1) A DL-based network is learned in a co-training manner to mine consensus information from 2D patches extracted from multiple planes. The DL-based network is called a "teacher model" in their work. (2) The trained teacher model is then used to assign pseudo labels to the unlabeled data. Multi-planar fusion is applied to generate more reliable labels to alleviate the errors occurring in the pseudo labeling and thus can help train better segmentation networks. (3) An additional network, called a "student model", is trained on the union of the manual labeled data and automatically labeled data (called self-labeled samples) to enlarge the data variation of the training data (Table 7.7).

7.2.6.3 Discussion

Compared to CT and CBCT images, the superior soft-tissue contrast of sMRI improves the prostate segmentation accuracy and alleviate the issue of prostate volume overestimation when using CT images alone. However, in sMRI-aided segmentation methods, the registration between training MRI and CT or CBCT will affect the sMRI image quality and thus affect the segmentation network performance. In this sense, the registration error also affects the delineation accuracy ultimately. Thus, this kind of method relies on the accurate deformable image registration.

Hybrid methods can be practical for clinical applications since hybrid methods do not need a large number of training multi-modality data samples to provide comprehensive information or a large number of manual delineated contours for learning the target annotations (the annotation of multiple organs in 3D volumes requires massive labor from radiologists). Testing of hybrid methods' performance in segmentation of multiple complex anatomical structures, such as the 2017 AAPM Thoracic Auto-segmentation Challenge, will be a future research direction.

7.3 BENCHMARK

Benchmarking, such as the 2017 AAPM Thoracic Auto-segmentation Challenge [149], can be helpful to understand the advantages and disadvantages of each method and compare architectures. Figure 7.8 shows a visual result of one patient using a DL-based method from the challenge [12].

From the report of this challenge [149], there were seven participants who completed the online challenge. Five out of seven participants used DL-based methods. In addition to those reported DL-based methods participating in this challenge, a review of recent studies using this benchmark data was performed. The numerical results of different DL-based methods are listed as follows:

There was not a single method or architecture that outperforms others in every organ structure. This highlights the inherent challenge of multi-organ segmentations. From Table 7.8, it seems that U-net-GAN has the best performance in esophagus, lungs, and spinal cord, yet gives heart contours

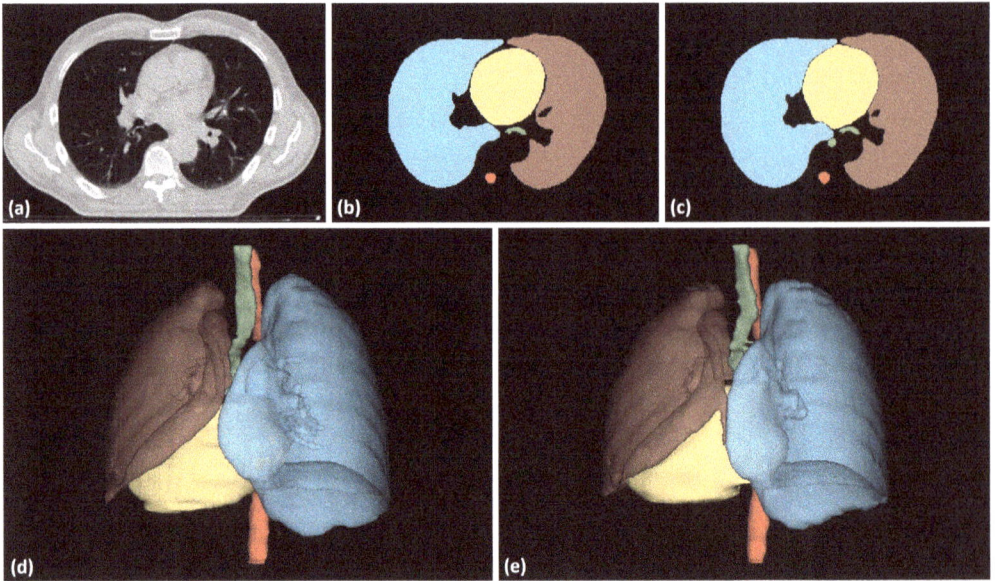

FIGURE 7.8 Visual result of one patient's data via DL-based method: (a) shows the CT image in axial view, (b) and (c) show the manual contour and segmented contour in axial view, respectively; (d) and (e) show the manual contour and segmented contour in 3D view.

trailing behind other methods. The results in this table also seem to indicate that 3D methods tend to perform better than the 2D counterpart. This is studied further in Chapter 8. It is worth noting that, except for the esophagus, most methods reach the level of accuracy equivalent to interobserver variability within clinical use.

7.4 CONCLUSION

This chapter covered the current state-of-the-art DL-based auto-segmentation architecture used for auto-delineation. Detailed discussion was given to cover all proposed DL-based methods found in the literature. Comparisons were made to highlight the novelty of each architectural approach and contrasted the pros and cons of them. In general, multi-organ segmentation is a challenging topic and currently there is not a clear architectural solution. The 2017 AAPM Thoracic Auto-segmentation Challenge can be used as a benchmark to show the relative performance of recent DL-based methods. It further demonstrates the difficulty of this problem. Although there is no single architecture that outperformed others for all organs, the data do provide some insights towards the superiority of certain methods.

Judging from the statistics of the cited works, there is a clear trend of using FCNs to perform end-to-end semantic segmentation for multi-organ automatic segmentation. Recently, GAN-based methods have been used to enhance the reliability of segmented contours. R-CNNs and hybrid methods have started to gain popularity in medical image segmentation, although only a few methods surveyed were applied to multi-organ segmentation.

DL-based multi-organ segmentation techniques represent a significant innovation in daily practices of radiation therapy workflow, expediting the segmentation process, enhancing contour consistency, and promoting compliance to delineation guidelines [4, 12, 52, 86, 88, 91, 96]. Furthermore, DL-based multi-organ segmentation could facilitate online adaptive radiotherapy to improve clinical outcomes.

TABLE 7.8

Comparison of the Results from DL-Based Methods Using Datasets from the 2017 AAPM Thoracic Auto-segmentation Challenge

Metric	Method	Esophagus	Heart	Left Lung	Right Lung	Spinal Cord
DSC	DCNN Team Elekta	0.72 ± 0.10	0.93 ± 0.02	0.97 ± 0.02	0.97 ± 0.02	0.88 ± 0.037
	3D U-net [150]	0.72 ± 0.10	0.93 ± 0.02	0.97 ± 0.02	0.97 ± 0.02	0.89 ± 0.04
	Multi-class CNN Team Mirada	0.71 ± 0.12	0.91 ± 0.02	0.98 ± 0.02	0.97 ± 0.02	0.87 ± 0.110
	2D ResNet Team Beaumont	0.61 ± 0.11	0.92 ± 0.02	0.96 ± 0.03	0.95 ± 0.05	0.85 ± 0.035
	3D and 2D U-net Team WUSTL	0.55 ± 0.20	0.85 ± 0.04	0.95 ± 0.03	0.96 ± 0.02	0.83 ± 0.080
	U-net-GAN [12]	0.75 ± 0.08	0.87 ± 0.05	0.97 ± 0.01	0.97 ± 0.01	0.90 ± 0.04
MSD (mm)	DCNN Team Elekta	2.23 ± 2.82	2.05 ± 0.62	0.74 ± 0.31	1.08 ± 0.54	0.73 ± 0.21
	3D U-net [150]	2.34 ± 2.38	2.30 ± 0.49	0.59 ± 0.29	0.93 ± 0.57	0.66 ± 0.25
	Multi-class CNN Team Mirada	2.08 ± 1.94	2.98 ± 0.93	0.62 ± 0.35	0.91 ± 0.52	0.76 ± 0.60
	2D ResNet Team Beaumont	2.48 ± 1.15	2.61 ± 0.69	2.90 ± 6.94	2.70 ± 4.84	1.03 ± 0.84
	3D and 2D U-net Team WUSTL	13.10 ± 10.39	4.55 ± 1.59	1.22 ± 0.61	1.13 ± 0.49	2.10 ± 2.49
	U-Net-GAN [12]	1.05 ± 0.66	1.49 ± 0.85	0.61 ± 0.73	0.65 ± 0.53	0.38 ± 0.27
HD95 (mm)	DCNN Team Elekta	$7.3 + 10.31$	5.8 ± 1.98	2.9 ± 1.32	4.7 ± 2.50	2.0 ± 0.37
	3D U-net [150]	$8.71 + 10.59$	6.57 ± 1.50	2.10 ± 0.94	3.96 ± 2.85	1.89 ± 0.63
	Multi-class CNN Team Mirada	7.8 ± 8.17	9.0 ± 4.29	2.3 ± 1.30	3.7 ± 2.08	2.0 ± 1.15
	2D ResNet Team Beaumont	8.0 ± 3.80	8.8 ± 5.31	7.8 ± 19.13	14.5 ± 34.4	2.3 ± 0.50
	3D and 2D U-net Team WUSTL	37.0 ± 26.88	13.8 ± 5.49	4.4 ± 3.41	4.1 ± 2.11	8.10 ± 10.72
	U-net-GAN [12]	4.52 ± 3.81	4.58 ± 3.67	2.07 ± 1.93	2.50 ± 3.34	1.19 ± 0.46

*Note: Methods not followed by a reference were from the 2017 AAPM Thoracic Auto-segmentation Challenge report [149].

ACKNOWLEDGMENTS

This chapter was supported in part by the National Cancer Institute of the National Institutes of Health under Award Number R01CA215718, and Dunwoody Golf Club Prostate Cancer Research Award, a philanthropic award provided by the Winship Cancer Institute of Emory University.

REFERENCES

1. MH Hesamian, W Jia, XJ He, and P Kennedy, "Deep learning techniques for medical image segmentation: achievements and challenges," (in English), *J Digit Imaging*, vol. 32, no. 4, pp. 582–596, Aug 2019, doi:10.1007/s10278-019-00227-x.
2. T Zhou, S Ruan, and S Canu, "A review: deep learning for medical image segmentation using multi-modality fusion," *Array*, vol. 3–4, p. 100004, Sep 1 2019, https://doi.org/10.1016/j.array.2019.100004.

3. Y Lei et al., "Male pelvic multi-organ segmentation aided by CBCT-based synthetic MRI," (in English), *Phys Med Biol,* vol. in press, Dec 18 2019, doi:10.1088/1361-6560/ab63bb.

4. X Dong et al., "Synthetic MRI-aided multi-organ segmentation on male pelvic CT using cycle consistent deep attention network," (in English), *Radiother Oncol*, vol. 141, pp. 192–199, Dec 2019, doi:10.1016/j.radonc.2019.09.028.

5. Y Lei et al., "CT prostate segmentation based on synthetic MRI-aided deep attention fully convolution network," (in English), *Med Phys,* vol. in press, Nov 20 2019, doi:10.1002/mp.13933.

6. Y Lei et al., "Whole-body PET estimation from low count statistics using cycle-consistent generative adversarial networks," (in English), *Phys Med Biol*, vol. 64, no. 21, p. 215017, Nov 4 2019, doi:10.1088/1361-6560/ab4891.

7. B van der Heyden et al., "Dual-energy CT for automatic organs-at-risk segmentation in brain-tumor patients using a multi-atlas and deep-learning approach," (in English), *Sci Rep*, vol. 9, Mar 11 2019, doi:ARTN 412610.1038/s41598-019-40584-9.

8. YB Fu et al., "A novel MRI segmentation method using CNN-based correction network for MRI-guided adaptive radiotherapy," (in English), *Med Phys*, vol. 45, no. 11, pp. 5129–5137, Nov 2018, doi:10.1002/mp.13221.

9. YB Fu, S Liu, HH Li, and DS Yang, "Automatic and hierarchical segmentation of the human skeleton in CT images," (in English), *Phys Med Biol*, vol. 62, no. 7, pp. 2812–2833, Apr 7 2017, doi:10.1088/1361-6560/aa6055.

10. Y Lei et al., "Ultrasound prostate segmentation based on multidirectional deeply supervised V-Net," (in English), *Med Phys*, vol. 46, no. 7, pp. 3194–3206, Jul 2019, doi:10.1002/mp.13577.

11. T Wang et al., "Learning-based automatic segmentation of arteriovenous malformations on contrast CT images in brain stereotactic radiosurgery," (in English), *Med Phys*, vol. 46, no. 7, pp. 3133–3141, Jul 2019, doi:10.1002/mp.13560.

12. X Dong et al., "Automatic multiorgan segmentation in thorax CT images using U-net-GAN," (in English), *Med Phys*, vol. 46, no. 5, pp. 2157–2168, May 2019, doi:10.1002/mp.13458.

13. B Wang et al., "Deeply supervised 3D fully convolutional networks with group dilated convolution for automatic MRI prostate segmentation," (in English), *Med Phys*, vol. 46, no. 4, pp. 1707–1718, Apr 2019, doi:10.1002/mp.13416.

14. T Wang et al., "A learning-based automatic segmentation and quantification method on left ventricle in gated myocardial perfusion SPECT imaging: a feasibility study," (in English), *J Nucl Cardiol,* vol. in press, Jan 28 2019, doi:10.1007/s12350-019-01594-2.

15. J Wu et al., "Deep morphology aided diagnosis network for segmentation of carotid artery vessel wall and diagnosis of carotid atherosclerosis on black-blood vessel wall MRI," (in English), *Med Phys*, vol. 46, no. 12, pp. 5544–5561, Dec 2019, doi:10.1002/mp.13739.

16. Y Liu et al., "Head and neck multi-organ auto-segmentation on CT images aided by synthetic MRI," *Med Phys*, doi:10.1002/mp.14378.

17. Y Liu et al., "CT-based multi-organ segmentation using a 3D self-attention U-net network for pancreatic radiotherapy," *Med Phys*, doi:10.1002/mp.14386.

18. BJun Guo et al., "Automated left ventricular myocardium segmentation using 3D deeply supervised attention U-net for coronary computed tomography angiography; CT myocardium segmentation," *Med Phys*, vol. 47, no. 4, pp. 1775–1785, 2020, doi:10.1002/mp.14066.

19. X He et al., "Automatic segmentation and quantification of epicardial adipose tissue from coronary computed tomography angiography," *Phys Med Biol*, vol. 65, no. 9, p. 095012, Nov 5 2020, doi:10.1088/1361-6560/ab8077.

20. Y Fu et al., "Pelvic multi-organ segmentation on cone-beam CT for prostate adaptive radiotherapy," *Med Phys*, vol. 47, no. 8, pp. 3415–3422, 2020, doi:10.1002/mp.14196.

21. K Raza, and NKJA Singh, "A tour of unsupervised deep learning for medical image analysis," *ArXiv*, vol. abs/1812.07715, 2018.

22. HC Shin, MR Orton, DJ Collins, SJ Doran, and MO Leach, "Stacked autoencoders for unsupervised feature learning and multiple organ detection in a pilot study using 4D patient data," (in English), *IEEE T Pattern Anal*, vol. 35, no. 8, pp. 1930–1943, Aug 2013, doi:10.1109/Tpami.2012.277.

23. V Alex, K Vaidhya, S Thirunavukkarasu, C Kesavadas, and G Krishnamurthi, "Semisupervised learning using denoising autoencoders for brain lesion detection and segmentation," (in English), *J Med Imaging (Bellingham, Wash.)*, vol. 4, no. 4, p. 041311, Oct 2017, doi:10.1117/1.Jmi.4.4.041311.

24. P Vincent, H Larochelle, I Lajoie, Y Bengio, and P-AJJMLR Manzagol, "Stacked denoising autoencoders: learning useful representations in a deep network with a local denoising criterion," *AC Med*, vol. 11, pp. 3371–3408, 2010.

25. K Vaidhya, S Thirunavukkarasu, A Varghese, and G Krishnamurthi, "Multi-modal brain tumor segmentation using stacked denoising autoencoders," *BrainLes*, 2015, Cham: Springer, pp. 181–194.

26. SF Qadri, Z Zhao, D Ai, M Ahmad, and Y Wang, *Vertebrae Segmentation Via Stacked Sparse Autoencoder from Computed Tomography Images* (Eleventh International Conference on Digital Image Processing (ICDIP 2019)). SPIE, 2019.

27. F Li, H Qiao, B Zhang, and X Xi, "Discriminatively boosted image clustering with fully convolutional auto-encoders," *arXiv Pat Recog*, vol. 83, pp. 161–173, 2017.

28. X Guo, X Liu, E Zhu, and J Yin, "Deep clustering with convolutional autoencoders," in *Neural Information Processing*, D Liu, S Xie, Y Li, D Zhao, and E-SM El-Alfy, Eds., 2017, Cham: Springer International Publishing, pp. 373–382.

29. C Wang, A Elazab, F Jia, J Wu, and Q Hu, "Automated chest screening based on a hybrid model of transfer learning and convolutional sparse denoising autoencoder," (in English), *Biomed Eng Online*, vol. 17, no. 1, p. 63, May 23 2018, doi:10.1186/s12938-018-0496-2.

30. M Ahmad, J Yang, D Ai, SF Qadri, and Y Wang, "Deep-stacked auto encoder for liver segmentation," in *Advances in Image and Graphics Technologies*, Y Wang et al., Eds., 2018, Singapore: Springer, pp. 243–251.

31. X Wang, S Zhai, and Y Niu, "Automatic vertebrae localization and identification by combining deep SSAE contextual features and structured regression forest," (in English), *J Digit Imaging*, vol. 32, no. 2, pp. 336–348, Apr 2019, doi:10.1007/s10278-018-0140-5.

32. E Tappeiner et al., "Multi-organ segmentation of the head and neck area: an efficient hierarchical neural networks approach," *Int J Comput Assist Radiol Surg*, vol. 14, no. 5, pp. 745–754, May 2019, doi:10.1007/s11548-019-01922-4.

33. BH Menze et al., "The multimodal brain tumor image segmentation Benchmark (BRATS)," *IEEE Trans Med Imaging*, vol. 34, no. 10, pp. 1993–2024, Oct 2015, doi:10.1109/TMI.2014.2377694.

34. HR Roth et al., "DeepOrgan: multi-level deep convolutional networks for automated pancreas segmentation," (in English), *Lect Notes Comput Sci*, vol. 9349, pp. 556–564, 2015, doi:10.1007/978-3-319-24553-9_68.

35. KM He, XY Zhang, SQ Ren, and J Sun, "Delving deep into rectifiers: surpassing human-level performance on ImageNet classification," (in English), *2015 IEEE International Conference on Computer Vision (ICCV)*, pp. 1026–1034, 2015, doi:10.1109/Iccv.2015.123.

36. Y LeCun, L Bottou, Y Bengio, and P Haffner, "Gradient-based learning applied to document recognition," *Proc IEEE*, vol. 86, no. 11, pp. 2278–2324, 1998, doi:10.1109/5.726791.

37. O Russakovsky et al., "ImageNet large scale visual recognition challenge," *Int J Comput Vision*, vol. 115, no. 3, pp. 211–252, Dec 1 2015, doi:10.1007/s11263-015-0816-y.

38. A Krizhevsky, I Sutskever, and GE Hinton, "ImageNet classification with deep convolutional neural networks," (in English), *Commun ACM*, vol. 60, no. 6, pp. 84–90, Jun 2012, doi:10.1145/3065386.

39. MD Zeiler, and R Fergus, "Visualizing and understanding convolutional networks," in *ECCV*, D Fleet, T Pajdla, B Schiele, and T Tuytelaars, Eds., 2014, Cham: Springer International Publishing, pp. 818–833.

40. K Simonyan, and A Zisserman, "Very deep convolutional networks for large-scale image recognition," *Clin Orthop Relat Res,* vol. abs/1409.1556, 2014.

41. C Szegedy et al., "Going deeper with convolutions," in *2015 IEEE Conference on Computer Vision and Pattern Recognition (CVPR)*, June 7–12 2015, pp. 1–9, doi:10.1109/CVPR.2015.7298594.

42. K He, X Zhang, S Ren, and J Sun, "Deep residual learning for image recognition," in *2016 IEEE Conference on Computer Vision and Pattern Recognition (CVPR)*, June 27–30 2016, pp. 770–778, doi:10.1109/CVPR.2016.90.

43. G Huang, Z Liu, Lvd Maaten, and KQ Weinberger, "Densely connected convolutional networks," in *2017 IEEE Conference on Computer Vision and Pattern Recognition (CVPR)*, July 21–26 2017, pp. 2261–2269, doi:10.1109/CVPR.2017.243.

44. A Krizhevsky, I Sutskever, and GE Hinton, "ImageNet classification with deep convolutional neural networks," *Proceedings of the 25th International Conference on Neural Information Processing Systems (NIPS) – Volume 1*, 2012, Lake Tahoe, NV, pp. 1097–1105.

45. S Hamidian, B Sahiner, N Petrick, and A Pezeshk, "3D convolutional neural network for automatic detection of lung nodules in chest CT," (in English), *Proc SPIE Int Soc Optic Eng*, vol. 10134, 2017, doi:10.1117/12.2255795.

46. SG Armato, III et al., "The lung image database consortium (LIDC) and image database resource initiative (IDRI): a completed reference database of lung nodules on CT scans," *Med Phys*, vol. 38, no. 2, pp. 915–931, 2011/02/01 2011, doi:10.1118/1.3528204.

47. J Chmelik et al., "Deep convolutional neural network-based segmentation and classification of difficult to define metastatic spinal lesions in 3D CT data," *Med Image Anal*, vol. 49, pp. 76–88, Oct 2018, doi:10.1016/j.media.2018.07.008.

48. T Zhong, X Huang, F Tang, SJ Liang, XG Deng, and Y Zhang, "Boosting-based cascaded convolutional neural networks for the segmentation of CT organs-at-risk in nasopharyngeal carcinoma," (in English), *Med Phys*, Oct 10 2019, doi:10.1002/mp.13825.

49. LDv Harten, JMH Noothout, J Verhoeff, JM Wolterink, and I Išgum, "Automatic segmentation of organs at risk in thoracic CT scans by combining 2D and 3D convolutional neural networks," *Proceedings of the 2019 Challenge on Segmentation of Thoracic Organs at Risk in CT Images – SegTHOR@ISBI*, 2019, electronic proceedings.

50. JM Johnson, A Alahi, and L Fei-Fei, "Perceptual losses for real-time style transfer and super-resolution," *European Conference on Computer Vision – ECCV*, 2016, Cham: Springer, pp. 694–711.

51. JM Wolterink, T Leiner, MA Viergever, and I Išgum, "Dilated convolutional neural networks for cardio-vascular MR segmentation in congenital heart disease," in *Reconstruction, Segmentation, and Analysis of Medical Images*, MA Zuluaga, K Bhatia, B Kainz, MH Moghari, and DF Pace, Eds., 2017, Cham: Springer International Publishing, pp. 95–102.

52. K Men et al., "Deep deconvolutional neural network for target segmentation of nasopharyngeal cancer in planning computed tomography images," (in English), *Front Oncol*, vol. 7, Dec 20 2017, doi: ARTN 31510.3389/fonc.2017.00315.

53. HR Roth, L Lu, A Farag, A Sohn, and RM Summers, "Spatial aggregation of holistically-nested networks for automated pancreas segmentation," in *Medical Image Computing and Computer-Assisted Intervention – MICCAI 2016*, S Ourselin, L Joskowicz, MR Sabuncu, G Unal, and W Wells, Eds., 2016, Cham: Springer International Publishing, pp. 451–459.

54. P Hu, F Wu, J Peng, Y Bao, F Chen, and D Kong, "Automatic abdominal multi-organ segmentation using deep convolutional neural network and time-implicit level sets," *Int J Comput Ass Rad*, vol. 12, no. 3, pp. 399–411, Mar 1 2017, doi:10.1007/s11548-016-1501-5.

55. B Ibragimov and L Xing, "Segmentation of organs-at-risks in head and neck CT images using convolutional neural networks," (in English), *Med Phys*, vol. 44, no. 2, pp. 547–557, Feb 2017, doi:10.1002/mp.12045.

56. DK Jalal, R Ganesan, and A Merline, "Fuzzy-C-means clustering based segmentation and CNN-classification for accurate segmentation of lung nodules," *Asian Pac J Cancer Prev*, vol. 18, no. 7, pp. 1869–1874, Jul 27 2017, doi:10.22034/APJCP.2017.18.7.1869.

57. XR Zhou, R Takayama, S Wang, XX Zhou, T Hara, and H Fujita, "Automated segmentation of 3D anatomical structures on CT images by using a deep convolutional network based on end-to-end learning approach," (in English), *Proc Spie*, vol. 10133, 2017, doi: Unsp 101332410.1117/12.2254201.

58. HJ Bae et al., "A Perlin noise-based augmentation strategy for deep learning with small data samples of HRCT images," *Sci. Rep.*, vol. 8, no. 1, p. 17687, Dec 6 2018, doi:10.1038/s41598-018-36047-2.

59. E Gudmundsson, CM Straus, and SG Armato, 3rd, "Deep convolutional neural networks for the automated segmentation of malignant pleural mesothelioma on computed tomography scans," (in English), *J Med Imaging (Bellingham, Wash.)*, vol. 5, no. 3, p. 034503, Jul 2018, doi:10.1117/1.jmi.5.3.034503.

60. P Nardelli et al., "Pulmonary artery-vein classification in CT images using deep learning," *IEEE Trans Med Imaging*, vol. 37, no. 11, pp. 2428–2440, Nov 2018, doi:10.1109/TMI.2018.2833385.

61. B Thyreau, K Sato, H Fukuda, and Y Taki, "Segmentation of the hippocampus by transferring algorithmic knowledge for large cohort processing," *Med Image Anal*, vol. 43, pp. 214–228, Jan 2018, doi:10.1016/j.media.2017.11.004.

62. GT Wang et al., "Interactive medical image segmentation using deep learning with image-specific fine tuning," (in English), *IEEE Trans Med Imaging*, vol. 37, no. 7, pp. 1562–1573, Jul 2018, doi:10.1109/Tmi.2018.2791721.

63. XR Zhou et al., "Performance evaluation of 2D and 3D deep learning approaches for automatic segmentation of multiple organs on CT images," (in English), *Med Imaging 2018: Comput-Aid Diag*, vol. 10575, 2018, doi: Unsp 105752c10.1117/12.2295178.

64. H Liu, L Wang, Y Nan, F Jin, Q Wang, and J Pu, "SDFN: segmentation-based deep fusion network for thoracic disease classification in chest X-ray images," *Comput Med Imaging Graph*, vol. 75, pp. 66–73, Jul 2019, doi:10.1016/j.compmedimag.2019.05.005.

65. YC Tang et al., "Improving splenomegaly segmentation by learning from heterogeneous multi-source labels," (in English), *Med Imaging 2019: Image Process*, vol. 10949, 2019, doi: Artn 109490810.1117/12.2512842.

66. J Yun et al., "Improvement of fully automated airway segmentation on volumetric computed tomographic images using a 2.5 dimensional convolutional neural net," (in English), *Med Image Anal*, vol. 51, pp. 13–20, Jan 2019, doi:10.1016/j.media.2018.10.006.

67. J Zhu, J Zhang, B Qiu, Y Liu, X Liu, and L Chen, "Comparison of the automatic segmentation of multiple organs at risk in CT images of lung cancer between deep convolutional neural network-based and atlas-based techniques," *Acta Oncol*, vol. 58, no. 2, pp. 257–264, Feb 2019, doi:10.1080/02841 86X.2018.1529421.

68. O Ronneberger, P Fischer, and T Brox, "U-Net: convolutional networks for biomedical image segmentation," (in English), *Med Image Comput Comput-Assis Inter, Pt III*, vol. 9351, pp. 234–241, 2015, doi:10.1007/978-3-319-24574-4_28.

69. G Litjens et al., "A survey on deep learning in medical image analysis," *Med Image Anal*, vol. 42, pp. 60–88, Dec 2017, doi:10.1016/j.media.2017.07.005.

70. E Shelhamer, J Long, and T Darrell, "Fully convolutional networks for semantic segmentation," (in English), *Ieee T Pattern Anal*, vol. 39, no. 4, pp. 640–651, Apr 2017, doi:10.1109/Tpami.2016.2572683.

71. F Milletari, N Navab, and S-A Ahmadi, "V-Net: fully convolutional neural networks for volumetric medical image segmentation," *Fourth International Conference on 3D Vision*, pp. 565–571, 2016.

72. PF Christ et al., "Automatic liver and lesion segmentation in CT using cascaded fully convolutional neural networks and 3D conditional random fields," *International Conference on Medical Image Computing and Computer-Assisted Intervention – MICCAI*, 2016, Cham: Springer, pp. 415–423.

73. J Schlemper et al., "Attention gated networks: learning to leverage salient regions in medical images," *Med Image Anal*, vol. 53, pp. 197–207, Apr 2019, doi:10.1016/j.media.2019.01.012.

74. XR Zhou, R Takayama, S Wang, T Hara, and H Fujita, "Deep learning of the sectional appearances of 3D CT images for anatomical structure segmentation based on an FCN voting method," (in English), *Med Phys*, vol. 44, no. 10, pp. 5221–5233, Oct 2017, doi:10.1002/mp.12480.

75. L Wu, Y Xin, S Li, T Wang, P Heng, and D Ni, "Cascaded fully convolutional networks for automatic prenatal ultrasound image segmentation," in *2017 IEEE 14th International Symposium on Biomedical Imaging (ISBI 2017)*, April 18–21 2017, pp. 663–666, doi:10.1109/ISBI.2017.7950607.

76. T Binder, E Tantaoui, P Pati, R Catena, A Set-Aghayan, and M Gabrani, "Multi-organ gland segmentation using deep learning," (in English), *Front Med-Lausanne*, vol. 6, Aug 5 2019, doi: ARTN 173 10.3389/fmed.2019.00173.

77. Ö Çiçek, A Abdulkadir, SS Lienkamp, T Brox, and O Ronneberger, "3D U-Net: learning dense volumetric segmentation from sparse annotation," *International Conference on Medical Image Computing and Computer-Assisted Intervention – MICCAI*, 2016, Cham: Springer, pp. 424–432.

78. K Men, JR Dai, and YX Li, "Automatic segmentation of the clinical target volume and organs at risk in the planning CT for rectal cancer using deep dilated convolutional neural networks," (in English), *Med Phys*, vol. 44, no. 12, pp. 6377–6389, Dec 2017, doi:10.1002/mp.12602.

79. M Oda et al., "3D FCN feature driven regression forest-based pancreas localization and segmentation," (in English), *Deep Learning in Medical Image Analysis and Multimodal Learning for Clinical Decision Support*, vol. 10553, pp. 222–230, 2017, doi:10.1007/978-3-319-67558-9_26.

80. T Brosch and A Saalbach, "Foveal fully convolutional nets for multi-organ segmentation," (in English), *Med Imaging 2018: Image Process*, vol. 10574, 2018, doi:Unsp 105740u 10.1117/12.2293528.

81. L Chen, P Bentley, K Mori, K Misawa, M Fujiwara, and D Rueckert, "DRINet for medical image segmentation," *IEEE Trans Med Imaging*, vol. 37, no. 11, pp. 2453–2462, Nov 2018, doi:10.1109/TMI.2018.2835303.

82. Ad Gelder, and HJA Huisman, "Autoencoders for multi-label prostate MR segmentation," *ArXiv*, vol. abs/1806.08216, 2018.

83. E Gibson et al., "Automatic multi-organ segmentation on abdominal CT with dense V-networks," *IEEE Trans Med Imaging*, vol. 37, no. 8, pp. 1822–1834, Aug 2018, doi:10.1109/TMI.2018.2806309.

84. E Gibson et al., "NiftyNet: a deep-learning platform for medical imaging," (in English), *Comput Meth Prog Bio*, vol. 158, pp. 113–122, May 2018, doi:10.1016/j.cmpb.2018.01.025.

85. G Gonzalez, GR Washko, and RS Estepar, "Multi-structure segmentation from partially labeled datasets. application to body composition measurements on CT scans," (in English), *Image Analysis for Moving Organ, Breast, and Thoracic Images*, vol. 11040, pp. 215–224, 2018, doi:10.1007/978-3-030-00946-5_22.

86. U Javaid, D Dasnoy, and JA Lee, "Multi-organ segmentation of chest CT images in radiation oncology: comparison of standard and dilated UNet," (in English), *Advanced Concepts for Intelligent Vision Systems, Acivs 2018*, vol. 11182, pp. 188–199, 2018, doi:10.1007/978-3-030-01449-0_16.

87. H Kakeya, T Okada, and Y Oshiro, "3D U-JAPA-Net: mixture of convolutional networks for abdominal multi-organ CT segmentation," (in English), *Medical Image Computing and Computer-Assisted Intervention – MICCAI 2018, Pt IV*, vol. 11073, pp. 426–433, 2018, doi:10.1007/978-3-030-00937-3_49.

88. S Kazemifar et al., "Segmentation of the prostate and organs at risk in male pelvic CT images using deep learning," (in English), *Biomed Phys Eng Expr*, vol. 4, no. 5, Sep 2018, doi: UNSP 055003 10.1088/2057-1976/aad100.

89. HR Roth et al., "An application of cascaded 3D fully convolutional networks for medical image segmentation," (in English), *Comput Med Imag Grap*, vol. 66, pp. 90–99, Jun 2018, doi:10.1016/j.compmedimag.2018.03.001.

90. HR Roth et al., "A multi-scale pyramid of 3D fully convolutional networks for abdominal multi-organ segmentation," (in English), *Medical Image Computing and Computer-Assisted Intervention MICCAI 2018, Pt IV*, vol. 11073, pp. 417–425, 2018, doi:10.1007/978-3-030-00937-3_48.

91. N Tong, S Gou, S Yang, D Ruan, and K Sheng, "Fully automatic multi-organ segmentation for head and neck cancer radiotherapy using shape representation model constrained fully convolutional neural networks," (in English), *Med Phys*, vol. 45, no. 10, pp. 4558–4567, Oct 2018, doi:10.1002/mp.13147.

92. SH Zhou et al., "Fine-grained segmentation using hierarchical dilated neural networks," (in English), *Medical Image Computing and Computer-Assisted Intervention – MICCAI 2018, Pt IV*, vol. 11073, pp. 488–496, 2018, doi:10.1007/978-3-030-00937-3_56.

93. M Anthimopoulos, S Christodoulidis, L Ebner, T Geiser, A Christe, and S Mougiakakou, "Semantic segmentation of pathological lung tissue with dilated fully convolutional networks," *IEEE J Biomed Health Inform*, vol. 23, no. 2, pp. 714–722, Mar 2019, doi:10.1109/JBHI.2018.2818620.

94. G Chen, J Zhang, D Zhuo, Y Pan, and C Pang, "Identification of pulmonary nodules via CT images with hierarchical fully convolutional networks," (in English), *Med Biol Eng Comput*, vol. 57, no. 7, pp. 1567–1580, Jul 2019, doi:10.1007/s11517-019-01976-1.

95. S Chen et al., "Automatic multi-organ segmentation in dual-energy CT (DECT) with dedicated 3D fully convolutional DECT networks," *Med Phys,* Dec 9 2019, doi:10.1002/mp.13950.

96. S Elguindi et al., "Deep learning-based auto-segmentation of targets and organs-at-risk for magnetic resonance imaging only planning of prostate radiotherapy," *Phys Imaging Radiation Oncol*, vol. 12, pp. 80–86, Oct 1 2019, https://doi.org/10.1016/j.phro.2019.11.006.

97. X Gu, J Wang, J Zhao, and Q Li, "Segmentation and suppression of pulmonary vessels in low-dose chest CT scans," *Med Phys*, vol. 46, no. 8, pp. 3603–3614, Aug 2019, doi:10.1002/mp.13648.

98. XL Li, YY Wang, QS Tang, Z Fan, and JH Yu, "Dual U-Net for the segmentation of overlapping glioma nuclei," (in English), *IEEE Access*, vol. 7, pp. 84040–84052, 2019, doi:10.1109/Access.2019.2924744.

99. NQ Nguyen and SW Lee, "Robust boundary segmentation in medical images using a consecutive deep encoder-decoder network," (in English), *IEEE Access*, vol. 7, pp. 33795–33808, 2019, doi:10.1109/Access.2019.2904094.

100. B Park, H Park, SM Lee, JB Seo, and N Kim, "Lung segmentation on HRCT and volumetric CT for diffuse interstitial lung disease using deep convolutional neural networks," (in English), *J Digit Imaging*, vol. 32, no. 6, pp. 1019–1026, Dec 2019, doi:10.1007/s10278-019-00254-8.

101. J Park et al., "Fully automated lung lobe segmentation in volumetric chest CT with 3D U-Net: validation with intra- and extra-datasets," (in English), *J Digit Imaging*, May 31 2019, doi:10.1007/s10278-019-00223-1.

102. XAN Xu, FG Zhou, B Liu, and XZ Bai, "Multiple organ localization in CT image using triple-branch fully convolutional networks," (in English), *IEEE Access*, vol. 7, pp. 98083–98093, 2019, doi:10.1109/Access.2019.2930417.

103. W van Rooij, M Dahele, HR Brandao, AR Delaney, BJ Slotman, and WF Verbakel, "Deep learning-based delineation of head and neck organs at risk: geometric and dosimetric evaluation," (in English), *Int J Radiat Oncol*, vol. 104, no. 3, pp. 677–684, Jul 1 2019, doi:10.1016/j.ijrobp.2019.02.040.

104. MP Heinrich, O Oktay, and N Bouteldja, "OBELISK-Net: fewer layers to solve 3D multi-organ segmentation with sparse deformable convolutions," (in English), *Med Image Anal*, vol. 54, pp. 1–9, May 2019, doi:10.1016/j.media.2019.02.006.

105. R Trullo, C Petitjean, D Nie, DG Shen, and S Ruan, "Joint segmentation of multiple thoracic organs in CT images with two collaborative deep architectures," (in English), *Deep Learning in Medical Image Analysis and Multimodal Learning for Clinical Decision Support*, vol. 10553, pp. 21–29, 2017, doi:10.1007/978-3-319-67558-9_3.

106. X Ying, H Guo, K Ma, JY Wu, Z Weng, and Y Zheng, "X2CT-GAN: reconstructing CT from biplanar X-Rays with generative adversarial networks," *Proceedings of the IEEE Conference on Computer Vision and Pattern Recognition – CVPR*, 2019, pp. 10619–10628.

107. X Dong et al., "Deep learning-based attenuation correction in the absence of structural information for whole-body PET imaging," (in English), *Phys Med Biol,* vol. in press, Dec 23 2019, doi:10.1088/1361-6560/ab652c.

108. J Harms et al., "Paired cycle-GAN-based image correction for quantitative cone-beam computed tomography," (in English), *Med Phys*, vol. 46, no. 9, pp. 3998–4009, Sep 2019, doi:10.1002/mp.13656.

109. W Dai, N Dong, Z Wang, X Liang, H Zhang, and EP Xing, "SCAN: structure correcting adversarial network for organ segmentation in chest X-rays," *Deep Learning in Medical Image Analysis and Multimodal Learning for Clinical Decision Support – DLMIA/ML-CDS@MICCAI*, 2018, Cham: Springer, pp. 263–273.

110. Q Zhang, H Wang, H Lu, D Won, and SW Yoon, "Medical image synthesis with generative adversarial networks for tissue recognition," in *2018 IEEE International Conference on Healthcare Informatics (ICHI)*, June 4–7 2018, pp. 199–207, doi:10.1109/ICHI.2018.00030.

111. C Han, K Murao, Si Satoh, and H Nakayama, "Learning more with less: GAN-based medical image augmentation," *ArXiv*, vol. abs/1904.00838, 2019.

112. Y Lei et al., "MRI-only based synthetic CT generation using dense cycle consistent generative adversarial networks," (in English), *Med Phys*, vol. 46, no. 8, pp. 3565–3581, Aug 2019, doi:10.1002/mp.13617.

113. IJ Goodfellow et al., "Generative adversarial nets," *Advances in Neural Information Processing Systems – NIPS*, 2014, pp. 2672–2680.

114. X Yi, E Walia, and P Babyn, "Generative adversarial network in medical imaging: a review," *Med Image Anal*, vol. 58, p. 101552, 2019.

115. D Yang et al., "Automatic liver segmentation using an adversarial image-to-image network," *International Conference on Medical Image Computing and Computer-Assisted Intervention – MICCAI*, 2017, Cham: Springer, pp. 507–515.

116. AK Mondal, J Dolz, and C Desrosiers, "Few-shot 3D multi-modal medical image segmentation using generative adversarial learning," *ArXiv*, vol. abs/1810.12241, 2018.

117. K Kamnitsas et al., "Unsupervised domain adaptation in brain lesion segmentation with adversarial networks," in *Information Processing in Medical Imaging*, M Niethammer et al., Eds., 2017, Cham: Springer International Publishing, pp. 597–609.

118. P Moeskops, M Veta, MW Lafarge, KAJ Eppenhof, and JPW Pluim, "Adversarial training and dilated convolutions for brain MRI segmentation," in *Deep Learning in Medical Image Analysis and Multimodal Learning for Clinical Decision Support*, MJ Cardoso et al., Eds., 2017, Cham: Springer International Publishing, pp. 56–64.

119. M Rezaei et al., "A conditional adversarial network for semantic segmentation of brain tumor," *ArXiv*, vol. abs/1708.05227, 2017.

120. J Son, SJ Park, and K-H Jung, "Retinal vessel segmentation in fundoscopic images with generative adversarial networks," *ArXiv*, vol. abs/1706.09318, 2017.

121. W Zhu, X Xiang, TD Tran, GD Hager, and X Xie, "Adversarial deep structured nets for mass segmentation from mammograms," *IEEE 15th International Symposium on Biomedical Imaging*, pp. 847–850, 2017.

122. Z Li, Y Wang, and J Yu, "Brain tumor segmentation using an adversarial network," in *Brainlesion: Glioma, Multiple Sclerosis, Stroke and Traumatic Brain Injuries*, A Crimi, S Bakas, H Kuijf, B Menze, and M Reyes, Eds., 2018, Cham: Springer International Publishing, pp. 123–132.

123. M Rezaei, H Yang, and C Meinel, "Whole heart and great vessel segmentation with context-aware of generative adversarial networks," in *Bildverarbeitung für die Medizin 2018*, A Maier, TM Deserno, H Handels, KH Maier-Hein, C Palm, and T Tolxdorff, Eds., 2018, Berlin: Springer Berlin Heidelberg, pp. 353–358.

124. M Rezaei, H Yang, and C Meinel, "Conditional generative refinement adversarial networks for unbalanced medical image semantic segmentation," *ArXiv*, vol. abs/1810.03871, 2018.

125. Y Xue, T Xu, H Zhang, LR Long, and X Huang, "SegAN: adversarial network with multi-scale L1 loss for medical image segmentation," *Neuroinformatics*, vol. 16, no. 3–4, pp. 383–392, Oct 2018, doi:10.1007/s12021-018-9377-x.

126. L Zhang, M Pereañez, SK Piechnik, S Neubauer, SE Petersen, and AF Frangi, "Multi-input and dataset-invariant adversarial learning (MDAL) for left and right-ventricular coverage estimation in cardiac MRI," in *Medical Image Computing and Computer-Assisted Intervention – MICCAI 2018*, AF Frangi, JA Schnabel, C Davatzikos, C Alberola-López, and G Fichtinger, Eds., 2018, Cham: Springer International Publishing, pp. 481–489.

127. Y Zhang, S Miao, T Mansi, and R Liao, "Task driven generative modeling for unsupervised domain adaptation: application to X-ray image segmentation," (in English), *Medical Image Computing and Computer-Assisted Intervention – MICCAI 2018, Pt II*, vol. 11071, pp. 599–607, 2018, doi:10.1007/978-3-030-00934-2_67.

128. F Mahmood et al., "Deep adversarial training for multi-organ nuclei segmentation in histopathology images," *IEEE Trans Med Imaging*, Jul 5 2019, doi:10.1109/TMI.2019.2927182.

129. R Trullo, C Petitjean, B Dubray, and S Ruan, "Multiorgan segmentation using distance-aware adversarial networks," (in English), *J Med Imaging*, vol. 6, no. 1, Jan 2019, doi: Artn 014001 10.1117/1. Jmi.6.1.014001.

130. RB Girshick, J Donahue, T Darrell, and J Malik, "Rich feature hierarchies for accurate object detection and semantic segmentation," *IEEE Conference on Computer Vision Pattern Recognition*, pp. 580–587, 2013.

131. JRR Uijlings, Kvd Sande, T Gevers, and AWM Smeulders, "Selective search for object recognition," *Int J Comput Vision*, vol. 104, pp. 154–171, 2013.

132. RB Girshick, "Fast R-CNN," *IEEE International Conference on Computer Vision*, pp. 1440–1448, 2015.

133. S Ren, K He, RB Girshick, and J Sun, "Faster R-CNN: towards real-time object detection with region proposal networks," *IEEE Trans Pattern Anal Machine Intel*, vol. 39, pp. 1137–1149, 2015.

134. X Huang, W Sun, TB Tseng, C Li, and W Qian, "Fast and fully-automated detection and segmentation of pulmonary nodules in thoracic CT scans using deep convolutional neural networks," *Comput Med Imaging Graph*, vol. 74, pp. 25–36, Jun 2019, doi:10.1016/j.compmedimag.2019.02.003.

135. K He, G Gkioxari, P Dollár, and RB Girshick, "Mask R-CNN," *IEEE International Conference on Computer Vision*, pp. 2980–2988, 2017.

136. M Liu, J Dong, X Dong H Yu, L Qi, "Segmentation of lung nodule in CT images based on mask R-CNN," *9th International Conference on Awareness Science and Technology*, pp. 1–6, 2018.

137. E Kopelowitz and G Engelhard, "Lung nodules detection and segmentation using 3D mask-RCNN," *ArXiv*, vol. abs/1907.07676, 2019.

138. Z Xu, Z Wu, and J Feng, "CFUN: combining faster R-CNN and U-net network for efficient whole heart segmentation," *ArXiv*, vol. abs/1812.04914, 2018.

139. D Bouget, A Jorgensen, G Kiss, HO Leira, and T Lango, "Semantic segmentation and detection of mediastinal lymph nodes and anatomical structures in CT data for lung cancer staging," *Int J Comput Assist Radiol Surg*, vol. 14, no. 6, pp. 977–986, Jun 2019, doi:10.1007/s11548-019-01948-8.

140. Y Li, L Zhang, H Chen, and N Yang, "Lung nodule detection with deep learning in 3D thoracic MR images," *IEEE Access*, vol. 7, pp. 37822–37832, 2019, doi:10.1109/ACCESS.2019.2905574.

141. X Xu, F Zhou, B Liu, D Fu, and X Bai, "Efficient multiple organ localization in CT image using 3D region proposal network," (in English), *IEEE Trans Med Imaging*, Jan 24 2019, doi:10.1109/tmi.2019.2894854.

142. J Wessel, MP Heinrich, Jv Berg, A Franz, and A Saalbach, "Sequential rib labeling and segmentation in chest X-ray using mask R-CNN," *ArXiv*, 2019.

143. R Zhang, C Cheng, X Zhao, and X Li, "Multiscale mask R-CNN-based lung tumor detection using PET imaging," (in English), *Mol Imaging*, vol. 18, p. 1536012119863531, Jan-Dec 2019, doi:10.1177/1536012119863531.

144. Y Lei et al., "MRI-based pseudo CT generation using classification and regression random forest," in *SPIE Medical Imaging*, 2019, vol. 10948, no. 1094843

145. X Yang et al., "MRI-based synthetic CT for radiation treatment of prostate cancer," in *International Journal of Radiation Oncology • Biology • Physics ASTRO*, 2018, vol. 102, no. 3: Elsevier, pp. S193–S194, https://doi.org/10.1016/j.ijrobp.2018.07.086.

146. Y Lei et al., "MRI-based synthetic CT generation using semantic random forest with iterative refinement," (in English), *Phys Med Biol*, vol. 64, no. 8, p. 085001, Apr 5 2019, doi:10.1088/1361-6560/ab0b66.

147. Y Lei et al., "MRI-based pseudo CT synthesis using anatomical signature and alternating random forest with iterative refinement model," (in English), *J Med Imaging*, vol. 5, no. 4, p. 043504, Oct 2018, doi:10.1117/1.Jmi.5.4.043504.

148. YY Zhou et al., "Semi-supervised 3D abdominal multi-organ segmentation via deep multi-planar co-training," (in English), *IEEE Wacv*, pp. 121–140, 2019, doi:10.1109/Wacv.2019.00020.

149. J Yang et al., "Autosegmentation for thoracic radiation treatment planning: a grand challenge at AAPM 2017," (in English), *Med Phys*, vol. 45, no. 10, pp. 4568–4581, Oct 2018, doi:10.1002/mp.13141.

150. X Feng, K Qing, NJ Tustison, CH Meyer, and Q Chen, "Deep convolutional neural network for segmentation of thoracic organs-at-risk using cropped 3D images," *Med Phys*, vol. 46, no. 5, pp. 2169–2180, May 1 2019, doi:10.1002/mp.13466.

8 Comparison of 2D and 3D U-Nets for Organ Segmentation

Dongdong Gu and Zhong Xue

CONTENTS

8.1 INTRODUCTION

Convolutional neural networks (CNNs) have become one of the most successful tools in the field of medical image processing. In recent years, typical CNNs originating from the field of computer vision [1–5] have found their application in segmenting medical images for quantitative analysis [6–9]. Among these methods, fully convolutional networks (FCNs) [10, 11] have been used for pixel-wise image segmentation. FCNs are only built using convolution, pooling, and activation layers, and use a patch-wise sliding window on the original input images. A detailed exploration of deep-learning architectures is given in Chapter 7. One of the drawbacks of FCNs is that segmentation must be performed for each pixel for a given position (center) of the sliding window. This was resolved by the U-net architecture [12], which modifies the FCN using an encoder and decoder structure that yields better segmentation in both digital natural image and medical image segmentation. The U-net has a symmetric multi-resolution structure and skips connections between the down-sampling path and the up-sampling path at each resolution. These skip connections combine local information at each resolution with the global information from the up-sampling blocks. The network has a large number of feature maps in the up-sampling path because of its symmetry. Therefore, it is sometimes necessary to reduce the number of channels in subsequent convolution layers to reduce memory usage.

The original U-net is entirely implemented for 2D or multi-channel 2D images (such as color images with RGB channels). However, in most biomedical image segmentation tasks, 3D or multiple volumes need to be processed. The extension of 2D U-net to 3D is straightforward, and almost all the basic functions and network building blocks required have been included in popular deep learning packages.

The purpose of this chapter is to explore and evaluate the performance of organ segmentation by comparing 2D and 3D U-nets based on performance on the lung CT segmentation challenge (LCTSC) datasets [13]. Such a dataset makes it possible to develop and benchmark deep learning segmentation algorithms for multi-organ segmentation.

In this chapter, the architectures of the 2D U-net and 3D U-net are presented and implemented. Then, these models are trained with a range of image resolutions, and the respective results are compared using the Dice similarity coefficient and the Hausdorff distance. The 2D networks are trained using axial plane images at resolutions 2 mm and 1 mm, and 3D networks are trained using isotropic 5 mm, 2 mm, and 1 mm resolutions. The networks are trained first with five simultaneous outputs as there are five organs to be segmented, and then the parameters of the networks are used as an initialization for training each organ-based segmentation so that the final networks are separately trained for each organ. This chapter will share the experiences of challenges and present the reader with detailed comparison between 2D and 3D U-nets.

8.2 STRUCTURES OF 2D AND 3D U-Nets

8.2.1 2D U-NET

Many studies have been accomplished processing volumetric images slice by slice [14] to train networks for 3D image segmentation. The advantages of 2D networks are that there are more training samples and variability (each slice is treated as an independent sample), and the requirement for graphics processing unit (GPU) memory is less. Figure 8.1 illustrates the architecture of the 2D U-net used in this chapter. The input of the network is a 2D slice, and the output consists of one or more probability maps dependent on the number of organs to be segmented. The in-plane resolution of the input images may vary between cases, such as 1 mm × 1 mm, 2 mm × 2 mm, etc. Therefore, linear interpolation is performed for each input image slice to convert patient data to a standardized resolution, and the nearest neighbor interpolation is performed for the corresponding ground truth masks.

Each block (rectangular boxes in Figure 8.1) of the network consists of a convolution layer, a batch normalization layer, and a ReLU layer. The size and the number of the channels are denoted at the bottom of the blocks. Arrows in the figure represent a direct feeding operation or a convolution layer to match the number of channels of two connected blocks. The down-sampling layer halves the size of the feature map and doubles the number of channels. Similarly, the up-sampling layer doubles the size of the feature map and halves the number of channels. The up-sampled feature maps are concatenated with the skip connection maps. As a result, on the decoder side, the number of channels is reduced for the first convolution block. The final output of the network consists of two channels by using a softmax layer. The two channels correspond to the foreground and background

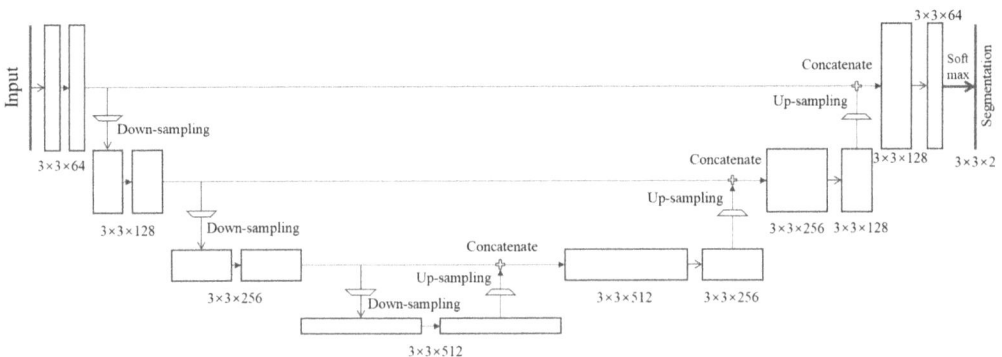

FIGURE 8.1   Illustration of the 2D U-net structure.

for organ segmentation. Compared to training a network that can simultaneously segment all the organs, experiments show that separate segmentation networks may perform better. Thus, the networks are trained for each type of organ in this chapter. Chapter 9 investigates the performance of multi-class and multi-label segmentation further.

The 2D U-net was implemented using PyTorch by following the initial design by Alexandre [15], with modifications to the number of channels, the skip connections, and the dataset class. The loss function used for training the network was the Dice loss. The training was performed in different resampled resolutions, i.e. the axial-plane resolutions used are 1 mm × 1 mm and 2 mm × 2 mm, respectively. For 2D segmentation, no interpolation was performed along the z-direction.

In the testing stage, each axial slice of the input CT volume was extracted and resampled according to the desired in-plane resolutions to match the model. The network yields a probability map through a final softmax layer for each organ. The output slices of probability maps were then resampled to the original input resolution based on linear interpolation. All the slices were finally stacked together to form a new volumetric probability map, with the same resolution of the input volume. A simple threshold was set to 0.8 empirically and applied to the probability map followed by morphological operations to extract the largest connected region as the segmentation result.

8.2.2   3D U-NET

The architecture of the 3D U-net (Figure 8.2) is similar to that of the 2D U-net (Figure 8.1). However, since the amount of GPU memory is limited, the numbers of channels at each level are set smaller than those in 2D U-net. The input of the 3D U-net is a 3D CT volume, and three spatial resolutions are used for training: 1 mm × 1 mm × 1 mm, 2 mm × 2 mm × 2 mm, and 5 mm × 5 mm × 5 mm. Trilinear interpolation was used for resampling input images, and nearest neighbor interpolation was used for resampling the ground truth masks. Each block of the network consists of a convolution layer, a batch normalization layer, and a RELU layer. In Figure 8.2, the size of kernels and number of channels are shown under each block. The down-sampling, up-sampling, pooling, and activation layers are the same as the 2D U-net except that 3D operations are used. The final output of the network is a 3D probability map of the softmax layer; $3 \times 3 \times 3$ kernels are used for all the networks, with stride of 1, and padding of 1. Dice loss was chosen for optimization of the network. The combined segmentation network was first trained for simultaneous five-organ segmentation and used as the initial network for each organ.

During the testing stage, the original CT volumes were first resampled to the desired resolution and fed to the 3D U-net to generate the probability map for each organ. These probability maps

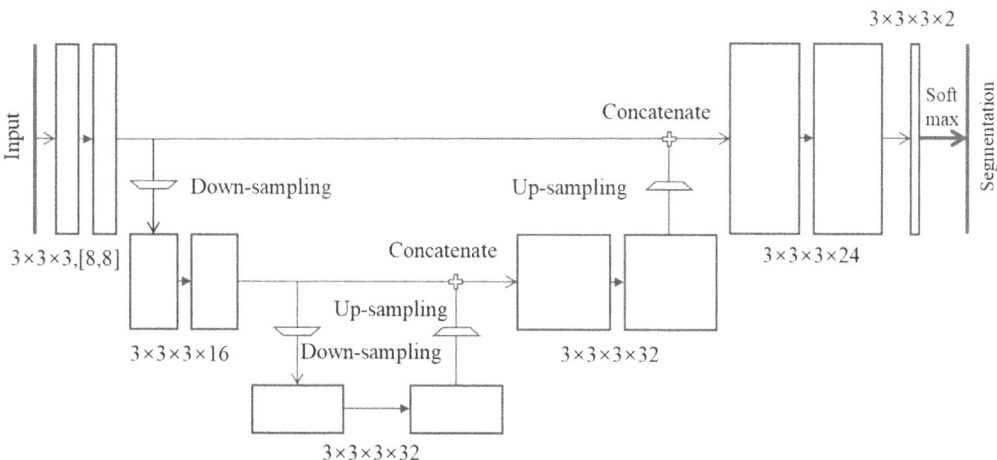

FIGURE 8.2   The structure of 3D U-net used in this chapter.

were then resampled to the original CT volume resolution for evaluation. Trilinear interpolation was used because the resampling was performed on probability maps rather than binary segmentation masks. Similarly, thresholding is performed, and the largest connected region is chosen as the final segmentation result.

8.3 EXPERIMENTAL RESULTS

8.3.1 DATASETS

The LCTSC datasets consist of 36 volumetric CT images for training and 24 images for testing. Each volume has five regions of interest (ROIs) to be segmented: the left and right lungs, the heart, the esophagus, and the spinal cord. All the ground truth masks are available for training and evaluation. Because some of the masks can slightly overlap each other (e.g. between the heart and the left lung masks), the ROI files are used separately and are not combined. Full details of this dataset are given in Chapter 1 and in Yang et al. [13].

The CT images from different patients do not have the same resolution and have different in-plane and z-direction pixel distances. Therefore, resampling is necessary before feeding to each network. Isotropic resolutions of 1 mm, 2 mm, and 5 mm were used for training and testing the 3D networks. For 2D U-net, resampling along the z-axis is not necessary. Thus, 36 training volumes were split into 5858 axial slices/images and resampled to desired in-plane resolutions to train the 2D U-net. These slices were similarly resampled into isotropic resolutions for training the 3D U-net.

8.3.2 EVALUATION METRICS

The Dice similarity coefficient and the 95% Hausdorff distance were used to evaluate the performance of segmentation on the original CT image spaces. Denoting the segmentation mask of an organ as S and the corresponding ground truth mask as T, the Dice coefficient is calculated by:

$$\text{Dice}\left(S,T\right) = \frac{2\left|S \cap T\right|}{\left|S\right| + \left|T\right|}, \tag{8.1}$$

where || represents the cardinalities or the number of voxels of the two masks.

The directed 95% Hausdorff distance is the point in S with distance to its closest point in T is greater or equal to exactly 95% of the other points in S. The undirected 95% Hausdorff distance is the average of two directed 95% Hausdorff distances. Mathematically, denoting the rth percentile as K_r, it is given as:

$$d_{H,r}\left(S,T\right) = \frac{\left[K_r\left(\min_{t \in T} d\left(s,t\right)\right), \forall s \in S\right] + \left[K_r\left(\min_{s \in S} d\left(t,s\right)\right), \forall t \in T\right]}{2}. \tag{8.2}$$

Besides the metrics of each subject, their average and standard deviation values are also computed. Chapter 15 further considers measures for the evaluation of auto-segmentation and alternative definitions and implementations are considered within that chapter.

8.3.3 IMPLEMENTATION DETAILS

Since CT intensities (corresponding to Hounsfield units) should be standard across different devices, they were not normalized but were cropped to the range of −1024 to 700 before training and testing. The pseudo code for the Pytorch Dataset class's __getitem__() function is summarized as follows:

```
Image = read_image(image_filename)
Image = intensity_clip(Image)
Segmentation = read_image(ground_truth_filename)
Image = resample(Image)
Segmentation = resample(Segmentation)
Index = generate_index(Segmentation)
Imaget = patch_sampler(Image, Index).cuda()
Segmentation = patch_sampler(Segmentation, Index).cuda()
Return {'image': Image, 'segmentation', Segmentation }
```

Note that the image patches used for training are randomly picked according to the ground truth segmentation. The sampling rule was that half of the patches picked would have their center points located within the region of interest of ground truth masks, and another half would not. In this way, the training data sampler was well balanced between the background regions and the ROIs.

The training procedure is illustrated by the following pseudo code:

```
for epoch in range(Number_Of_Epochs + 1):
for i, data in enumerate(dataLoader):
    image = data['image']
    segmentation = data['segmentation']
    ... (prepare tensor data)
    output = model(image)
    ... (prepare output)
    loss = dice_loss(output, segmentation)
    optimizer.zero_grad()
    loss.backward()
    optimizer.step()
```

The major parameters and configuration of the experiments are shown in Table 8.1. No validation set was used to choose which model will be used finally, since the number of samples available was limited. The model selected was the last trained model after reducing the learning rates from 0.0001 to 0.00001 and when the Dice loss plots are reasonably stable.

The training of all the five networks was fairly straightforward. However, due to limited GPU memory, only one 3D patch with size $256 \times 256 \times 256$ could be fed to the training per batch, resulting in bumpy Dice loss plots. Therefore, the maximum patch size chosen was set to $128 \times 128 \times 128$. For training the 3D U-net at resolution 1 mm, the results of 2 mm model on the training set were cropped to get smaller volumes, and the 1 mm 3D U-net model was trained only using the cropped images. It is in this context that the 1 mm 3D U-net is considered a refinement step for the corresponding 2 mm model.

Theoretically, the size of patches does not change network behavior. However, behavior may be dependent on the padding and how deep the network is. Each network has an effective field of view dependent on the network depth. Therefore, if one uses zero padding, a $128 \times 128 \times 128$ patch will yield a resultant probability map with size $80 \times 80 \times 80$ for the 3D U-net. Thus, the boundary effects or effective covering range for a voxel in the segmentation map could be 2.4 cm wider in radius for 1 mm resolution, which is fairly small. In other words, the network used here may not be deep enough to cover a large spatial region. Thus, the 3D U-net at 1 mm resolution could generate false positive regions outside the ROI. However, the U-net may detect image boundaries and detailed information with greater accuracy because of the fine resolution, and therefore improve performance.

Thus, to capture both global and local information using a U-net is difficult despite it being inherently multi-resolution. Increasing the depth so that the network has a wider effective field of view may better handle such situations and improve the performance. However, a consequence is that the computational cost and memory requirement become very high. An alternative solution is to apply the network on multi-resolution images. For example, a coarse-resolution network can

TABLE 8.1

List of Major Parameters and Configurations for Training

	3D U-net (1 mm)	3D U-net (2 mm)	3D U-net (5 mm)	2D U-net (1 mm)	2D U-net (2 mm)
Number of training samples	36 Volumes	36 Volumes	36 Volumes	5858 Slices	5858 Slices
Patch size	$128 \times 128 \times 128$	$128 \times 128 \times 128$	$64 \times 64 \times 64$	256×256	128×128
Batch size/minimal epochs	6 / 700	6 / 1000	6 / 1000	12 / 3000	30 /3000
Learning rate	$0.0001 \rightarrow 0.00001$	$0.0001 \rightarrow 0.00001$	$0.0001 \rightarrow 0.00001$	$0.0001 \rightarrow 0.00001$	$0.0001 \rightarrow 0.00001$
Loss	Dice (3D)	Dice (3D)	Dice (3D)	Dice (2D)	Dice (2D)
Optimizer	Adam	Adam	Adam	Adam	Adam
Machine and GPU	Tesla V100 DGXS	Tesla V100 DGXS	Tesla V100 DGXS	Tesla V100 DGXS	Tesla V100 DGXS

be first used to segment the ROI so that the ROI region can be cropped at high-resolution, and the fine-resolution network can be applied in a subsequent refinement step. Such an implementation actually works like a cascade model – the first stage has a wider field of view, and the last stage can be considered as fine tuning the results within a smaller spatial domain.

8.3.4 RESULTS

Before reporting the quantitative metrics, segmentation results of a subject produced using the 2D network are visualized in Figure 8.3. Overall, the two segmentations at resolutions of 2 mm and 1 mm were similar, but the segmentation of the high-resolution model is more accurate. For example, the 1 mm model yielded a more detailed segmentation shape and could match the boundaries better. It is also noted from the sagittal views that esophagus masks were not smooth along the z-direction for either 2D U-net models.

When comparing different organ segmentations, the overall segmentations of the lungs and the heart were much better than those of the esophagus and the spinal cord. This can be quantifiably confirmed from Figure 8.4, which plots the average and standard deviation values of (a) the Dice similarity coefficient and (b) the 95% Hausdorff distances between the segmentation results and the ground truth. For convenience, the results of 3D U-net are also plotted in this figure. By comparing all the 2D cases, especially for the esophagus, the spinal cord, and the heart, the high-resolution model outperformed the low-resolution one. The 95% Hausdorff distances of the left and right lungs were much smaller in the high-resolution model.

Figure 8.5 shows the segmentation results for the same subject for the 3D U-net segmentation. For convenience, the same ground truth is shown. Overall, the 1 mm U-net model yielded a more

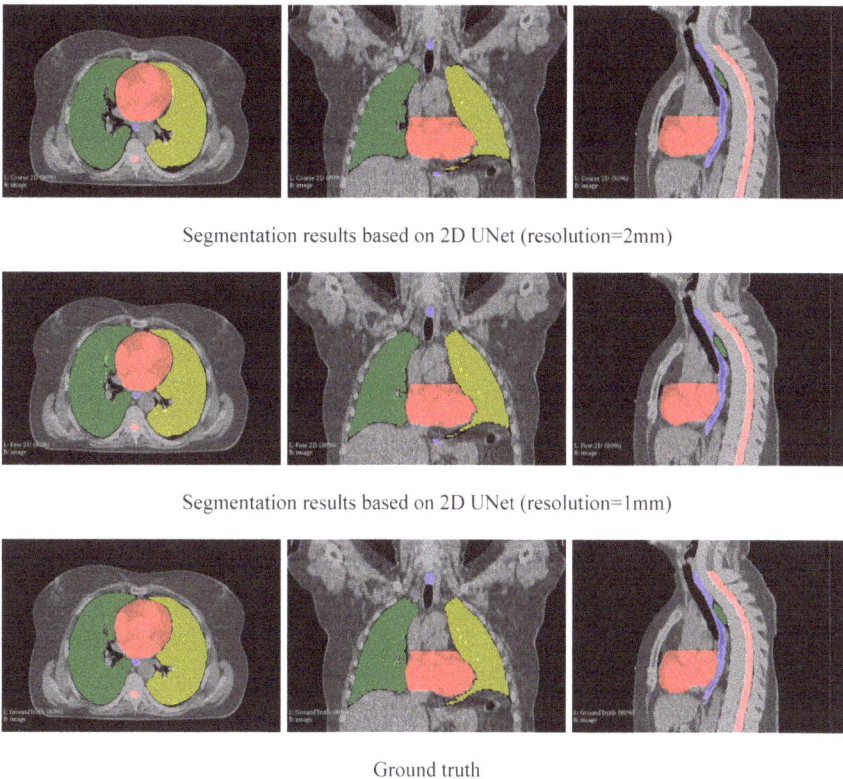

Segmentation results based on 2D UNet (resolution=2mm)

Segmentation results based on 2D UNet (resolution=1mm)

Ground truth

FIGURE 8.3 Comparison of low-resolution (2 mm, top) and high-resolution (1 mm, bottom) 2D U-net segmentation results. The last row shows the ground truth.

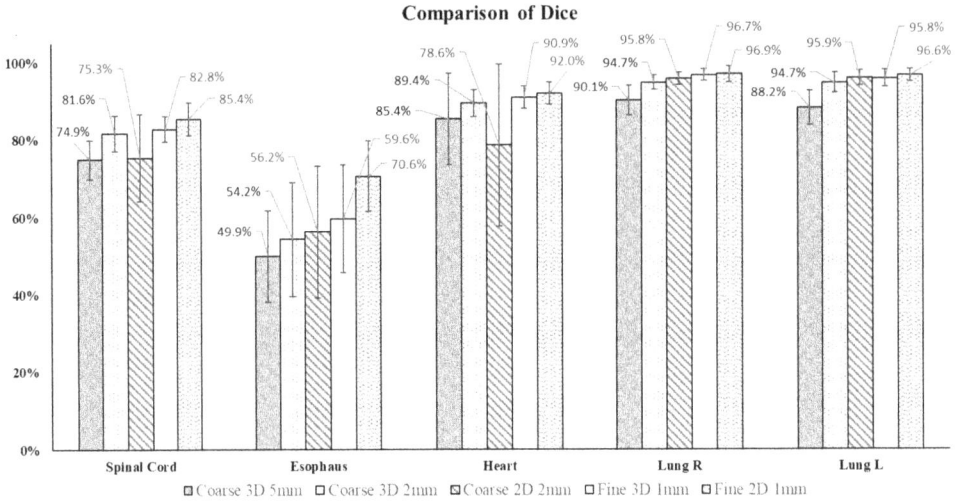

(a) Dice coefficients for 2D and 3D UNet results at different resolutions.

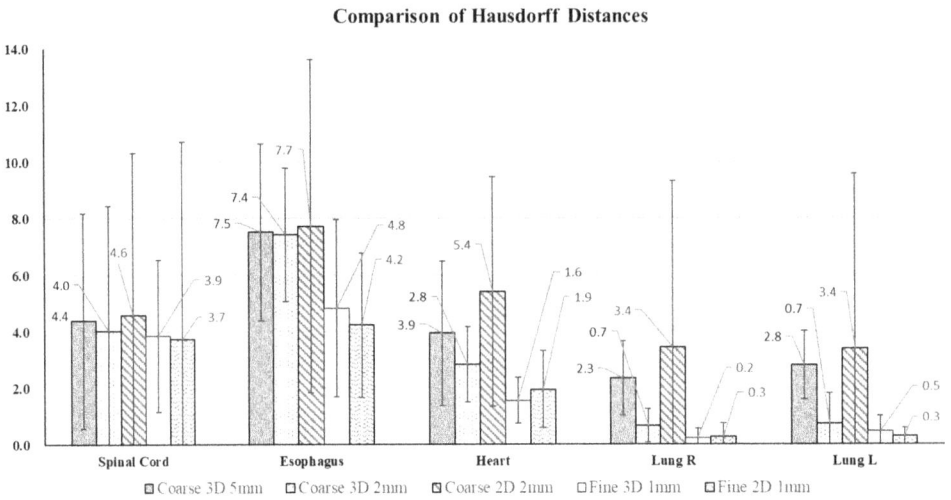

(b) Hausdorff distances for 2D and 3D UNet results at different resolutions.

FIGURE 8.4 Comparison of the mean and the standard deviation values of Dice coefficients of 2D and 3D U-nets at different resolutions. For each organ, the segmentation model from left to right are 3D coarse (5 mm), 3D coarse (2 mm), 2D coarse (2 mm), 3D fine (1 mm), and 2D fine (1 mm), respectively.

detailed segmentation, and compared to the 5 mm model, both the 2 mm and 1 mm models better matched the organ boundaries. The 2 mm model may have 1~2 voxel shifts because of the resampling procedure. Notice that there were some occasional disconnected spots for the 1 mm model. As discussed previously, this is a result of the selection of smaller patch sizes. Network inference must be performed by cropping the images into pieces and then stitching the results together as a consequence of the graphic card memory limitation and the size of the image. This would not be a problem with larger GPU memory if the images could be input as a whole.

The average and standard deviation of Dice coefficients and 95% Hausdorff distances between the segmentation results and the ground truth for the 3D U-net models are shown in Figure 8.4. In

Segmentation results based on 3D UNet (resolution=5mm)

Segmentation results based on 3D UNet (resolution=2mm)

Segmentation results based on 3D UNet (resolution=1mm)

Ground truth

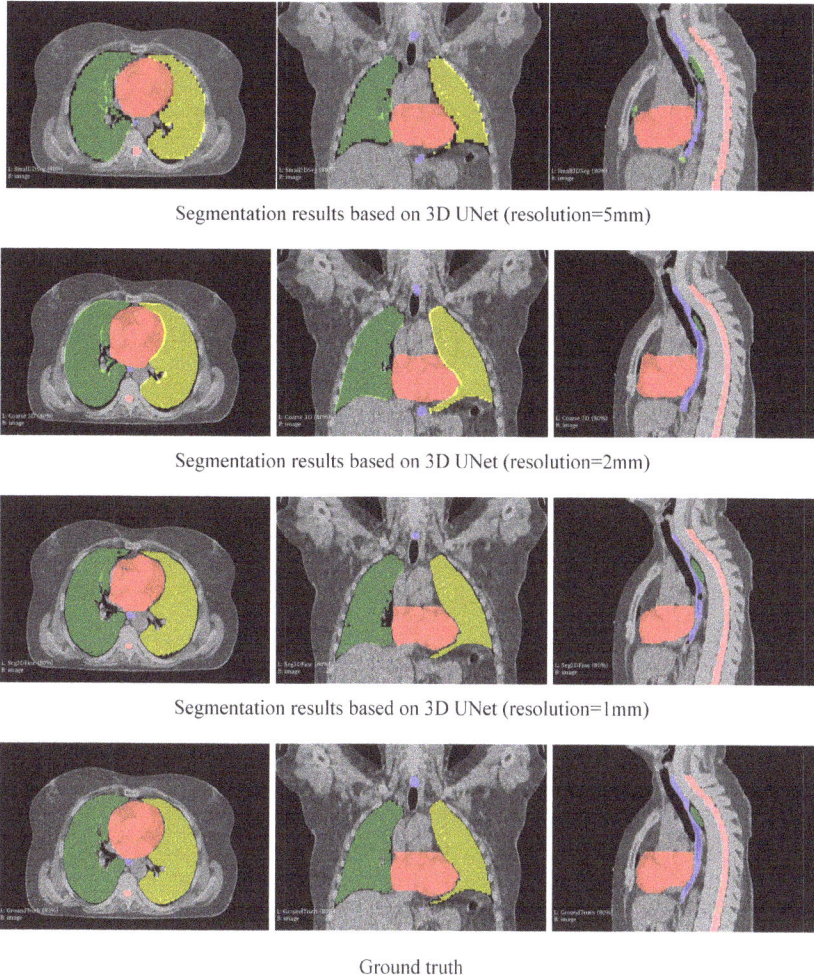

FIGURE 8.5   Comparison of 3D U-net results using three resolutions: 5 mm, 2 mm, and 1 mm.

all the five organs, the high-resolution 1 mm model outperformed the 2 mm and 5 mm low-resolution models for both Dice and 95% Hausdorff distances. The segmentation performance increased along with the resolution. The segmentation results of the lungs and the heart were much better than those of the esophagus and the spinal cord.

When comparing the 2D and 3D U-nets, Dice of the spinal cord and the heart in the 2 mm 3D model was higher than those in 2 mm resolution 2D model. However, the 2 mm 3D model achieved a better 95% Hausdorff distance across all the five organs compared to the 2 mm 2D model. When comparing the Dice values for the 1 mm resolution, the 3D model did not perform as well as the 2D model. This may have been caused by the cropping patch operation and padding effect. Overall, the results of 2D model and 3D model were comparable, and a significant improvement was not observed by using a 3D U-net for the dataset. This may seem counter-intuitive, however there were limited volumetric samples for the 3D model. Data augmentation could improve the performance of both models and are discussed in Chapter 11.

8.3.5 DISCUSSIONS

There are many variations for U-nets in the literature [6–9, 11, 14, 16–22], and it should be noted that this chapter does not intend to fully evaluate every configuration of the network structures or compare them. Rather, this chapter seeks to evaluate two basic 2D and 3D U-nets so that their effectiveness can be better understood. Similarly, it is also an open question about how to choose comparable network structures and blocks so that a 2D U-net and a 3D U-net have similar capacity. In the work of this chapter, 2D and 3D U-nets have been configured to be as similar as possible in terms of network structures, but with a different number of channels to fully use the capacity of the GPU memory. An alternative would be to keep the number of channels the same and reduce the capacity of the 2D network according to the memory constraint of the 3D network.

Selection of batch size could also affect the learning ability. One cannot use as many batches for 3D U-net as for the 2D U-net on account of the GPU memory limit. Moreover, the size of patches must be restricted for 3D U-net for the same reason. In this chapter, the patch size was set to $128 \times 128 \times 128$ for 3D models. This limits the training of 3D U-net by using a relatively smaller patch size compared to the size of the image (typically 500 pixels along each axis after resampling to 1 mm isotropic resolution). The selection of patch size is also closely related to the padding in the convolution blocks as previously mentioned. Ideally, one should not use padding and only compute the loss for the reduced size patches after network output to eliminate padding discrepancies. However, as the network becomes deeper and deeper, the resultant feature map becomes very small in size. An effective way to overcome this problem is to use a cascade model whereby a coarse resolution segmentation is used to crop the fine-resolution images so that segmentation on isotropic 1 mm resolution can be performed with a fine resolution network.

8.4   SUMMARY

By using the LCTSC datasets, the performance of 2D and 3D U-net segmentation were compared for five organs from CT volumes. The comparison was performed by training the image data in different spatial resolutions. For 2D U-net, only axial plane images were used and resampled, and for 3D U-net, this was performed by isotopically resampling all the images into desired resolutions. The advantage of 2D U-net is that training is faster, more channels can be used in the network hierarchy, and more training samples can be provided with the same size dataset. The disadvantage is that it has less continuity along the z-direction. On the other hand, the anatomical structures are smoother in 3D space. The drawback for 3D U-net is that the network may need more epochs to train, and there is less variability in the training samples due to a limited number of training volumes. The network also requires more GPU memory. If one looks into the details of the 3D U-net results, although Dice for low resolution models is lower than that of high resolution models, low resolution models capture global image information better and generate less ambiguity in terms of localization of ROIs. A multi-stage segmentation approach is therefore recommended so that low-resolution models first capture the shape of the structure of interest, and then a high-resolution model can further refine the segmentation.

REFERENCES

1. Y LeCun, L Bottou, Y Bengio, and P Haffner, "Gradient-based learning applied to document recognition," *Proceedings of the IEEE*, pp. 2278–2324, 1998.
2. A Krizhevsky, I Sutskever, and GE Hinton, "ImageNet classification with deep convolutional neural networks," in *Proceedings of the 25th International Conference on Neural Information Processing Systems – Volume 1*, pp. 1097–1105, 2012, Lake Tahoe, NV.
3. C Szegedy et al., "Going deeper with convolutions," in *The IEEE Conference on Computer Vision and Pattern Recognition*, pp. 1–9, 2015.
4. K He, X Zhang, S Ren, and J Sun, "Identity mappings in deep residual networks," in *European Conference on Computer Vision*, pp. 630–645, 2016, Springer.
5. G Huang, Z Liu, L van Der Maaten, and K Weinberger, "Densely connected convolutional networks," in *The IEEE Conference on Computer Vision and Pattern Recognition*, pp. 4700–4708, 2017.

6. Q Wang et al., "Segmentation of lung nodules in computed tomography images using dynamic programming and multidirection fusion techniques," *Acad Radiol*, vol. 16, no. 6, pp. 678–688, Jun 2009.

7. T Weikert, T Akinci D'Antonoli, J Bremerich, B Stieltjes, G Sommer, and AW Sauter, "Evaluation of an AI-powered lung nodule algorithm for detection and 3D segmentation of primary lung tumors," *Contrast Media Mol Imaging*, vol. 2019, p. 1545747, 2019.

8. K Edwards et al., "Abdominal muscle segmentation from CT using a convolutional neural network," *Proc SPIE Int Soc Opt Eng*, vol. 11317, Feb 2020.

9. J Chmelik et al., "Deep convolutional neural network-based segmentation and classification of difficult to define metastatic spinal lesions in 3D CT data," *Med Image Anal*, vol. 49, pp. 76–88, Oct 2018.

10. J Long, E Shelhamer, and T Darrell, "Fully convolutional networks for semantic segmentation," in *2015 IEEE Conference on Computer Vision and Pattern Recognition (CVPR)*, pp. 3431–3440, 2015, Boston, MA.

11. T Fechter, S Adebahr, D Baltas, I Ben Ayed, C Desrosiers, and J Dolz, "Esophagus segmentation in CT via 3D fully convolutional neural network and random walk," *Med Phys*, vol. 44, no. 12, pp. 6341–6352, Dec 2017.

12. O Ronneberger, P Fischer, and T Brox, "U-Net: convolutional networks for biomedical image segmentation," in *Medical Image Computing and Computer-Assisted Intervention – MICCAI 2015*, vol. Part III, pp. 234–241, 2015, Munich, Germany, Springer.

13. J Yang et al., "Autosegmentation for thoracic radiation treatment planning: a grand challenge at AAPM 2017," *Med Phys*, vol. 45, no. 10, pp. 4568–4581, Oct 2018.

14. HJ Bae et al., "Fully automated 3D segmentation and separation of multiple cervical vertebrae in CT images using a 2D convolutional neural network," *Comput Methods Programs Biomed*, vol. 184, p. 105119, Feb 2020.

15. M Alexandre, *UNet: Semantic Segmentation with PyTorch*. May 1 2020. https://github.com/milesial/Pytorch-UNet

16. S Gou, N Tong, SX Qi, S Yang, RK Chin, and K Sheng, "Self-channel-and-spatial-attention neural network for automated multi-organ segmentation on head and neck CT images," *Phys Med Biol*, vol. 65, no. 245034, Feb 25 2020. https://doi.org/10.1088/1361-6560/ab79c3

17. X Li, Z Gong, H Yin, H Zhang, Z Wang, and L Zhuo, "A 3D deep supervised densely network for small organs of human temporal bone segmentation in CT images," *Neural Netw*, vol. 124, pp. 75–85, Apr 2020.

18. H Liu et al., "A cascaded dual-pathway residual network for lung nodule segmentation in CT images," *Phys Med*, vol. 63, pp. 112–121, Jul 2019.

19. Z Liu et al., "Segmentation of organs-at-risk in cervical cancer CT images with a convolutional neural network," *Phys Med*, vol. 69, pp. 184–191, Jan 2020.

20. S Noguchi, M Nishio, M Yakami, K Nakagomi, and K Togashi, "Bone segmentation on whole-body CT using convolutional neural network with novel data augmentation techniques," *Comput Biol Med*, vol. 121, p. 103767, Jun 2020.

21. Y Yang, H Jiang, and Q Sun, "A multiorgan segmentation model for CT volumes via full convolution-deconvolution network," *Biomed Res Int*, vol. 2017, p. 6941306, 2017.

22. J Zhu, J Zhang, B Qiu, Y Liu, X Liu, and L Chen, "Comparison of the automatic segmentation of multiple organs at risk in CT images of lung cancer between deep convolutional neural network-based and atlas-based techniques," *Acta Oncol*, vol. 58, no. 2, pp. 257–264, Feb 2019.

9 Organ-Specific Segmentation Versus Multi-Class Segmentation Using U-Net

Xue Feng and Quan Chen

CONTENTS

9.1 INTRODUCTION

In clinical practice of radiation treatment planning, multiple organs need to be segmented from the CT to calculate the dose distribution on each organ. Deep convolutional neural networks (DCNN) represented by U-net have been widely used in this application and demonstrated far superior performance than all traditional methods [1]. To achieve the multi-organ segmentation task, there can be two design choices: a single network can be designed and trained with direct multi-class segmentation output or multiple organ-specific networks can be trained with each one performing a binary class segmentation. This study aims to perform a comparison of these two options using the data from the 2017 AAPM Thoracic Auto-segmentation Challenge and evaluate the advantages and disadvantages of each method.

9.2 MATERIALS AND METHODS

9.2.1 DATASETS

The 2017 AAPM Thoracic Auto-segmentation Challenge dataset was used in this study. Details of this dataset are provided in Chapter 1 and in Yang et al. [1]. The organs to be segmented included the esophagus, heart, left and right lungs, and spinal cord. The Radiation Therapy Oncology Group (RTOG) contouring guideline [2–3] was used for the ground truth labeling and evaluation of the output.

9.2.2 Network Structure

As CT images are often acquired in three dimensions (3D), a 3D U-net was used as the backbone network structure to more effectively exploit the full 3D spatial information directly from the image volume. Chapter 8 evaluates the performance difference between two-dimensional (2D) and 3D networks. All 2D operations were replaced with their 3D counterparts [4]. Padding was used during convolution to maintain the spatial dimension of each layer so that the output labels would have the same size as the input images. The dimensions of the input image were set to be $72 \times 208 \times 208$ determined by the graphic processing unit (GPU) memory available and the average aspect ratio of the chest CT. The 72 corresponds to the craniocaudal direction and the number of features at the first layer was set to 24. Two sets of 3D convolution filters of $3 \times 3 \times 3$, batch normalization layer, and the rectified linear activation function were used for each encoding block. With each pooling step, the spatial dimension was reduced in all directions and the number of features was doubled. The final segmentation map contained p classes. For the organ-specific networks, p is 2 as it only contains the target organ and background; for multi-organ networks, p is 6 to represent five organs and the background. Figure 9.1 shows the network structure.

9.2.3 Pre-Processing and Downsampling

As the CT imaging protocol including pixel spacing, axial slice thickness, and field-of-view in the z-direction can vary from different scans, to reduce the variability within the dataset including both training and testing cases and to fit the same input matrix size to the network, image pre-processing was performed. In addition, limited by the GPU memory, downsampling of the original images was performed. As the axial field-of-view (FOV) of CT images is often larger than the body, for all axial slices, it was first resampled to have the in-plane resolution of 1.9512 mm $\times$ 1.9512 mm^2 and center-cropped the resulting images to 208×208. The number of slices was resampled to be 72 without cropping or padding as some organs extend to the very top or bottom slices. Thus. the resulting 3D input image size was $72 \times 208 \times 208$, matching the network architecture input. The ground truth labels were processed using the same workflow to be consistent with the CT images. Finally, to normalize the image intensity, the voxel values outside of -1000 to 600 Hounsfield units (HU) were set to -1000 and 600, respectively. The resulting images were then normalized to the range $[0, 1]$.

9.2.4 Quantitative Evaluation Metrics

When ground truth contours are available, the automatic segmentation results can be evaluated using quantitative measures. A detailed discussion of quantitative evaluation measure is given in Chapter 15. The measures used in the 2017 AAPM Thoracic Auto-segmentation Challenge, the Dice coefficient, mean surface distance, and 95% Hausdorff distance, were used in this study. The definitions of these measures follow those used in the challenge and may vary from the definition or implementation given in Chapter 15, therefore the definitions are provided here.

The Dice coefficient (D) is calculated as:

$$D = \frac{2|X \cap Y|}{|X| + |Y|}$$

where X and Y are the ground truth and the algorithm segmented contours, respectively. The directed average Hausdorff measure is the average distance of a point in X to its closest point in Y, given as:

$$\vec{d}_{H,avg}(X,Y) = \frac{1}{|X|} \sum_{x \in |X|} min_{y \in |Y|} d(x,y)$$

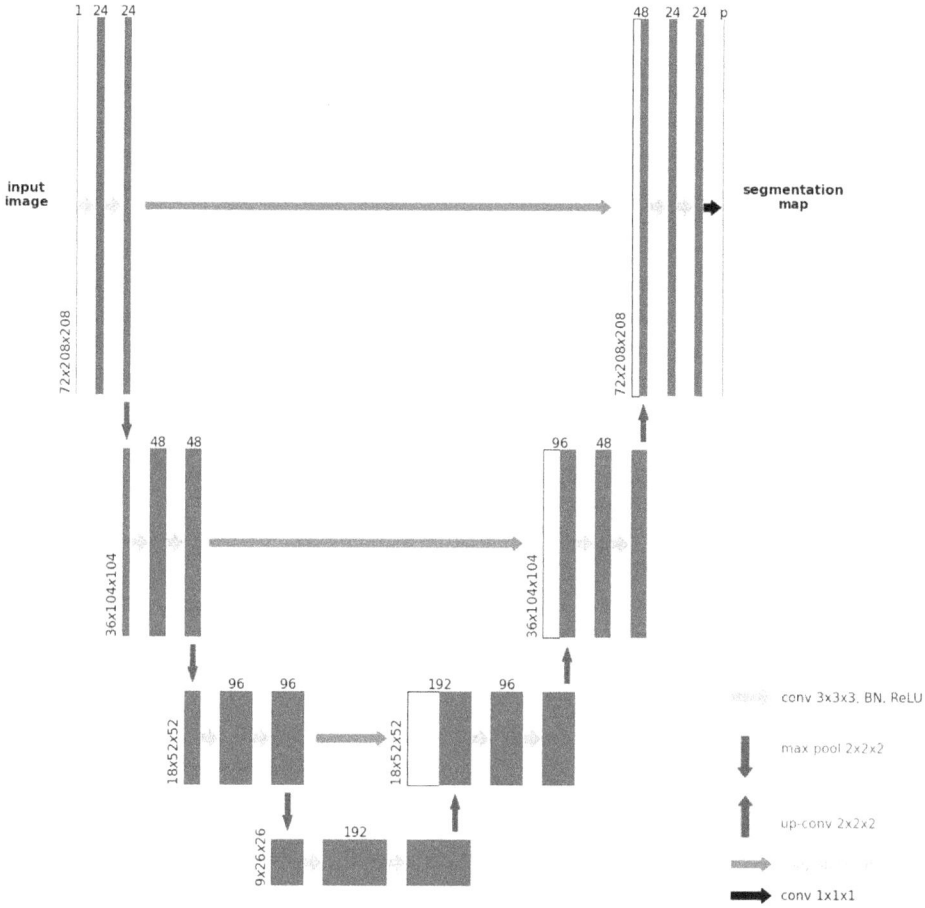

FIGURE 9.1 General structure for 3D U-net used in this application. Each encoding block consists of two sets of consecutive convolutions: batch norm and rectifier linear activation layers. Padding was used to maintain the spatial dimension during convolution. The number of features was doubled after each pooling layer. Long range connections were used by concatenating the outputs from the corresponding encoding blocks with the decoding blocks. For multi-organ segmentation network, $p = 1 +$ number of organs; for organ-specific networks, $p = 2$.

The mean surface distance (MSD) is then defined as the average of the two directed average Hausdorff measures:

$$MSD = \frac{\vec{d}_{H,avg}\left(X,Y\right) + \vec{d}_{H,avg}\left(Y,X\right)}{2}$$

The 95% directed percent Hausdorff measure is the 95th percentile distance over all distances from points in X to their closest point in Y. Denoting the 95th percentile as K_{95}, this is given as:

$$\vec{d}_{H,95}\left(X,Y\right) = K_{95}\left(min_{y \in |Y|}d(x,y)\right) \forall x \in X$$

The undirected 95% Hausdorff distance (HD95) is then defined as the average of the two directed distances:

$$HD95 = \frac{\vec{d}_{H,95}\left(X,Y\right) + \vec{d}_{H,95}\left(Y,X\right)}{2}$$

9.2.5 Implementation and Comparison Experiments

Both the multi-organ network and each organ-specific network model were implemented using the TensorFlow framework. To reduce the effects of the unbalanced voxels of different organs, the weighted cross-entropy loss was used. For multi-organ network, the relative weights for background and five organs were: background: 1.0, spinal cord: 2.0, right lung: 1.0, left lung: 1.0, heart: 1.0, esophagus: 3.0; for the organ-specific network, the same relative weights were used but the background was defined to include all voxels other than the specific organ. In practice, as smaller organs have much fewer corresponding voxels, their weights in the cross-entropy loss function are often increased to avoid gradient vanishing in learning these structures. However, a very large weight can also lead to increased false positives. The weights for the spinal cord were thus empirically set to be 2.0 and for the esophagus to be 3.0 without a very deep investigation into the actual effect on the performance. The Adam optimizer [5] with a learning rate of 0.0005 was used. The training process ran for 200 epochs in which each epoch looped through all the training cases once. During training, data augmentation was performed by applying random translations, rotations, and scaling to the input images and the corresponding ground truth label maps at each iteration. Further discussion of data augmentation is given in Chapter 11.

It is noted that the multi-label maps, although often using integer values from 0 to 5 to denote the different classes, they are not continuous numerically. Therefore, in order to avoid any interpolation errors, such as two neighboring voxels having labels of 0 and 6 may generate an intermediate voxel with value of 3, which corresponds to another class, the multi-label map to multiple binary maps for each organ were first converted and then applied the transformations to each one separately. Furthermore, after applying the random transformations to the binary maps, a threshold value of 0.5 was applied to each interpolated organ segmentation to convert back to binary values. The trained networks were then deployed to the testing dataset. As a simple post-processing step, isolated voxels labeled as pertaining to a specific organ which were not connected to the majority of the voxels belonging to that organ were regarded as false positives and removed during post-processing.

In the experiments, 24 cases were randomly selected for training both networks and the remaining 12 were used for performance evaluation and comparison. The Dice coefficient, MSD, and HD95 were calculated, respectively. Students' t-tests were used and $p < 0.05$ was used as the criteria for statistical significance.

9.3 RESULTS

Tables 9.1, 9.2, and 9.3 show the Dice scores, MSD, and HD95 of the multi-organ network and organ-specific networks. No statistically significant differences were observed for all organs, indicating that the multi-organ and organ-specific networks yielded comparable results. Furthermore, although the organ-specific networks showed slightly better Dice scores for the spinal cord and esophagus, the MSD and HD95 of the spinal cord were larger, confirming that it is difficult to claim the superiority of one method over another.

Figure 9.2 shows the Dice scores for training and validation datasets during the training process. For the multi-organ network, the mean values of all organs are shown. Comparing different organs

TABLE 9.1

Dice Scores of Multi-Organ Networks and Organ-Specific Networks

	Spinal Cord	Lung_R	Lung_L	Heart	Esophagus
Multi-organ	0.807 ± 0.056	0.972 ± 0.007	0.965 ± 0.012	0.899 ± 0.032	0.512 ± 0.141
Organ specific	0.816 ± 0.044	0.973 ± 0.007	0.964 ± 0.009	0.892 ± 0.027	0.543 ± 0.125
p-value	0.449	0.207	0.819	0.310	0.054

TABLE 9.2

MSD (mm) of Multi-Organ Networks and Organ-Specific Networks

	Spinal Cord	Lung_R	Lung_L	Heart	Esophagus
Multi-organ	1.832 ± 0.903	0.996 ± 0.236	1.102 ± 0.471	3.380 ± 1.373	6.693 ± 5.934
Organ specific	1.964 ± 1.119	0.936 ± 0.236	1.158 ± 0.449	3.653 ± 1.214	5.681 ± 4.664
p-value	0.705	0.267	0.534	0.277	0.152

TABLE 9.3

HD95 (mm) of Multi-Organ Networks and Organ-Specific Networks

	Spinal Cord	Lung_R	Lung_L	Heart	Esophagus
Multi-organ	8.268 ± 7.717	3.798 ± 0.897	4.498 ± 3.084	9.677 ± 3.921	20.884 ± 18.00
Organ-specific	9.801 ± 9.158	3.525 ± 0.875	4.737 ± 3.091	11.44 ± 5.508	19.44 ± 15.94
p-value	0.585	0.416	0.730	0.136	0.121

with the organ-specific networks, the spinal cord and esophagus are the two organs that are more difficult to learn as the training and testing Dice scores showed sudden jumps around iteration 1000 while the jumps happened at a much earlier stage for the other three organs. This is due to the fact that the spinal cord and esophagus are relatively smaller organs and the network learned to segment background first. The multi-organ networks also showed a stepwise increase with the second jump during iteration 1300–1800, which are assumed to be due to spinal the cord and esophagus. Comparing with organ-specific networks, the time it took to reach the final performance was also longer due to the reduced gradients as only two organs were not learned, which contributed to the gradient descent process. Despite the differences in convergence speed, the plots showed that all networks were able to converge to the optimal performance with less than half of the total number of iterations.

9.4 DISCUSSION

In this study the performances of the multi-organ segmentation network and organ-specific network were compared when all other settings such as input images, network structure, and training and testing augmentations were kept the same. The only difference between these two strategies is the last layer, as the former aims to simultaneously segment all organs while the latter segments one organ every time by treating all voxels not pertaining to the target organ as background. Intuitively, the multi-organ network has more information provided during training as a multi-label segmentation map is provided; therefore, ideally it is possible to utilize the label information from other organs to help the segmentation of the target organ. On the contrary, in the organ-specific network, the detailed information of other organs is not available but is merged with the background. This is similar to the concept of multi-task learning vs single-task learning [6–7]. It is generally regarded that multi-task learning can yield better performance than single-task learning if the parallel tasks are strongly correlated, or share common features, so that one task can benefit from other tasks. However, it is often challenging to perform a rigorous proof in CNN, as although there are many shared parameters of the different tasks, the detailed correlations are hard to analyze. For this specific application of thoracic organ segmentation, although the voxels of each organ are mutually exclusive and segmenting one organ may benefit from the contours of another organ, the segmentation criteria mostly relies on the target organ itself, or a combination with some key anatomical

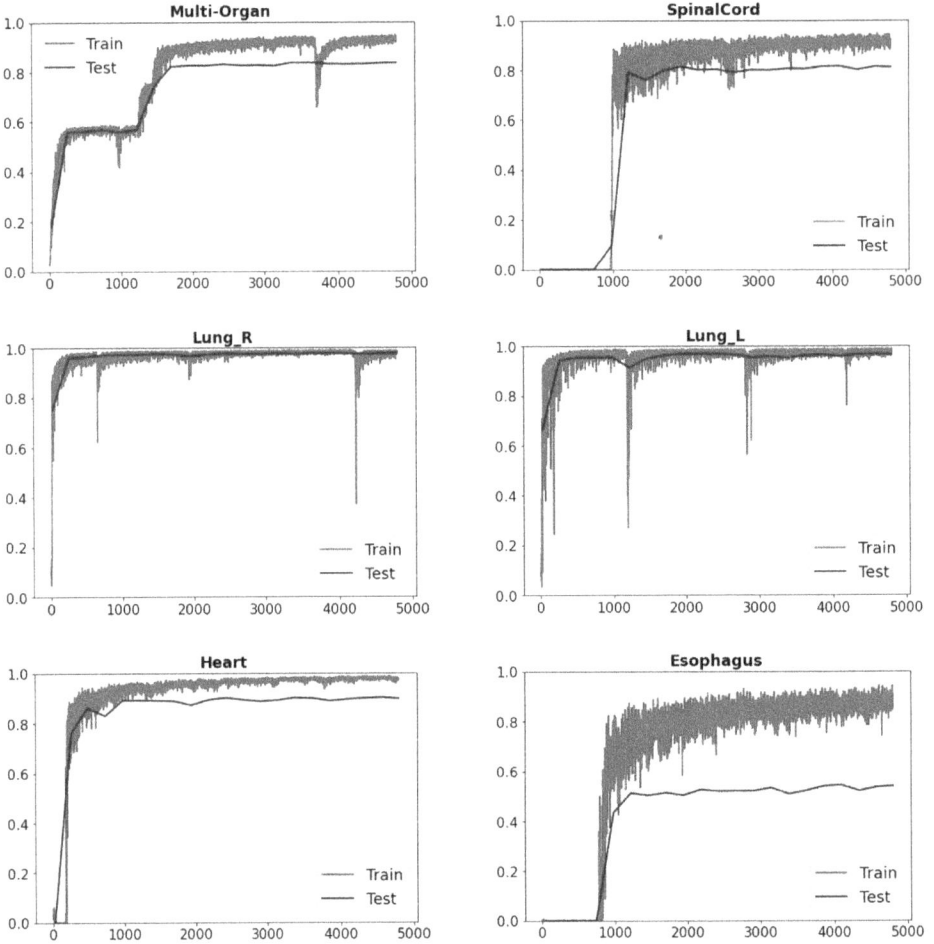

FIGURE 9.2   Dice scores of the training cases and testing cases during training for the multi-organ network and organ-specific networks. The Dice scores of the multi-organ network showed a stepwise increase, meaning that the segmentation of different organs was learned consecutively rather than simultaneously. The spinal cord and esophagus were more difficult to learn as the performance only increased after iteration 1000.

biomarkers. Therefore, it is hard to justify how much benefit the segmentation of one organ can get from the availability of the segmentation of other organs. The experiments in this study suggest that there is negligible benefit, as both strategies yield almost the same results on the validation datasets.

Comparing the two strategies, the multi-organ segmentation network ensures that all output labels are mutually exclusive, as the class with the highest probability is assigned to each voxel; therefore, no potential conflict exists. For organ-specific networks, as each network yields a binary label map, it is possible that one voxel is classified as belonging to the specific organ by multiple networks so that conflict resolving is needed. This often happens for boundary voxels between two organs. In this case, the heart shares some boundaries with both lungs so that conflicts may happen, as shown in Figure 9.3 of a final segmentation. One solution is to record the probabilistic output of each organ-specific network and compare among different networks to follow the network yielding the highest foreground probability for a given voxel. In addition, although organs are very unlikely to overlap, the organ-specific networks provide a convenient way to handle overlap regions of interest (ROIs) such as organ and lesion within the organ as separate networks can be trained for organ and lesion segmentations. However, one significant disadvantage of the organ-specific network is

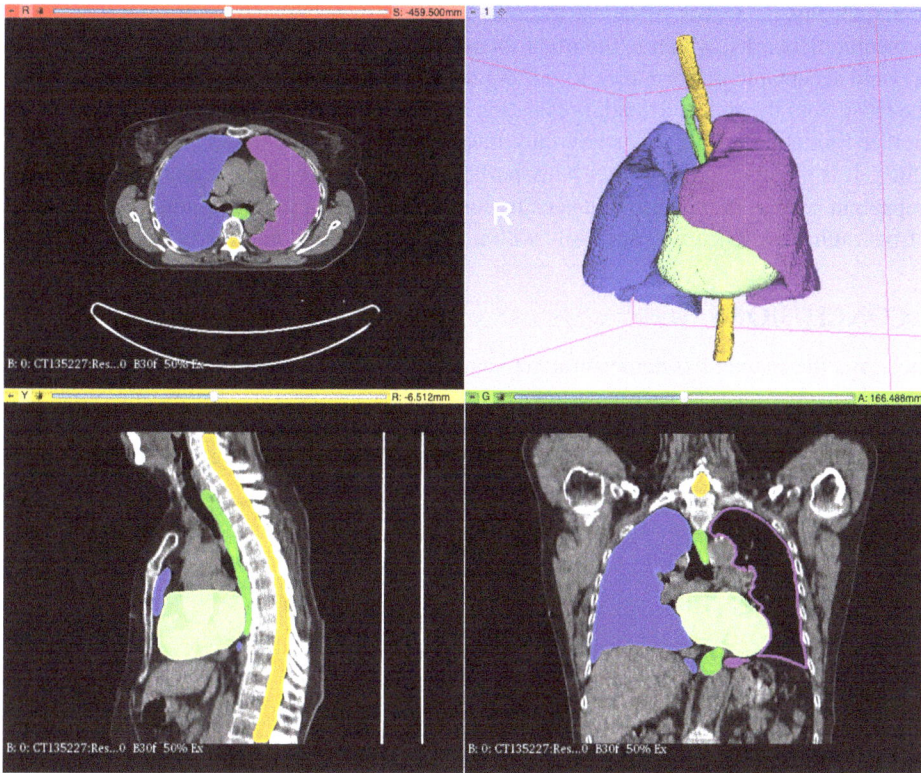

FIGURE 9.3 Five organs automatically segmented on a validation case. All OARs have satisfactory contours with no obvious mistakes. Minimal human interaction is needed.

the prolonged training and testing time as multiple models need to be trained and deployed to accomplish the same task.

Although the multi-organ and organ-specific networks did not show any differences in performance, a key factor that can affect the performance is the spatial resolution of the input images. In a previous study [8], the output of a multi-organ network was used to crop the images to only contain one organ and trained organ-specific networks on each cropped image; the performance was significantly improved compared to the multi-organ network. It is expected as more spatial details become available to make organ segmentation more accurate. This strategy is only possible with the organ-specific networks as many organs will be missing in the cropped images so that the multi-organ network will not be able to make correct segmentations. Alternatively, although it is possible to use patch-based strategy [9] to extract small patches and slide over the whole volume to train the multi-organ network, the spatial dimension is limited by the patch size so that no global information can be learned.

In this study, the Dice curves were analyzed during the training process to investigate the convergence of the networks. As expected, the smaller organs are more difficult to learn as they converge slower, even with organ-specific networks. The curves on the multi-organ networks also showed that instead of gradual learning for all organs, it first learned to segment the easy organs with a sudden increase, during which the segmentation of hard organs is learned. A detailed comparison did show that the organ-specific networks can learn the segmentation more rapidly with fewer iterations between the "not learned" and "learned" states. However, as all models converge rather rapidly, these differences are not likely to make a significant impact in practice.

One significant limitation of this study is the relatively small datasets, as only 36 cases were used in the experiments. While such a small training set may be insufficient to perform high quality segmentation, this dataset is sufficient to make a comparison between multi-organ networks and organ-specific networks, which is the main focus of this chapter. Furthermore, both strategies are able to yield acceptable performance for all organs with the exception of the esophagus which suffers the most from the reduced spatial resolution.

Another limitation is that this study only investigated the comparisons for thoracic organ segmentation. It is possible that for other body parts, where the inter-connection of different organs is more apparent such as in the head and neck region, a multi-organ segmentation network can benefit more from multi-task learning, especially with a greatly increased number of organs.

9.5   CONCLUSIONS

In conclusion, the chapter has demonstrated that in thoracic organ segmentation, there are no differences between a multi-organ segmentation network and organ-specific networks in terms of performance. This is likely due to the fact that the organ segmentation is largely independent. However, organ-specific networks are more attractive as they can be used to take high resolution images that only contain the specific organ as the input to improve the performance, but at the cost of prolonged training and testing times.

ACKNOWLEDGMENTS

The authors would like to thank NVIDIA Corporation for providing the GPU grant support for Dr. Quan Chen's lab.

REFERENCES

1. Yang J, Veeraraghavan H, Aramato SG III, et al. Autosegmentation for thoracic radiation treatment planning: a grand challenge at AAPM 2017. *Med Phys* 2018;45(10):4568–4581. doi: 10.1002/mp.13141.
2. Kong FM, Ten Haken RK, Schipper M, et al. Effect of midtreatment PET/CT-adapted radiation therapy with concurrent chemotherapy in patients with locally advanced non-small-cell lung cancer: a phase 2 clinical trial. *JAMA Oncol.* 2017;3(10):1358–1365. doi:10.1001/jamaoncol.2017.0982.
3. Kong FM, Ritter T, Quint DJ, et al. Consideration of dose limits for organs at risk of thoracic radiotherapy: atlas for lung, proximal bronchial tree, esophagus, spinal cord, ribs, and brachial plexus. *Int J Radiat Oncol Biol Phys* 2011;81(5):1442–1457.doi:10.1016/j.ijrobp.2010.07.1977.
4. Cicek O, Abdulkadir A, LienKamp SS, Brox T, Ronneberger O. 3D U-Net: learning dense volumetric segmentation from sparse annotation. arXiv:1606.06650 [cs.CV].
5. Kingma DP, Ba J. Adam: a method for stochastic optimization. arXiv:1412.6980 [cs.LG].
6. Ruder S. An overview of multi-task learning in deep neural networks. arXiv:1706.05098 [cs.LG].
7. Zhang Y, Yang Q. A survey on multi-task learning. arXiv:1707.08114 [cs.LG].
8. Feng X, Qing K, Tustison NJ, Meyer CH, Chen Q. Deep convolutional neural network for segmentation of thoracic organs-at-risk using cropped 3D images. *Med Phys* 2019;46(5):2169–2180. doi:10.1002/mp.13466.
9. Kim H, Jung J, Kim J, et al. Abdominal multi-organ auto-segmentation using 3D-patch-based deep convolutional neural network. *Sci Rep.* 2020;10(1):6204. Published 2020 Apr 10. doi:10.1038/s41598-020-63285-0.

10 Effect of Loss Functions in Deep Learning-Based Segmentation

Evan Porter, David Solis, Payton Bruckmeier, Zaid A. Siddiqui, Leonid Zamdborg, and Thomas Guerrero

CONTENTS

10.1 INTRODUCTION

Traditional problem-solving algorithms define a problem and a specific set of steps required to arrive at a solution. In contrast, a deep learning model is a statistical framework, which, when trained stochastically, arrives at a solution. For the model to effectively converge to a solution, it must be able to evaluate the quality of candidate solutions as it learns. Loss functions, also called objective functions or cost functions, quantify the quality of a candidate solution during the model training process. For each step during training, the model's weights are progressively updated to yield predictions which minimize the loss function. Because the loss function dictates the model's measure of success and the degree to which the weights are updated, choosing the proper loss function for a given task is vital.

At the beginning of training a model, the weights are randomly initialized and generally incapable of making any useful predictions. However, through backpropagation training, models can learn to solve tasks across many divergent domains. Take, for example, the simple problem of segmenting the skull on a CT image, as shown in Figure 10.1.

The backpropagation training process is broken into three steps: prediction, evaluation, and backpropagation. During the first step, the training input data flows through the model which is simply a series of mathematical operations, most commonly convolutional operations. The data which is returned from the model is referred to as a prediction. In the skull segmentation example, the model is provided with a two-dimensional CT image slice as input, from which it generates a prediction for a segmentation mask. From the example in Figure 10.1, the current model's skull prediction is non-ideal and further training, or updates to the model's weights, is warranted. Next, the error of the prediction, in relation to the ground truth, is calculated using the loss function. In the final training step, the gradient of the error is calculated with respect to each model weight. Then every weight is updated by the scaled gradient of the error, with the intent of minimizing each weight's contribution to the error in subsequent predictions. The scaling factor, commonly called the learning rate, is represented by λ in Figure 10.1. Therefore, to allow for backpropagation training, a loss function must have scalar-valued output and be differentiable with respect to the model weights. A complete training process repeats these three steps until the output of the loss function, or prediction error, is minimized. Ideally, upon finishing training, the model weights should converge upon a state capable of robustly solving the given task.

In addition to dictating what is learned, a loss function can influence how easily a model converges upon a solution. Like many optimization problems, the training of deep learning models utilizes a multi-dimensional gradient descent. A simple visual representation of the training process would be the act of navigating to the lowest point on an uneven plane, such as those shown in Figure 10.2. If the plane possesses many depressions in addition to the true lowest point, it would be difficult to detect the lowest point globally or merely locally; after all, the only knowledge is of the local surroundings, not if there is a deeper depression elsewhere on the plane.

To adapt this to deep learning terminology, the x–y axis of the surface represents all potential model weight combinations, and the z-axis indicates the loss function performance of the current weight combination. During training, the model is initialized randomly within the weight possibility space. Then, as the model trains, it explores the space of its possible weight combinations to minimize the loss function. Optimal loss functions therefore have an easily computed gradient path towards the global minimum.

The set of weights which minimize the loss function are referred to as the global minimum, and the other sets of weights which produce loss functions lower than their surroundings as the local

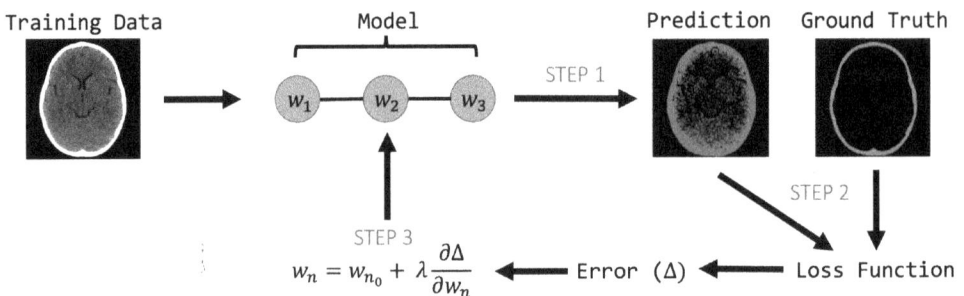

FIGURE 10.1 The steps in training a deep learning model. Step 1, from the training data, a prediction is made. Step 2, using the loss function, the ground truth and prediction are compared, and an error is determined. Step 3, each weight is updated proportionally to the gradient of the error.

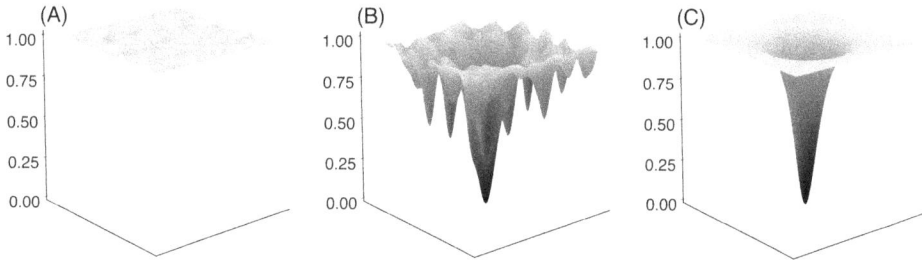

FIGURE 10.2 A visual depiction of loss functions where the x-y axis is model weight combinations, and the z-axis is the loss function. With an incorrectly chosen loss function (A), a poorly suited loss function (B) and an easily trainable loss function (C).

minima. If a loss function completely unsuited to the data is selected, it is unlikely the model will train at all. Such a visualized loss space [1] example is given in Figure 10.2a. If a poorly suited, but trainable, loss function is chosen instead, there will be both a global minimum and local minima, as in Figure 10.2b. However, if a carefully chosen a loss function well suited for the task is used, finding the global minimum will be both simple and efficient, as seen in Figure 10.2c.

A well-chosen loss function has a significant role in reaching an optimal solution for a given deep learning task. This chapter will cover the necessary elements of a loss function, the challenge of segmentation tasks for loss functions, common loss functions, and their applications, dealing with imperfect data, choosing a starting loss function, and troubleshooting methods to help overcome frequent challenges in medical image segmentation.

10.2 ADMISSIBILITY OF A LOSS FUNCTION

To understand the importance of admissibility, imagine that two people are bidding to build a fence enclosure for a farmer's sheep. The farmer only tells both designers that whoever designs the fence with the shortest length will be hired. The first designer, using his knowledge of geometry, designs a circular fence, large enough to encircle the flock. On the other hand, the second designer proposes to build a fence only around himself, declaring himself 'outside' the fence. Clearly, this second solution fails to enclose the flock, which is the original purpose of a building fence. However, the farmer presented the ideal solution as that which minimized fence distance, not that which minimized the danger to the sheep. In a deep learning context, the farmer's loss function, length of fence, was not admissible to his true intentions behind building the fence.

While the second solution may seem outlandish, deep learning models are inherently prone to converging upon these lazy solutions. For segmentation tasks, common lazy solutions are models which do not predict every structure, predict highly smoothed structures, or models which uniformly predict a single structure. To prevent these lazy solutions, a loss function must be carefully chosen which defines the ideal solution to the task, minimizes the risk of unintended results, and ensures effective convergence to a robust solution.

10.3 PRESENTING THE PROBLEM

The remainder of this chapter covers the proper combination of ground truth data and loss functions and presents a selection of different losses useful for image segmentation. For discussion, a segmentation task is considered where a ground truth label mask is available in which each voxel is designated as either a member of the class or not. These ground truth label masks can be organized as either a multi-label or multi-class segmentation tasks, both of which can be used to train a deep learning model. Multi-label and multi-class segmentation are discussed further in Chapter 9, but a brief discussion is included here, since the definition of the problem impacts the choice of loss function.

A multi-label segmentation allows for each voxel to be a member of multiple classes, as well as not a member of any class. An example of a multi-label segmentation is a patient with multiple thoracic structures and a body contour. In this case, every voxel classified as "heart" would also be member of the "body" class. And, for any voxel exterior to the body, class membership would not be required.

A multi-class segmentation is a restriction of a multi-label segmentation task, where each voxel is a mutually exclusive classification. This means that each voxel must, and can only, be a member of a single segmentation class. For example, when contouring the left and right lung, each voxel will be one of three classes: left lung, right lung, or neither lung. Through the inclusion of the "neither", also referred to as the "background" class, the problem allows for every voxel to be a member of a class. To restrict voxels from having membership of multiple classes, or likewise to reduce a multi-label to a multi-class segmentation problem, binary operators (i.e. AND, OR, and NOT) can be utilized.

Strict adherence to the multi-class labeling rules is important because any mislabeled voxels will interfere with the model's training. Take, for example, a voxel which was not assigned any of left lung, right lung, or neither. During the training process, a prediction of any class membership will falsely be evaluated as an error and will be backpropagated into the model weights, potentially interfering with the otherwise properly trained parameters.

Although multi-class labeling restricts the preparation and data organization of the ground truth labels, doing so also restricts the complexity of any prediction. By reducing the degrees of freedom possible in a solution, the overall solution space is restricted, and the gradient descent is simplified. This means that, for most tasks, preparing the ground truth as a multi-class problem will result in quicker convergence to a solution.

As a depiction of both label types, Figure 10.3 demonstrates different representations of an arbitrary 2D image composed of a partially overlapping circle and triangle. Figure 10.3b shows a "one-hot encoded" multi-label data set representation of the original image, Figure 10.3a. In this case, a third dimension is added to the 2D image, with each position along this dimension called a channel, where each channel represents membership of the pixel position to different categories, or classes, of data. A pixel value of 1 in channel 1, Figure 10.3b left, would indicate that the pixel belongs to the circle region, and a pixel value of 1 in channel 2 would indicate that the pixel belongs to a triangle

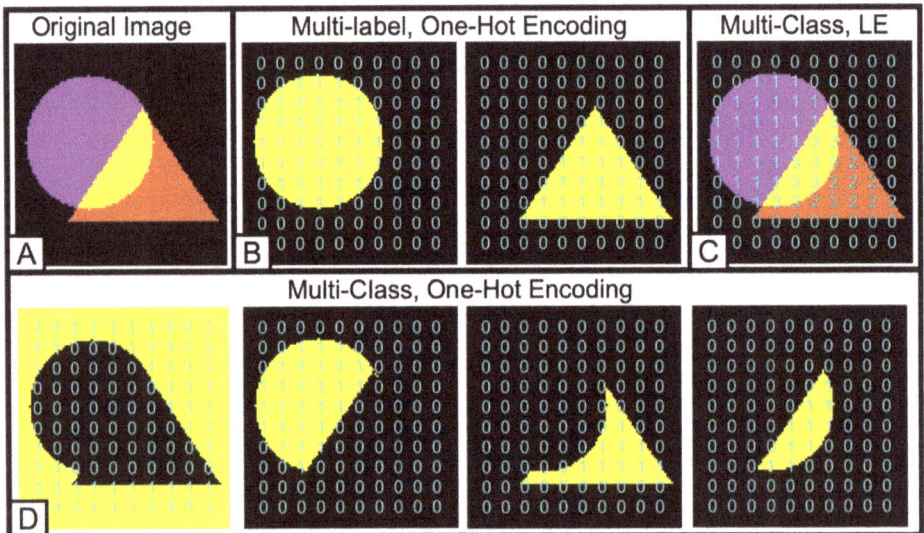

FIGURE 10.3 (A) The original image of a circle and triangle sharing an overlapped region is shown. (B) A one-hot encoded multi-label representation of image A. (C) A multi-class label encoding (LE) representation of image A. (D) A one-hot encoded multi-class representation of image A.

region. It is important to note that in a multi-label representation of the data, a given pixel position may hold a value of 1 in either channel, indicating that the pixel position belongs to both the circle region and triangle region. This contrasts the with multi-class representation of the dataset, which must hold mutually exclusive classifications. In Figure 10.3c, a multi-class label-encoded data representation of Figure 10.3a is shown. In this representation, a unique integer label is assigned to each pixel, which indicates to which classification category the pixel belongs: 0 – background, 1 – circle only, 2 – triangle only, 3 – intersection region of the circle and triangle. Because this is a multi-class representation, a new classification is needed to indicate membership of the pixel in the overlapping region. In Figure 10.3d, a one-hot encoded multi-class representation of Figure 10.3a is shown. In a similar fashion to Figure 10.3b, multiple channels are again utilized to indicate the category a given pixel belongs to (from left to right): channel 1 – background, channel 2 – circle only, channel 3 – triangle only, channel 4 – circle and triangle intersection. As will be discussed later, though similar in their composition, the use of either a multi-label or multi-class representation (Figure 10.3b vs Figure 10.3d) for a dataset may hold distinct advantages for loss functions and their application.

The output of a neural network needs to match the dimensionality of the target ground truth labels. For segmentation, this requires a special output layer to convert the regression from the network into class probabilities for each voxel in the input. Multi-class segmentation requires a softmax function, which is a scaled activation which maps the neural network to a normalized distribution function representing the per-channel estimation of class membership (the sum of the classes for a given voxel predication is equal to one). Despite the output of a softmax activation being normalized, the model output should not be confused with a probabilistic (i.e. Frequentist or Bayesian) output for class membership. This means that probabilistic statistical tests or utilizing a probabilistic determination to inform clinical decisions is not a valid interpretation of a network's output. Instead, in order to make an inference, each voxel has a class assigned to the channel with the highest value, typically by applying a maximum argument (argmax) function, ensuring each voxel is a member of only a single class. However, during model training, the loss is computed from the raw outputs (without the argmax function applied) to compute and backpropagate the gradient of the error with respect to all possible classes.

For a model to achieve multi-label segmentation, the model should conclude with a sigmoid function as the final activation. This ensures that the model outputs normalized, class-independent, per-voxel class membership predictions. Since the sigmoid function is independent for each output channel, a voxel having membership in multiple classes is a valid prediction. Then, during inference, a sigmoid-activated prediction is rounded to the nearest binary value, allowing each voxel the potential of being a member of multiple classes. And, similarly to multi-class segmentation training, the loss function should be computed on the raw, or unrounded, predictions.

10.4 COMMON LOSS FUNCTIONS

10.4.1 Mean Squared Error

Mean squared error (MSE) is one of the simplest loss functions used in deep learning. MSE is computed by the mean of the squared loss between the prediction and ground truth, given by Equation 10.1:

$$MSE = \frac{1}{n} \sum_{i=1}^{n} \left(X_i - Y_i \right)^2 \tag{10.1}$$

Throughout the chapter, X will denote the prediction and Y the ground truth. For each tensor, X and Y, there exist n classes, encoded as channels. Because it does not require multi-class or multi-label segmentation input, MSE is applicable to both prediction types. However, unlike other functions (e.g. Dice loss), the MSE loss scores during training are not correlative to common segmentation

comparison metrics, limiting MSE's overall interpretability. While MSE is an acceptable loss function for certain situations, more specialized loss functions are available for segmentation tasks.
Mean squared error has a built-in implementation in Keras with keras.losses.mean_squared_error,
in TensorFlow with tf.compat.v1.losses.mean_squared_error and in PyTorch using torch.nn.MSELo
ss.

10.4.1.1   Cross Entropy

The term "cross entropy" describes a family of logarithmic loss functions, typically referring to one
of two types: binary cross entropy and categorical cross entropy. For both functions, they follow
the same basic formula, as given by Equation 10.2, but differ by the expected input prediction type.

$$CE = -\sum_{i=1}^{n} Y_i \log(X_i)$$ (10.2)

10.4.1.2   Binary Cross Entropy

Binary cross entropy is a logarithmic loss function designed for multi-label problems, where the
data is limited to a binary value designating class membership. This is most commonly utilized
with ground truth data which has been one-hot encoded. This is paired with a model with a sigmoid
function as the final activation, providing an output vector with values from zero to one. To reiter-
ate, a multi-label problem would be a task which has volumetrically overlapping segmentations. An
example of this is the BraTS Challenge MRI dataset [2–4], a brain lesions dataset with structure for
the enhancing tumor (ET), tumor core (TC), and whole tumor (WT). To allow for predictions with
overlapping structures, the model should output a three channel mask, with each channel corresponding to one of ET, TC, or WT. Binary cross entropy has native implementations in TensorFlow
with tf.compat.v1.losses.sigmoid_cross_entropy, in Keras using keras.losses.binary_crossentropy,
and in PyTorch with torch.nn.BCELoss.

10.4.1.3   Categorical Cross Entropy

Like binary cross entropy, categorical cross entropy is a logarithmic function. Categorical cross
entropy is designed to work with multi-class problems and is compatible with models that have
softmax final activation. These models then predict the certainty that any given voxel belongs to
each class. Typically, multi-class problems are most useful for segmentation tasks which do not have
overlapping classes, such as segmentation of either left or right lung. To prepare the ground truth
and model for a softmax activation, the output should have $n + 1$ channels, where n corresponds
to the number of segmented structures. This leaves an additional channel to correspond to voxels
which are not a member of any class, referred to as the background. Categorical cross entropy has
native implementations in TensorFlow with tf.compat.v1.losses.softmax_cross_entropy, Keras with
keras.losses.categorical_crossentropy and in PyTorch with torch.nn.CrossEntropyLoss.

10.4.2   Dice Loss

The Sørensen-Dice coefficient, commonly referred to as the Dice similarity coefficient (DSC), was
developed for biostatisticians to determine the similarity between two populations [5]. Although it
was originally designed to work with tabular binary data, it has proven to be a useful tool for binary
segmentation analysis as well [6]. For a set of two contours, the prediction X and the ground truth Y,
the Dice similarity coefficient can be determined with Equation 10.3:

$$DSC = \frac{2|X \cap Y| + \epsilon}{|X| + |Y| + \epsilon}$$ (10.3)

where ϵ represents a small value to prevent from having zero division errors when both X and Y are empty and to ensure that DSC = 1 in that instance. Then, the Dice loss function is simply 1 – DSC, or the negative of the Dice similarity coefficient.

An implementation of the Dice similarity coefficient and Dice loss in Python code using Numpy, Keras, and PyTorch implementations are available at https://github.com/Auto-segmentation-in-Radiation-Oncology/Chapter-10. It is noted that each implementation looks somewhat different. This is partially because of the different functions and syntax of each library, but also because both the PyTorch and Keras implementations are designed to work with non-binary output data during the training process.

```
# Numpy compatible dice loss
import numpy as np

def dice_coef(output, labels):
    # Computes the dice coefficient of two numpy arrays
    eps = np.finfo(float).jpg
    intersection = np.sum(output * labels)
    denominator = np.sum(output) + np.sum(labels)
    return (2 * intersection + eps) / (denominator + eps)

def dice_loss(output, labels):
    # Computes the dice loss of two numpy arrays
    return 1 - dice_coef(output, label)
```

10.4.3   HAUSDORFF DISTANCE LOSS

With the previously discussed loss functions, the prediction agreement was determined by the relative similarity of the structures. This means a well performing prediction could be quite volumetrically accurate but have little penalization for discontinuities in the volume.

Take Figure 10.4 for example, where the prediction is generally in strong agreement with the ground truth, but the prediction also includes a small region with large spatial separation from the ground truth. To translate this to radiation oncology segmentation for treatment planning, while Figure 10.4 has a high DSC, if this were a target volume, the resulting treatment plan would differ

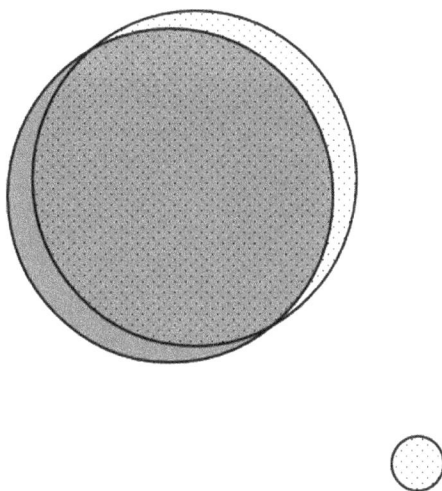

FIGURE 10.4 A visual comparison of two volumetrically similar contours with poor Hausdorff distance agreement. The dark region represents a ground truth contour whereas the light region indicates the prediction.

vastly from the ground truth's plan. For radiation oncology treatment planning, an accurate segmentation is primarily one with relatively minimal spatial difference from the ground truth.

To robustly determine the maximum spatial separation between two structures, Hausdorff distance (HD), as provided by Equation 10.4, can be computed:

$$HD = \max \left\{ \sup_{y \in Y} \inf_{x \in X} d(y, x), \sup_{x \in X} \inf_{y \in Y} d(y, x) \right\} \tag{10.4}$$

where *sup*, *inf* represent the supremum and infimum of the distances, and the distances represented as $d(y,x)$ are computed between a point from each contour set. Effectively, this metric computes the minimum distance between every point from surface one to two and from surface two to one. Then, the HD is the maximum distance separation from the mappings in either direction, which is the greatest spatial discrepancy between the two surfaces. Unlike previous loss functions such as Dice loss, HD is only dependent on the contour surfaces, meaning a ring and a filled contour could yield the same HD. Thus, HD is particularly sensitive to disjoint segmentations and, when used as a loss function, will reinforce accurate contour boundary predictions. The simplest way to create a HD loss is to compute the negative HD [7]. An implementation of HD is contained within the excellent MedPy library [8].

A shortcoming of the Hausdorff distance loss function is that it is spatially dependent and highly sensitive to outliers. If the individual image and segmentation masks used to train the model vary in field of view, or have inconsistent voxel dimensions, the HD will be non-uniform across the training set. Most commonly, image voxel dimensions vary in the z-axis, which, if uncorrected, could yield inconsistent results on the superior or inferior boundaries of a contour. Correction can be achieved by either resampling the image to uniform voxel dimensions or generating the training data sets with corresponding voxel dimensions and passing them into the loss function. Further, the Hausdorff distance metric is sensitive to outliers, which may be overcome by substituting a percentile or mean Hausdorff distance, instead of the traditional total maximum distance.

10.5 DEALING WITH CLASS IMBALANCE

Despite the amazing capabilities of deep learning models, they can also be lazy and will frequently take any available shortcuts to get nearest to the correct answer. Consider the lazy approach to the task of segmenting Figure 10.5 into one of three classes: square, circle, and triangle.

Between the structures in Figure 10.5, the triangle is 0.5% of the total area, the circle is 23.3% of the total area, and the background square is 76.2%. Now, if the deep learning model's performance

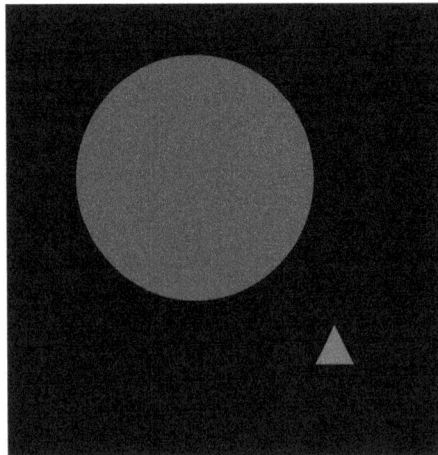

FIGURE 10.5   A representation of an unbalanced segmentation task.

is graded with an unbalanced loss function, the deep learning model could omit learning of the triangle completely and be within 99.5% accuracy of the ideal prediction. In this case, if each shape represented an anatomical structure, failing to segment any necessary structure would constitute a clinically unacceptable prediction, independent of the accuracy of the other structures.

When choosing a loss function, the task's inherent class balance should be considered to prevent overfitting to only the most dominant classes. While there is no rule-of-thumb for when to choose a balanced or unbalanced loss function, when a task is in fact balanced, the majority of imbalance adjusted loss functions asymptotically approach their unbalanced counterparts. It is recommended to begin with one of the following loss functions if the significance of class imbalance is unknown.

10.5.1 WEIGHTED CROSS ENTROPY

For tasks with a known and constant magnitude of class imbalance throughout the samples, the imbalance can be compensated for using weighted cross entropy [9]. In similar fashion to the standard cross entropy, weighted cross entropy is a logarithmic function compatible with either multi-label or multi-class data. However, the implementation differs by having a per-class scaling or weighting factor. A cross entropy a binary problem is given in Equation 10.5:

$$CE = -\sum_{i=1}^{n} w_i Y_i \log(X_i)$$

(10.5)

where w_i represents the per-class weighting to compensate for class imbalance. If $w_i > 1$, then the model will decrease false negatives and if $w_i < 1$ then the model will decrease false positives. To understand false negatives and false positives in the context of segmentation, see Section 10.5.5 on sensitivity specificity loss. Weighted cross entropy is natively implemented in TensorFlow, accessible at `tf.nn.weighted_cross_entropy_with_logits`, and in PyTorch, the standard `torch.nn.CrossEntropyLoss` function accepts weights as a parameter. A Keras-compatible implementation is below:

```
import tf.keras.backend as K

class weighted_cross_entropy:
    def __init__(self, weights):
        self.weights = weights
    def binary(y_true, y_pred):
        y_true = y_true * self.weights
        y_pred = y_pred * self.weights
        y_pred = K.clip(y_pred, K.jpgilon(), 1-K.jpgilon())
        logit = K.log(y_pred / (1. - y_pred))
        term0 = logit*y_true
        term1 = K.log(1 + K.exp(-1 * K.abs(logit)))
        bce = K.maximum(logit, 0) - term0 + term1
        return bce
    def categorical(y_true, y_pred):
        y_pred /= K.sum(y_pred, axis=-1, keepdims=True)
        y_pred = K.clip(y_pred, self.jpg, 1 - self.jpg)
        cce = -1 * K.sum(self.weights * y_true * K.log(y_pred))
        return cce
```

Using a similar implementation as weighted cross entropy, other weighted loss functions exist (e.g. weighted Hausdorff distance [10]). Furthermore, it is feasible that any multi-class loss function could be manually adapted to account for class imbalance by including defined class specific weightings.

10.5.2 GENERALIZED DICE LOSS

Dice loss is one of the most common loss functions, but it unfortunately is not entirely robust to class imbalances. To account for imbalances, the generalized Dice loss weights the per class Dice score with the inverse square of that class's ground truth volume [11]. This metric is then given by Equation 10.6:

$$GDL = 1 - \frac{2 * \sum_{l} w_l |X \cap Y| + \in}{\sum_{l} w_l |X + Y| + \in} \text{ where } w_l = \frac{1}{Y_l^2} \tag{10.6}$$

where the summation in numerator and denominator represents calculation on a per-class basis. Through inclusion of the weighting factor, the generalized Dice score biases towards classes with a smaller volume proportional to their under-representation.

The strength of the generalized Dice loss function is that it does not require user hyperparameter tuning to compensate for class imbalance. Generalized Dice loss is a good imbalance compensating loss function to test first due to this lack of required tuning. After experimenting with generalized Dice loss, other loss functions with greater tunability can be explored to determine if any potential performance gains exist. A Numpy, Keras, and Tensorflow compatible implementation of generalized Dice loss are provided.

```
# Numpy compatible generalized dice loss
import numpy as np

def generalized_dice_coef(output, labels):
    # Computes the generalized dice coefficient of two numpy arrays
    eps = np.finfo(float).jpg
    total = np.sum(labels)
    sum_dims = tuple(range(labels.ndim))
    w = 1 / (np.sum((labels/total), axis=sum_dims[:-1])**2 + eps)
    numerator = np.sum(w * np.sum(output * labels, axis=sum_dims))
    denominator = np.sum(w * np.sum(output + labels, axis=sum_dims))
    return (2 * numerator + eps) / (denominator + eps)

def generalized_dice_loss(output, labels):
    # Computes the generalized dice loss of two numpy arrays
    return 1 - generalized_dice_coef(output, labels)
```

10.5.3 NO-BACKGROUND DICE LOSS

The no-background Dice loss function is a boundary condition of the generalized Dice loss function for when only small structures and the background exist. To address the foreground-background class imbalance, the Dice loss can be calculated on only the structures of interest, excluding the background altogether. During training, the reported loss function is then simply one minus the average Dice coefficient of the structures. Considering that this utilizes the standard Dice similarity coefficient, the loss function only accounts for large imbalances between structures and the background class, not imbalances which may exist between classes. As such, this loss function is best suited for a model concluding with a softmax activation used to segment a single or paired small volume structure.

```
# Numpy compatible dice loss
# Requires the standard dice loss implementation as well
import numpy as np

def no_bkgd_dice_loss(output, labels):
    # Computes the dice loss from arrays with background channel at 0
    return 1 - dice_coef(output[..., 1:], label[..., 1:])
```

10.5.4 Focal Loss

Focal loss, as the name implies, adds a focusing mechanism into cross entropy loss which reduces the relative importance of high-confidence predictions [12]. This is particularly relevant for multi-class problems with a final softmax activation where the predictions are certainties of class membership. As the model trains and confidence increases for the membership of certain voxels, those highly confident predictions are down-weighted in the loss function. When applied to imbalanced problems, the model will quickly and confidently learn that much of the image volume is a part of the background class. Once this occurs, the focal loss function will shift significance away from the background and on to the accuracy of the remaining structures.

Focal loss achieves dynamic rebalancing by including a scaling factor which decays to zero as probability approaches one. In a simple n-class case, the focal loss is given by Equation 10.7:

$$FL = -\sum_{i=1}^{n} \alpha_i \left(1 - X_i\right)^{\gamma_i} \log\left(X_i\right)$$

(10.7)

where in the equation, X_i is the predicted class membership certainty, α_i is a user adjustable per-class weighting factor and γ_i is a user-adjustable per-class focusing parameter, although most implementations leave the focusing parameter equal across all classes. When the focusing parameter is $\gamma = 0$, the loss function is equivalent to the cross-entropy loss function. But, as γ becomes larger, the magnitude of focusing increases.

10.5.5 Sensitivity Specificity Loss

To understand sensitivity specificity loss, it is necessary to first understand how sensitivity (recall), specificity, and precision relate to medical image segmentation. During the calculation of each of these metrics, the voxel-wise accuracy of the segmentations is considered. With this in mind, four classes of predictions to foreground or background class can be assigned (see Table 10.1).

Table 10.2 shows how sensitivity, specificity, precision, and recall are defined to represent for the quality of segmentation from these classes.

The Dice loss could be represented as the product of recall and precision, as shown in Equation 10.8.

$$Dice = \frac{2|X \cap Y|}{|X| + |Y|} = \frac{TP}{TP + FP} * \frac{TP}{TP + FN}$$

(10.8)

TABLE 10.1

Classification Types for Binary Segmentations

Classification Type	Description
True Positive (TP)	Indicates a voxel correctly classified as a member of the class
True Negative (TN)	Indicates a voxel correctly classified as not a member of the class. Depending on encoding, this may represent the background
False Positive (FP)	Indicates a voxel incorrectly classified as a member of the class, when the ground truth designates it as not a member, or as the background
False Negative (FN)	Indicates a voxel incorrectly classified as not a member of the class, typically representing background over-prediction

TABLE 10.2

Metrics to Evaluate Binary Segmentations

Statistic	Equation	Description
Sensitivity or Recall	$TP/(TP + FN)$	Represents a model's ability to correctly segment the ROI, with score penalization due to structure under-segmentation, or the prediction of false negatives
Specificity	$TN/(TN + FN)$	Measures the background segmentation accuracy, with penalization due to ROI over-segmentation
Precision	$TP/(TP + FP)$	A measure of a model's capabilities to segment the ROI, with scoring penalization resulting from over-segmentation of the structure, or the prediction of false positives

To calculate sensitivity specificity loss (SSL) [13], the balance between the two terms can be adjusted with a factor, in this case r. This would then give the sensitivity specificity loss function as in Equation 10.9:

$$SSL = r\frac{TP}{TP + FP} * (1 - r)\frac{TN}{TN + FP} \tag{10.9}$$

which is computed as a combination of the mean squared errors between the prediction (sensitivity) and the background (specificity), which is provided in Equation 10.10:

$$SSL = r\frac{\sum_i (X_i - Y_i)^2 X_i}{\sum_i X_i} + (1 - r)\frac{\sum_i (X_i - Y_i)^2 (1 - X_i)}{\sum_i (1 - X_i)} \tag{10.10}$$

where the background to foreground weighting can be accounted for with the r factor, where a higher value of r places a larger emphasis on the sensitivity, or foreground voxels. Like weighted cross entropy, this loss function allows for user adjustable weighting to compensate for class imbalances present in the task.

10.5.6 TVERSKY LOSS

If the segmentation task requires higher sensitivity to either false negatives or false positives, a variable index, such as the Tversky loss can be utilized [14]. The Tversky index is given in Equation 10.11:

$$Tversky = \frac{|X \cap Y|}{|X \cap Y| + \alpha |\sim Y| + \beta |\sim X|} = \frac{TP}{TP + \alpha * FP + \beta * FN} \tag{10.11}$$

where the $\sim$ operator indicates the relative complement of the Boolean array and the values of α and β are hyperparameters corresponding to the magnitude of the penalization for FP and FN, respectively. Through adjusting the ratio of α/β, the performance of the loss function can be modified. In the instance that $\alpha = \beta = 0.5$, the Tversky loss function becomes equivalent to the Dice loss function.

The Tversky loss function's strength is, if the user is so inclined, it that can be adjusted to exactly counteract the task's class imbalance or segmentation needs. For example, segmentation tasks which prioritize ROI coverage could have a lower ratio of α/β, whereas tasks which require minimal over-expansion of segmentations would utilize a higher ratio.

```
# Numpy compatible Tversky coefficient and loss
class losses:
    def __init__(self, alpha=0.5, beta=0.5, loss=True):
        self.alpha = alpha
        self.beta = beta

    def tversky(self, output, labels):
        # Calculates the tversky coefficient or loss
        eps = np.finfo(float).jpg
        true_pos = np.sum(output * labels)
        a_false_pos = self.alpha * np.sum(labels * (1 - output))
        b_false_neg = self.beta * np.sum(output * (1 - labels))
        tversky = (true_pos + eps) / (true_pos + a_false_pos + b_false_neg
                  + eps)
        return tversky

    def tversky_loss(self, output, labels):
        return 1 - self.tversky(output, labels)
```

10.6 COMPOUND LOSS FUNCTIONS

Many of the loss functions discussed in this chapter exhibit unique properties which make them well suited for segmentation tasks. Occasionally, however, problems require properties at the intersection of multiple loss functions. Fortunately, different loss functions can be combined to span a larger set of properties.

10.6.1 DICE + CROSS ENTROPY

The combination of cross entropy and Dice loss is a popular pairing for loss functions [15]. Alone, the Dice loss is robust to minor class imbalances but does not allow for weighting of false positives or false negatives. The two terms within a weighted binary cross entropy function, however, can be modified to increase or decrease the penalty for false negative or false positive values. When Dice loss and cross entropy losses are combined, the result is a partially class imbalanced loss function with variable sensitivity for false predictions.

10.6.2 DICE + FOCAL LOSS

A further example of combined loss function is Dice loss and focal loss [16]. More precisely, this loss function implementation utilized the Tversky loss function with $\alpha = \beta = 0.5$, although these hyperparameters could have been tuned differently for this task. Through the combination, this joint loss function combines both the volumetric dependency of the Dice loss and the focal loss property of increased importance of highly uncertain predictions.

10.6.3 NON-LINEAR COMBINATIONS

To generate the most utility from a combined loss function, the balance between the terms should exhibit non-linear behavior. A strong loss function combination should choose loss functions which each possess unique properties. For some tasks, these behaviors can be more powerful at the early or late stages of training.

For example, take the Hausdorff loss function. Traditionally, the 100th percentile Hausdorff distance is highly sensitive to spatial outliers which limits the usefulness during early training. However, this becomes an asset during late training stages, as it can accurately discriminate against spatial outliers, thus fine-tuning performance.

Another example of a potential non-linear combination is Dice and focal loss. In the original loss function implementation, the Dice loss term dominates for epochs with poor validation set performance. Then, the importance of the focal loss term increases as the validation set performance improves. This gradual shift in balance allows the model to partially train on Dice loss before becoming dominated by focal loss and being penalized for high prediction uncertainty.

It should be noted that non-linear loss function combinations will require additional hyperparameter tuning and are more likely to train inconsistently. A suggested workflow is to begin training the model with only the initially dominant term. Then, once hyperparameter-tuned, the loss function can be expanded with the minor terms, before re-tuning the hyperparameters.

10.7   DEALING WITH IMPERFECT DATA

For most medical image segmentation tasks, the training data set must be large, diverse, and high quality. Unfortunately, particularly in medicine, creating such a training set is a time-consuming undertaking. This is particularly problematic when the generation of ground truth labels requires an expert, whose time is likely at a premium.

An ongoing field of research attempts to create methods and loss functions to train high quality models from imperfect data. In many clinical cases, only the relevant selection of all organs-at-risk are segmented. This means that the original clinical dataset may not be densely populated with all structures on all cases. For cases that lack a labeled structure, gradient backpropagation will penalize a model's potentially accurate prediction due to imperfections in the ground truth.

A few attempts to account for imperfect data, particularly sparsely labeled ground truths, have achieved success through modification of the loss function. For example, Bokhorst et al. [17] trained a U-net model from sparsely labeled histology images by only backpropagating the loss function from channels which had "valid" ground truth labels. Zhu et al. [16] extended this concept by not only masking for only "valid" ground truths but weighting each class at the inverse of their occurrence. In doing so, the loss function compensated for the inter-class imbalance deriving from the sparsely labeled ground truth. Although these are promising first steps, the further adaptation of loss functions to train robustly on imperfect data will continue to garner interest for medical image segmentation. For further discussion on data set preparation, see Chapter 14.

10.8   EVALUATING A LOSS FUNCTION

In the proceeding sections, many differing loss functions and their application were discussed. With the numerous loss function choices, picking a starting point can be overwhelming. A decision tree to help choose an initial loss function is provided in Figure 10.6. However, to get the most out of the chosen loss function, a user should understand how to evaluate and tune the loss function's performance.

Typical deep learning strategy dictates a dataset be separated into three unique subsets: training, validation, and testing. The training set, as the name implies, is used to train the model and is the largest of the three subsets. During the training process, predictions made from this data are used for backpropagation weight updates. Following every epoch, the training model makes predictions from a smaller subset of data, the validation set, where predictions are made without updating the model's weights. It should be repeated that deep learning models are lazy and will take whatever shortcuts are available. Commonly, this shortcut is overfitting by memorization. When a model memorizes, it begins to perform outstandingly on the training dataset without learning generalizable features, which means it cannot replicate this performance equally on an unlearned dataset, such as the validation set. The model's progress can be monitored in real time by frequently predicting the validation dataset, preventing time from being wasted when the training is non ideal. Typically, the relationship of training and validation loss falls into one of four categories, as shown in Figure 10.7.

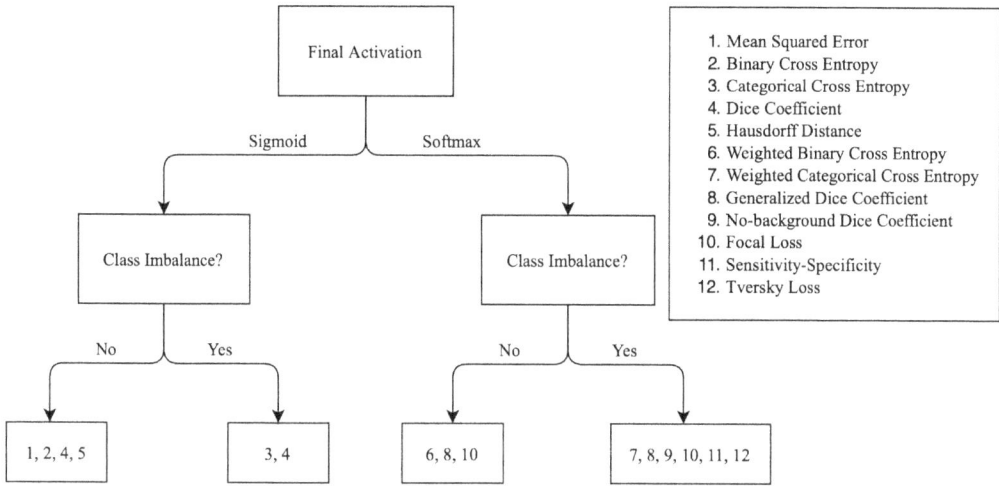

FIGURE 10.6   A flowchart to aid in determining the proper loss function for a given task.

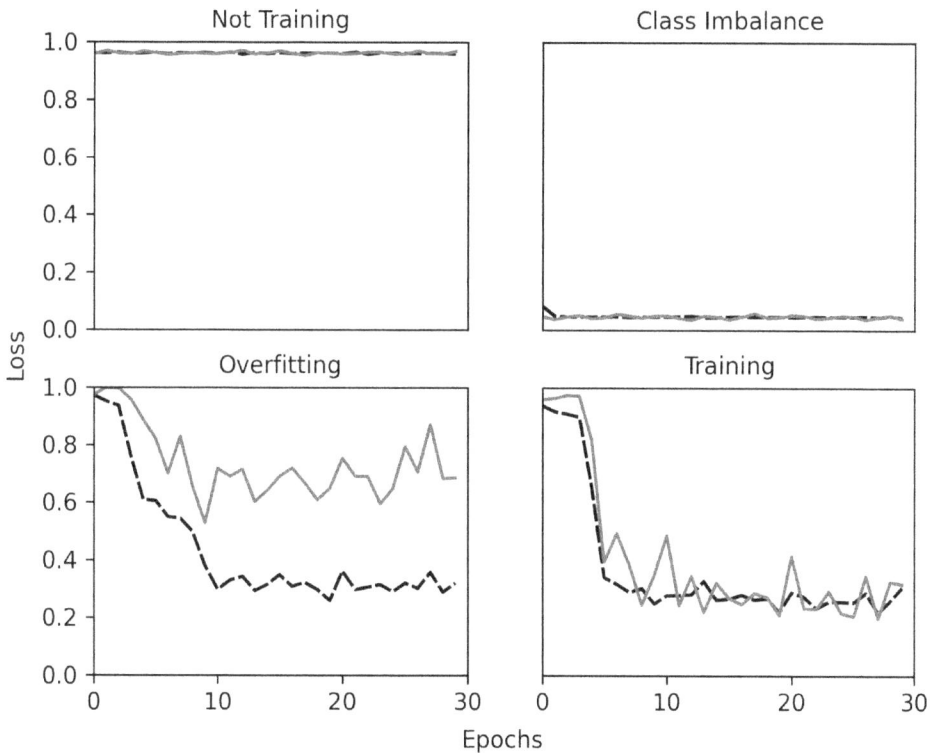

FIGURE 10.7   A representation of different types of relationships between the training loss (slightly lighter) and validation loss (slightly darker). Top-left: A model which does not train. Top-right: A highly imbalanced data set with a poorly suited loss function. Bottom-left: A model which overfits on the training set. Bottom-right: A model which trains.

A model that consistently performs poorly on both losses across all epochs, as seen in the top-left of Figure 10.7, is indicative of a model that is not training. Unfortunately, there is no clear-cut reason why a model does not train, but troubleshooting should progress through the training process. Beginning with the data, this issue may arise from training data or ground truth labels that are incorrectly formatted or not properly corresponding. Within the model, errant graph connections or incorrect final activation and loss function pairings can prevent the model from properly backpropagating the gradient. Finally, hyperparameters may be poorly selected, causing weights to change too quickly or coarsely to successfully converge to the minima.

A model which immediately produces outstanding and desirable results, like that shown in the top-right of Figure 10.7, is indicative of a highly unbalanced task paired with an unbalanced loss function. At the start of training, a model's weights are randomly initialized, and are never expected to perform perfectly after only a few iterations of the training cycle. This behavior is typically characterized by a model becoming trapped in an overwhelming local minimum, such as predicting one class for the entire volume. This can be troubleshot through experimentation with alternative loss functions.

An overfitting model, as given in the bottom-left of Figure 10.7, has a loss function that consistently decreases while the validation loss remains unchanged. To prevent overfitting, common techniques may be to introduce dropout into the model or utilizing optimizer regularization. Additionally, the training data can be augmented to simulate a more diverse dataset. Approaches to data augmentation are discussed in Chapter 11.

When everything comes together, and a deep learning model learns properly, both loss functions are expected to decrease relatively steadily and asymptotically to the same value, as shown in the bottom-right of Figure 10.7. It is important to note that the rate of convergence will vary based on task, model, and optimizer. In this instance, the model was able to learn a generalizable feature from the training data and perform equally well on the validation set. The possibility exists, however, that the chosen loss function is not indicative of desired performance. To check this, the model's predictions on the validation set should be compared to the ground truth with additional metrics. If these metrics also indicate strong performance, a final prediction on the test set can be made. A more detailed discussion of evaluation of model performance is presented in Chapter 15.

For a deep learning model to converge upon a generalizable solution, the method by which it gauges performance, the loss function, must be carefully chosen. The loss function dictates the backpropagation process, and in turn how a model learns, because the loss function quantifies the fitness of the model's predictions. While educated guessing may assist in selecting a loss function, finding the ideal function typically requires experimentation with different loss functions or combinations. The most popular loss functions were described within this chapter, but there exist many niche functions which were not discussed. As techniques for medical image segmentation evolve, pioneering individuals will continue to develop novel loss functions capable of greater admissibility and ease of trainability.

REFERENCES

1. Li H, Xu Z, Taylor G, Studer C, Goldstein T. Visualizing the Loss Landscape of Neural Nets. arXiv:171209913 [cs, stat] [Internet]. 2017 Dec 28 [cited 2019 Jul 9]. Available from: http://arxiv.org/abs/1712.09913

2. Menze BH, Jakab A, Bauer S, Kalpathy-Cramer J, Farahani K, Kirby J, et al. The Multimodal Brain Tumor Image Segmentation Benchmark (BRATS). *IEEE Trans Med Imaging*. 2015;34(10):1993–2024. PMID: 25494501

3. Bakas S, Akbari H, Sotiras A, Bilello M, Rozycki M, Kirby JS, et al. Advancing the Cancer Genome Atlas glioma MRI Collections with Expert Segmentation Labels and Radiomic Features. *Sci Data*. 2017;4:170117. PMID: 28872634

4. Bakas S, Reyes M, Jakab A, Bauer S, Rempfler M, Crimi A, et al. Identifying the Best Machine Learning Algorithms for Brain Tumor Segmentation, Progression Assessment, and Overall Survival Prediction in the BRATS Challenge. 2019 Apr 12 [cited 2020 Oct 2]. Available from: https://www.repository.cam.ac.uk/handle/1810/291597

5. Dice LR. Measures of the Amount of Ecologic Association Between Species. *Ecology* [Internet]. 1945;26(3):297–302 [cited 2019 Jun 24]. Available from: https://esajournals.onlinelibrary.wiley.com/doi/abs/10.2307/1932409

6. Milletari F, Navab N, Ahmadi S. V-Net: Fully Convolutional Neural Networks for Volumetric Medical Image Segmentation. In: *2016 Fourth International Conference on 3D Vision (3DV)*. 2016. pp. 565–571.

7. Karimi D, Salcudean SE. Reducing the Hausdorff Distance in Medical Image Segmentation with Convolutional Neural Networks. arXiv:190410030 [cs, eess, stat] [Internet]. 2019 Apr 22 [cited 2020 Mar 30]. Available from: http://arxiv.org/abs/1904.10030

8. Maier O. loli/medpy [Internet]. 2020 [cited 2020 Oct 2]. Available from: https://github.com/loli/medpy

9. Ronneberger O, Fischer P, Brox T. U-Net: Convolutional Networks for Biomedical Image Segmentation. In: Navab N, Hornegger J, Wells WM, Frangi AF, editors. *Medical Image Computing and Computer-Assisted Intervention – MICCAI 2015*. Springer International Publishing; 2015. pp. 234–241. (Lecture Notes in Computer Science).

10. Ribera J, Güera D, Chen Y, Delp EJ. Locating Objects Without Bounding Boxes. arXiv:180607564 [cs] [Internet]. 2019 Apr 3 [cited 2020 Mar 30]. Available from: http://arxiv.org/abs/1806.07564

11. Sudre CH, Li W, Vercauteren T, Ourselin S, Cardoso MJ. Generalised Dice Overlap as a Deep Learning Loss Function for Highly Unbalanced Segmentations. Deep Learning in Medical Image Analysis and Multimodal Learning for Clinical Decision Support Lecture Notes in Computer Science [Internet]. 2017;10553:240–248 [cited 2019 Jun 24]. Available from: http://arxiv.org/abs/1707.03237

12. Lin T-Y, Goyal P, Girshick R, He K, Dollár P. Focal Loss for Dense Object Detection. arXiv:170802002 [cs] [Internet]. 2018 Feb 7 [cited 2020 Mar 24]. Available from: http://arxiv.org/abs/1708.02002

13. Brosch T, Yoo Y, Tang LYW, Li DKB, Traboulsee A, Tam R. Deep Convolutional Encoder Networks for Multiple Sclerosis Lesion Segmentation. In: Navab N, Hornegger J, Wells WM, Frangi AF, editors. *Medical Image Computing and Computer-Assisted Intervention – MICCAI 2015*. Springer International Publishing; 2015. pp. 3–11. (Lecture Notes in Computer Science).

14. Salehi SSM, Erdogmus D, Gholipour A. Tversky Loss Function for Image Segmentation Using 3D Fully Convolutional Deep Networks. arXiv:170605721 [cs] [Internet]. 2017 Jun 18 [cited 2019 Aug 24]. Available from: http://arxiv.org/abs/1706.05721

15. Taghanaki SA, Zheng Y, Zhou SK, Georgescu B, Sharma P, Xu D, et al. Combo Loss: Handling Input and Output Imbalance in Multi-Organ Segmentation. arXiv:180502798 [cs] [Internet]. 2018 Oct 22 [cited 2020 Mar 23]. Available from: http://arxiv.org/abs/1805.02798

16. Zhu W, Huang Y, Zeng L, Chen X, Liu Y, Qian Z, et al. AnatomyNet: Deep Learning for Fast and Fully Automated Whole-Volume Segmentation of Head and Neck Anatomy. *Med Phys* [Internet]. 2019;46(2):576–89. [cited 2020 Mar 30]. Available from: https://aapm.onlinelibrary.wiley.com/doi/abs/10.1002/mp.13300

17. Bokhorst JM, Pinckaers H, van Zwam P, Nagtegaal I, Laak J van der, Ciompi F. Learning from Sparsely Annotated Data for Semantic Segmentation in Histopathology Images. 2018 [cited 2020 Mar 30]. Available from: https://openreview.net/forum?id=SkeBT7BxeV

11 Data Augmentation for Training Deep Neural Networks

Zhao Peng, Jieping Zhou, Xi Fang, Pingkun Yan,
Hongming Shan, Ge Wang, X. George Xu, and Xi Pei

CONTENTS

11.1 OVERVIEW

Data augmentation is a popular technique for reducing overfitting and improving the generalization capabilities of deep neural networks. Augmentation encompasses a suite of techniques that enhances the size and diversity of training datasets. It plays a critical role when the amount of high-quality ground truth data is limited, and acquiring new examples is costly and time-consuming, a very common problem in medical image analysis, including auto-segmentation for radiation therapy [1]. This chapter reviews current advances in data augmentation techniques applied to auto-segmentation in radiation oncology, including geometric transformations, intensity transformation, and artificial data generation. In addition, an example application of these data augmentation methods for training deep neural networks for segmentation in the domain of radiation therapy is provided.

11.2 INTRODUCTION AND LITERATURE REVIEW

With the support of big data, deep convolutional neural networks have performed remarkably well on many computer vision tasks [2–8]. However, many application domains do not have access to big data, such as medical image analysis. It is especially difficult to build big medical image datasets due to the rarity of diseases, patient privacy, the requirement of medical experts for labeling, and the expense and manual effort needed to conduct medical imaging processes. In order to successfully build well-generalizing deep models, a huge amount of ground truth data is needed to avoid the

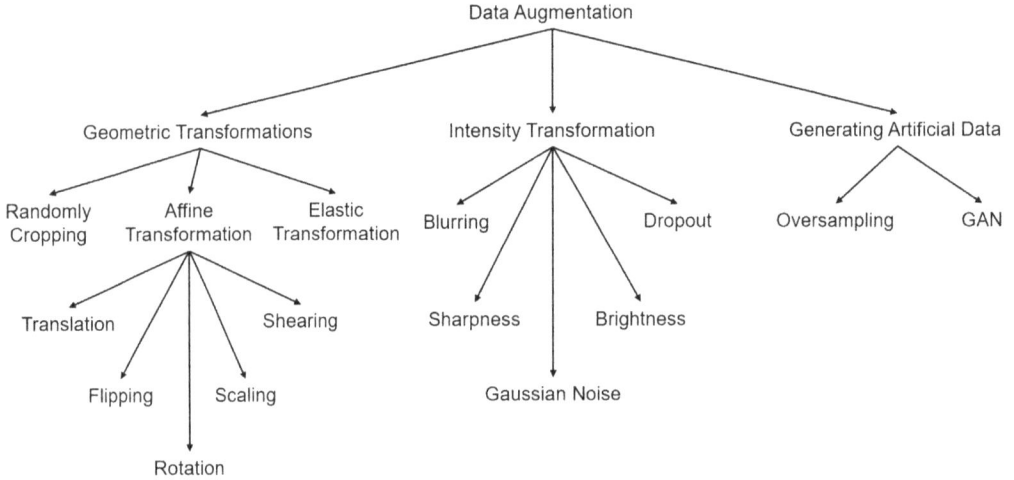

FIGURE 11.1 The taxonomy of data augmentation methods in auto-segmentation for radiation oncology applications.

overfitting of such a large-capacity neural network method, and "memorizing" the training set [9]. It is a generally accepted notion that bigger datasets result in better deep learning models [10–12]. To combat the problem of limited size medical training sets, data augmentation has been widely used in medical image analysis [13–16]. It encompasses a series of techniques that enhance the size and diversity of training datasets so that better deep learning models can be built using them.

Shorten and Khoshgoftaar [17] summarized general data augmentation algorithms in natural images, including geometric transformations, color space augmentations, kernel filters, mixing images, random erasing, feature space augmentation, adversarial training, generative adversarial networks, neural style transfer, and meta-learning. However, in radiation oncology, the images involved are medical images such as computed tomography (CT), magnetic resonance imaging (MRI), and positron emission tomography (PET), which are different from natural images. For example, CT is grayscale while natural images are color. In addition, there are also some differences in the data augmentation methods for different learning tasks. In this chapter, considering data augmentation for auto-segmentation in radiation oncology, the literature is reviewed, and several types of data augmentation methods are summarized. Figure 11.1 illustrates the range of methods that can be employed for data augmentation. Geometric transformations, intensity transformation, and artificial data generation are considered in this chapter.

11.3 GEOMETRIC TRANSFORMATIONS

The most commonly used geometric transformations for data augmentation include flipping, rotation, translation, scaling, shearing, and cropping. The flipping operation creates a mirror reflection of an original image along one or more selected axes. The rotation operation is done by rotating the image right or left on an axis between 1° and 359°. The translation operation shifts the entire image by a given number of pixels in a chosen direction, while applying padding accordingly. The scaling operation zooms in or out an image along one or more selected axes. The shear operation displaces each point in an image in a selected direction. This displacement is proportional to its distance from the line which goes through the origin and is parallel to this direction. The cropping operation can be used as a practical processing step for image data with mixed height and width dimensions by cropping a central patch of each image. Additionally, random cropping is usually adopted to increase the variety of training examples [18–20]. The arbitrary combinations of the flipping, rotation, translation, scaling, and shearing are usually called affine transformations [21].

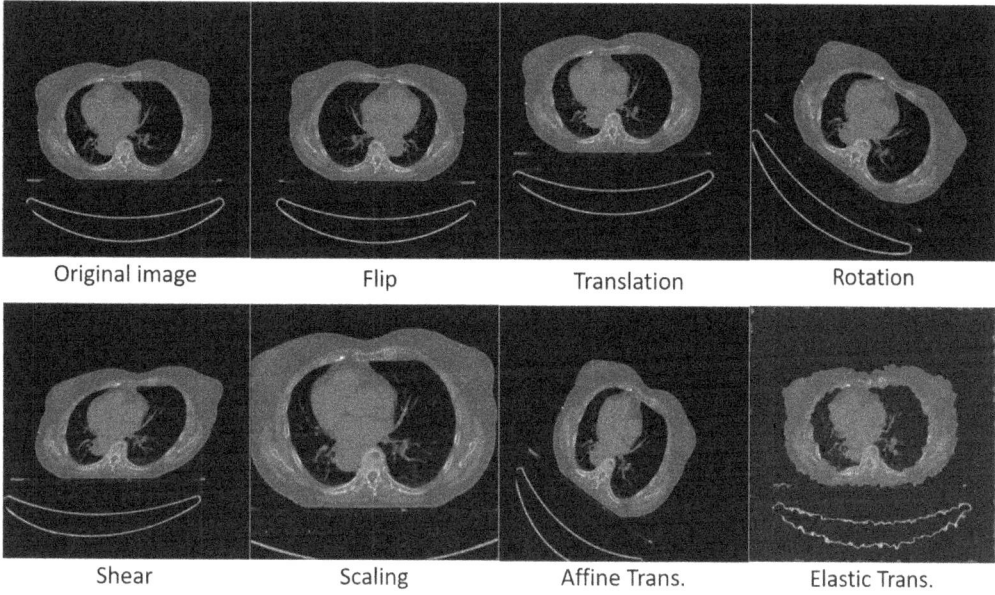

FIGURE 11.2 Applying geometric transformations to CT images.

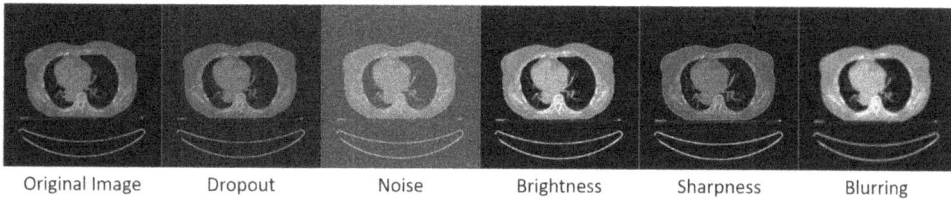

FIGURE 11.3 Applying intensity transformations to CT images.

Affine transforms preserve the parallelism of lines in the input and output images. Those operations are the easiest to implement and have proven useful on datasets such as ImageNet [3]. They are also widely used in the medical image segmentation task [22–27].

Another common geometric transformation is elastic transformation, which can lead to the distortion of shapes in the image and bring the different training examples from the affine transformation [28]. Considering the great variability in the shape and appearance of the tissue, organ, and tumor, this operation can be especially useful in medical image analysis. Elastic transformations are often used in combination with affine transformations [29–32], which can greatly increase the diversity of the training examples. Figure 11.2 presents examples by applying these geometric transformations to a CT image.

11.4 INTENSITY TRANSFORMATION

Intensity transformation refers to change of pixel intensity values, either locally or across the entire image. Methods include adding Gaussian noise, random dropout, shifting, and scaling of pixel-intensity values (for example modifying the image brightness), sharpening, blurring, and more [33–36]. Such operations can be especially useful in medical image analysis, where different training images are acquired in different locations and using different scanners, hence they can be intrinsically heterogeneous in the pixel intensities or intensity gradients. Figure 11.3 shows some examples of intensity transformation on a CT image.

11.5   ARTIFICIAL DATA GENERATION

Oversampling is a traditional method for data augmentation by synthesizing new samples using the existing training data. This approach primarily focuses on alleviating problems due to class imbalance. Random oversampling (ROS) is a naive approach which duplicates images randomly from the minority class until a desired class ratio is achieved. Intelligent oversampling techniques date back to the synthetic minority over-sampling technique (SMOTE), developed by Chawla et al. [37]. SMOTE created new instances by interpolating new points from existing instances via k-nearest neighbors. Later, Inoue [38] introduced a simple but surprisingly effective data augmentation technique named SamplePairing. A new sample was synthesized from one image by overlaying another image randomly chosen from the training data (i.e. taking an average of two images for each pixel). Zhang et al. [39] introduced a data oversampling routine, termed mixup, which blended two examples drawn at random from the training data by weighted summation. Their experiments showed that mixup improves the generalization of state-of-the-art neural network architectures.

Generative adversarial nets (GANs) are a method for synthesizing data using deep neural networks. GANs consist of a generator, which synthesizes samples, and a discriminator, which evaluates the reality of synthetic samples. GANs were first introduced in 2014 [40], and since then various works on GAN extensions, such as DCGANs [41], WGAN [42], and CycleGANs [43], were published. GANs have been widely used for data augmentation. Sandfort et al. [44] used a Cycle-GAN-based data augmentation to improve generalizability in CT segmentation tasks. Frid-Adar et al. [45] used GAN-based data augmentation for liver lesion classification. This improved classification performance from 78.6% sensitivity and 88.4% specificity using classic augmentations to 85.7% sensitivity and 92.4% specificity using GAN-based data augmentation. Tang et al. [46] used pix2pix-GAN-based data augmentation to enhance lymph node segmentation; the Dice score increased about 2.2% (from 80.3% to 82.5%). Zou et al. [47] used a CycleGAN-based framework to generate domain adaptive images to realize unsupervised segmentation of images in the target domain.

11.6   APPLICATIONS OF DATA AUGMENTATION

11.6.1   DATASETS AND IMAGE PREPROCESSING

In this chapter, two datasets were used: The 2017 Lung CT Segmentation Challenge (LCTSC) [48–50] detailed in Chapter 1, and a Pancreas-CT (PCT) dataset, which contains 43 abdominal contrast enhanced CT scan patients with eight segmented organs (the spleen, left kidney, gallbladder, esophagus, liver, stomach, pancreas, and duodenum) [22, 30, 49, 51]

For each patient in these datasets, the Hounsfield unit (HU) values were processed using a minimum threshold of −200 and a maximum threshold of 300 prior to being normalized to yield values between 0 to 1. In order to focus on organs and suppress the background information, the image was cropped to a region of interest according to the body contour in the original CT images and used as training data. Finally, to circumvent computer memory limitation, data resampling was performed using linear interpolation for CT images and using nearest interpolation for the labels. The resulting resolution after resampling was 2.0 mm × 2.0 mm × 2.5 mm for the LCTSC dataset, and 2.0 mm × 2.0 mm × 1.0 mm for the PCT dataset.

11.6.2   TRAINING, VALIDATION, AND TESTING FOR ORGAN SEGMENTATION

The network used in this study was based on the 3D U-net [52, 53] shown in Figure 11.4; the network consists of an encoder and a decoder. The role of the decoder network is to map the low-resolution encoder feature maps to full input resolution feature maps for pixel-wise classification [54]. The encoder contains four repeated residual blocks. Each block consists of four

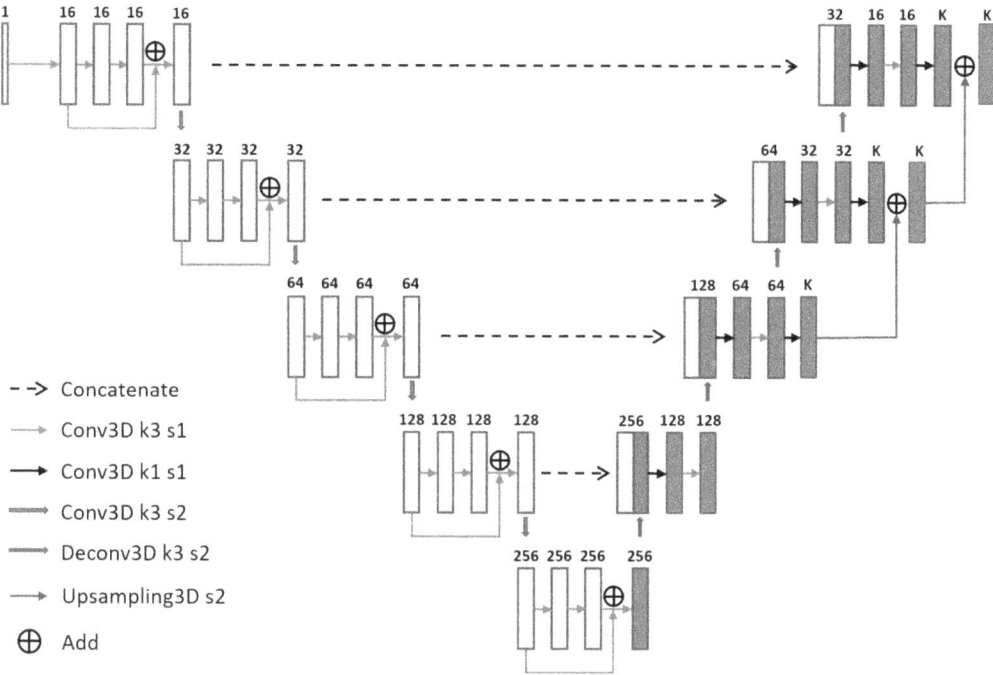

FIGURE 11.4   The network architecture used in the chapter.

convolutional modules. Each convolutional module is composed by a convolution layer with the kernel of $3 \times 3 \times 3$, an instance normalization, and a leaky rectified linear unit with coefficient of 0.3. For each residual block, the stride of convolution layer in the convolutional modules is $1 \times 1 \times 1$ with the exception of the last convolutional module in which the stride is $2 \times 2 \times 2$ to achieve downsampling. There is a spatial dropout layer between the early two convolutional modules to prevent the network from overfitting. The decoder contains four repeated segmentation blocks. Each block consists of two convolutional modules and one deconvolutional module. The four dashed arrows in the figure indicate four skipping connections that copy and reuse early feature-maps as the input to later layers that have the same feature-map size by a concatenation operation to preserve high-resolution features. In the final three segmentation blocks, a $1 \times 1 \times 1$ convolution layer is used to map the feature tensor to a probability tensor with the channels of the desired number of classes before all the results are merged by the upsampling operation to enhance the precision of segmentation results. Finally, a softmax activation is used to output a probability of each class for every voxel [55].

A five-fold cross-validation method was adopted for this work [56]. The entire dataset is randomly split, using the "random.shuffle()" function in Python, into five non-overlapping subsets for training, validation, and testing in the ratio of 3:1:1 (i.e. three subsets for training, one subset for validation, and one subset for testing). The validation process is used to monitor the training process and to prevent overfitting. To reduce the potential for bias, the five randomly split subsets were rotated five times to report the average performance over these five different hold-out testing subsets, as illustrated in Figure 11.5. The five-fold cross validation strategy is key to ensuring the independence of the testing data, i.e. each sample is used in the testing subsets only once.

To assess the value of data augmentation, the segmentation model is trained with and without data augmentation. Geometric transformations, including flipping, rotation, and random cropping were used for data augmentation. Patches are first randomly extracted from the resampled CT

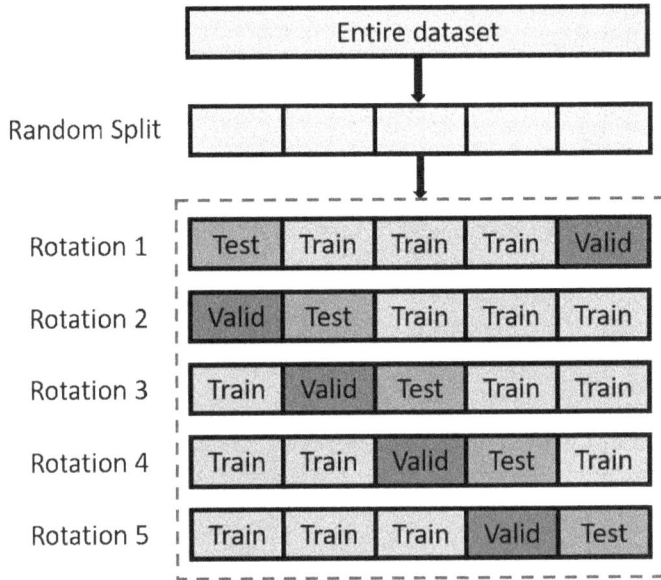

FIGURE 11.5 Example of splitting and rotation using the five-fold cross-validation method for the dataset involving five subsets [53].

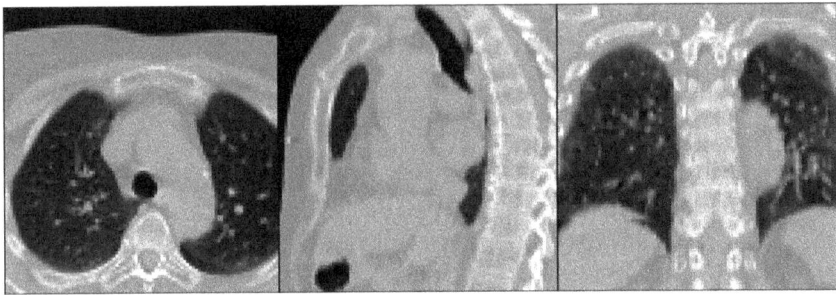

FIGURE 11.6 An example to illustrate patches from LCTSC, the database used in the training in terms of axial, sagittal, and coronal views [53].

images, and then flipped or rotated along some axes. In this experiment, the rotation angles are 0°, 90°, 180°, or 270°, and the patch size is 96 × 96 × 96. Figure 11.6 shows an example of such patches from LCTSC used in the training in terms of axial, sagittal, and coronal views. Finally, the network was trained with the patches and their corresponding labels. The weighted Dice similarity coefficient was used as the loss function, defined as:

$$\text{Loss} = -\frac{1}{N*K}\sum_{i=1}^{N}\sum_{k=1}^{K}\frac{2*\sum_{v=1}^{V}(p_{i,k,v}*y_{i,k,v})+\varepsilon}{\sum_{v=1}^{V}p_{i,k,v}+\sum_{v=1}^{V}y_{i,k,v}+\varepsilon},$$

where $p_{i,k,v}$ is the predicted probability of the voxel v of the sample i belonging to the class k, $y_{i,k,v}$ is the ground truth label (0 or 1), N is the number of samples, K is the number of classes, V is the number of voxels in one sample, and ε is a smooth factor (set to be 1 in this study). The initial learning rate was 0.0005, and the Adam algorithm [57] was used to optimize the parameters of the network. The validation loss was calculated for every epoch, and the learning rate was halved when the validation loss no longer decreased after 30 consecutive epochs. To prevent overfitting, the training process was terminated when the validation loss no longer decreased after 50 consecutive epochs.

At the testing stage, patches were first extracted from each CT image with a moving window. The window size was $96 \times 96 \times 96$ and the stride was 48 in each direction. In other words, multiple patches are extracted from one patient and fed into the network. The output of the network was a probability tensor for each patch. Then all probability tensors were merged from the same patient with a mean operator in the overlapping area to obtain the final probability tensor. Next, the class of each voxel was determined by the largest probability. This resulted in preliminary results of organ segmentation. Using the nearest neighbor interpolation, the preliminary segmentation results were resampled to the size of original CT images to obtain the final organ segmentation result.

All experiments described above were performed on a Linux computer system. Keras with TensorFlow as the backend was used as the platform for designing and training the neural network [58]. The hardware includes (1) GPU – NVIDIA GeForce Titan X Graphics Card with 12 GB memory, and (2) CPU – Intel Xeon Processor X5650 with 16 GB memory.

11.7 EVALUATION CRITERIA

The Dice similarity coefficient (DSC) was used to evaluate the performance of organ segmentation [59]:

$$\text{DSC} = \frac{2|A \cap B|}{|A| + |B|}$$

where A is the manually segmented organ (i.e. the ground truth) and B is the automatically segmented organ by the network. The DSC ranges from 0 to 1 with the latter indicating a perfect performance. Chapter 15 gives alternative measures that could be used for this evaluation.

11.8 RESULTS

The performance of the network with and without data augmentation in organ segmentation was evaluated in terms of the DSC. The segmentation results of all organs are summarized in Figure 11.7 and Table 11.1. The Dice scores showed significant improvement after data augmentation, especially for small or indistinguishable organs. A visual comparison of manual segmentation and automatic segmentation (with and without data augmentation) for all organs from both LCTSC and PCT is shown in Figure 11.8a and b. The results indicate that data augmentation can boost the segmentation performance for deep neural networks.

11.9 DISCUSSION

In this chapter, a range of data augmentation methods are summarized. It was found that the affine transformations are still the most widely used in practice, because they are easy to implement and operate in real time due to low time complexity. In addition, an interesting characteristic of these augmentation methods is their ability to be combined. For example, samples taken from GANs can be augmented with geometry transformation such as flip, rotation, and translation to create more samples. Such hybridizing techniques from various data augmentation algorithmic groups have the potential to further boost the performance of large-capacity deep learning model.

Data augmentation can be applied not only to the training stage but also the test stage. Test-stage augmentation is analogous to ensemble learning in the data space. Instead of aggregating the predictions of different learning algorithms, predictions are aggregated across augmented images. For example, all segmentation results can be predicted after the CT image is flipped, rotated, and scaled, then the segmentation results are averaged to form the final segmentation result of the CT

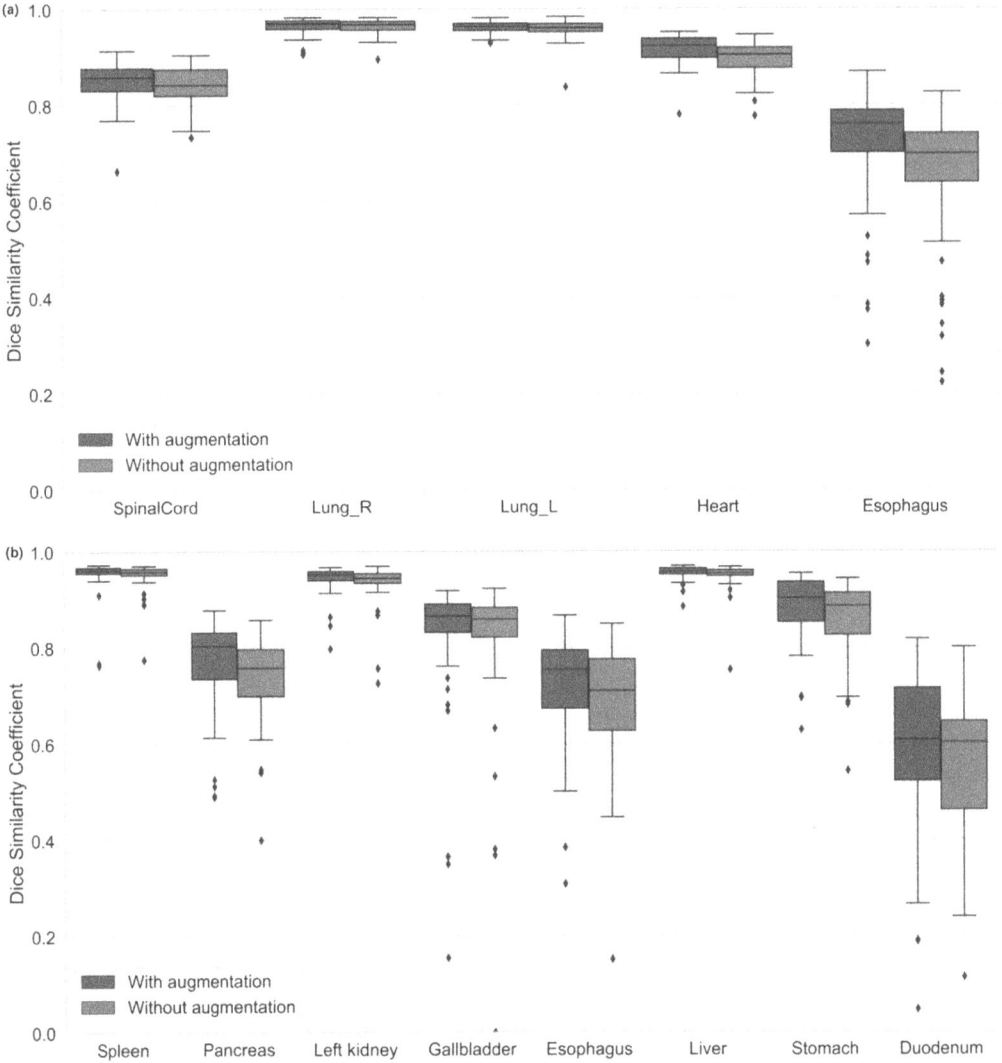

FIGURE 11.7 Evaluation of organ segmentation performance in terms of DSC. (a) Data based on 60 patients from the LCTSC database. (b) Data based on 43 patients from the PCT database.

image. Such an approach focuses more on the high segmentation accuracy instead of the prediction speed.

Combined with self-supervised learning, data augmentation can effectively improve the performance of a deep learning model. A common practice for data augmentation is to assign the same label to all augmented samples of the same source. However, if the augmentation results in large distributional discrepancy among them (e.g. rotations), forcing their label invariance may be too difficult to solve and often reduces the performance. To solve this challenge, Lee et al. [60] proposed a simple yet effective idea of learning the joint distribution of the original and self-supervised labels of augmented samples. The joint learning framework is easier to train, and enables an aggregated inference combining the predictions from different augmented samples for improving the performance.

In the two application examples, the data augmentations such as flipping, rotation, and random cropping were all used to train a 3D CNN model to automatically segment multiple organs in

TABLE 11.1

Comparison of the Segmentation Results (Mean ± Standard Deviation) between Models with and without Data Augmentation

Dataset	Organ	Dice (with Augmentation)	Dice (without Augmentation)
LCTSC	Spinal cord	0.853 ± 0.043	0.842 ± 0.041
	Right lung	0.965 ± 0.017	0.963 ± 0.015
	Left lung	0.961 ± 0.014	0.957 ± 0.020
	Heart	0.916 ± 0.030	0.895 ± 0.035
	Esophagus	0.723 ± 0.118	0.657 ± 0.139
PCT	Spleen	0.951 ± 0.043	0.950 ± 0.032
	Pancreas	0.764 ± 0.105	0.736 ± 0.091
	Left kidney	0.941 ± 0.038	0.932 ± 0.047
	Gall bladder	0.812 ± 0.161	0.784 ± 0.217
	Esophagus	0.713 ± 0.121	0.676 ± 0.133
	Liver	0.954 ± 0.016	0.946 ± 0.033
	Stomach	0.876 ± 0.074	0.857 ± 0.085
	Duodenum	0.586 ± 0.175	0.565 ± 0.143

patient-specific CT images using two datasets. The segmentation performance of most of organs was acceptable – the Dice was over 0.8 – but for some organs such as the duodenum, esophagus, pituitary, the segmentation performance of the network was found to be relatively poor, despite augmentation, because the organ and its surrounding tissues have similar pixel values in CT image, making the boundary difficult to detect by the CNN model. The results indicated that the data augmentation may not be sufficient to solve this problem. To improve the segmentation performance of these organs, more high-quality data or multi-modality data may be needed.

Although data augmentation has become a key part in training deep neural networks for auto-segmentation, there are still promising and unexplored research pathways in the literature. For example, there is no consensus about the best strategy for combining various data augmentation methods. One important consideration is the intrinsic bias in the initial, limited dataset. There are no existing augmentation techniques that can correct a dataset that has very poor diversity with respect to the testing data. All these data augmentation algorithms perform best under the assumption that the training data and testing data are both drawn from the same data distribution. If this is not true, it is very unlikely that these methods will be useful. Additionally, an interesting question for practical data augmentation is how to determine augmented dataset size. There is no consensus as to which ratio of original-to-final dataset size will result in the best performing model.

11.10 SUMMARY

Deep learning models rely on big datasets to avoid overfitting. Data augmentation is a very useful technique for constructing bigger datasets. In this chapter, the state-of-the-art data augmentation methods applied to automatic segmentation for radiation oncology were reviewed, including geometric transformation, intensity transformation, and artificial data generation. In addition, an example of data augmentation in segmentation was demonstrated on two datasets. The results indicated that data augmentation can be very important for training the organ segmentation model. Data augmentation has become a critical part of the deep learning-powered methods.

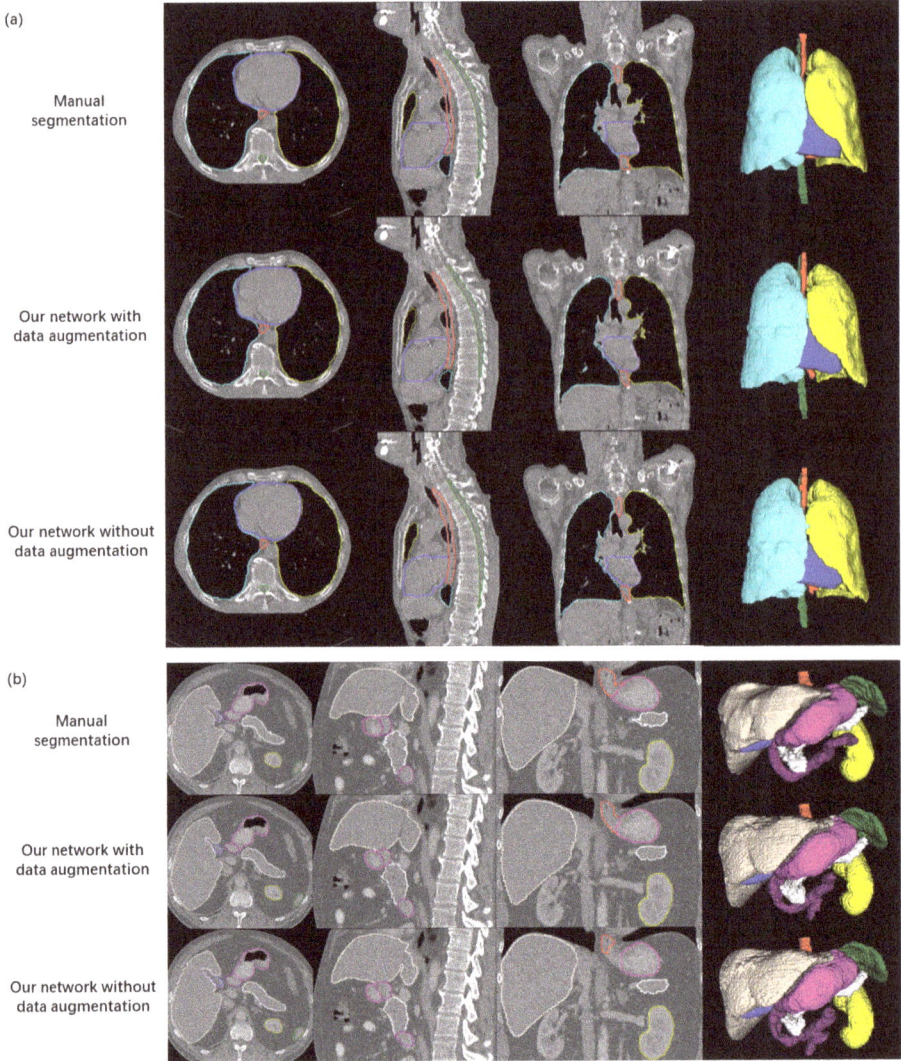

FIGURE 11.8 Examples for visual comparison of organ segmentation between manual methods from LCTSC or PCT database and the automatic methods, in terms of axial, sagittal, coronal, and 3D views (from left to right). (a) LCTSC database showing left lung (yellow), right lung (cyan), heart (blue), spinal cord (green), and esophagus (red). (b) PCT database showing spleen (green), pancreas (white), left kidney (yellow), gallbladder (blue), esophagus (red), liver (bisque), stomach (magenta), and duodenum (purple).

ACKNOWLEDGMENTS

This work was supported in part by Wisdom Tech, in part by an International Training Grant from the American Association of Physicists in Medicine (AAPM), and in part by various grants: NIH/NIBIB (R42EB019265-01A1, U01EB017140, R01EB026646), NIH/NCI (R01CA233888 and R01CA237267), and National Natural Science Foundation of China (11575180).

REFERENCES

1. H Shan, X Jia, P Yan, Y Li, H Paganetti, and G Wang, "Synergizing medical imaging and radiotherapy with deep learning," *Machine Learning: Science and Technology*, vol. 1, no. 021001, 2020. https://io pscience.iop.org/article/10.1088/2632-2153/ab869f/meta

2. Y LeCun, L Bottou, Y Bengio, and P Haffner, "Gradient-based learning applied to document recognition," *Proceedings of the IEEE*, vol. 86, no. 11, pp. 2278–2324, 1998.

3. A Krizhevsky, I Sutskever, and GE Hinton, "ImageNet classification with deep convolutional neural networks," *Communications of the ACM*, vol. 60, no. 6, pp. 84–90, 2017.

4. J Redmon, S Divvala, R Girshick, and A Farhadi, "You only look once: unified, real-time object detection," presented at the Computer Vision and Pattern Recognition, 6/27/2016.

5. K He, G Gkioxari, P Dollar, and R Girshick, "Mask R-CNN," *IEEE Transactions on Pattern Analysis*, vol. 42, no. 2, pp. 386–397, 2/1/2020.

6. J Donahue et al., "Long-term recurrent convolutional networks for visual recognition and description," *IEEE Transactions on Pattern Analysis*, vol. 39, no. 4, pp. 677–691, 4/1/2017.

7. L-C Chen, G Papandreou, I Kokkinos, K Murphy, and AL Yuille, "DeepLab: semantic image segmentation with deep convolutional nets, atrous convolution, and fully connected CRFs," *IEEE Transactions on Pattern Analysis*, vol. 40, no. 4, pp. 834–848, 4/1/2018.

8. X Fang, B Du, S Xu, BJ Wood, and P Yan, "Unified multi-scale feature abstraction for medical image segmentation," in *Medical Imaging 2020: Image Processing*, 2020, vol. 11313: International Society for Optics and Photonics, p. 1131319.

9. Y LeCun, Y Bengio, and G Hinton, "Deep learning," *Nature*, vol. 521, no. 7553, pp. 436–444, 5/28/2015.

10. A Halevy, P Norvig, and F Pereira, "The unreasonable effectiveness of data," *IEEE Intelligent Systems*, vol. 24, no. 2, pp. 8–12, 2009.

11. C Sun, A Shrivastava, S Singh, and A Gupta, "Revisiting unreasonable effectiveness of data in deep learning era," in *Proceedings of the IEEE International Conference on Computer Vision*, 2017, pp. 843–852.

12. G Wang, "A perspective on deep imaging," *IEEE Access*, vol. 4, pp. 8914–8924, 2016.

13. Z Hussain, F Gimenez, D Yi, and DL Rubin, "Differential data augmentation techniques for medical imaging classification tasks," presented at the American Medical Informatics Association Annual Symposium, 1/1/2017.

14. S-C Park, JH Cha, S Lee, W Jang, CS Lee, and JK Lee, "Deep learning-based deep brain stimulation targeting and clinical applications," *Frontiers in Neuroscience-Switz*, vol. 13, 10/24/2019.

15. E Gibson et al., "NiftyNet: a deep-learning platform for medical imaging," *Computer Methods and Programs in Biomedicine*, vol. 158, pp. 113–122, 5/1/2018.

16. H Shan et al., "3-D convolutional encoder-decoder network for low-dose CT via transfer learning from a 2-D trained network," *IEEE Transactions on Medical Imaging*, vol. 37, no. 6, pp. 1522–1534, 2018.

17. C Shorten and TM Khoshgoftaar, "A survey on image data augmentation for deep learning," *Journal of Big Data*, vol. 6, no. 1, p. 60, 2019. https://link.springer.com/article/10.1186/s40537-019-0197-0

18. R Takahashi, T Matsubara, and K Uehara, "Data augmentation using random image cropping and patching for deep CNNs," *IEEE Transactions on Circuits and Systems for Video*, pp. 1–1, 1/1/2020.

19. X Dong et al., "Automatic multiorgan segmentation in thorax CT images using U-net-GAN," (in English), *Medical Physics*, vol. 46, no. 5, pp. 2157–2168, May 2019.

20. K Men, JR Dai, and YX Li, "Automatic segmentation of the clinical target volume and organs at risk in the planning CT for rectal cancer using deep dilated convolutional neural networks," (in English), *Medical Physics*, vol. 44, no. 12, pp. 6377–6389, Dec 2017.

21. CC Stearns, and K Kannappan, "Method for 2-D affine transformation of images," Ed: Google Patents, 1995.

22. E Gibson et al., "Automatic multi-organ segmentation on abdominal CT with dense v-networks," *IEEE Transactions on Medical Imaging*, vol. 37, no. 8, pp. 1822–1834, 2018.

23. Q Dou et al., "3D deeply supervised network for automated segmentation of volumetric medical images," *Medical Image Analysis*, vol. 41, pp. 40–54, 2017.

24. XM Li, H Chen, XJ Qi, Q Dou, CW Fu, and PA Heng, "H-DenseUNet: hybrid densely connected UNet for liver and tumor segmentation from CT volumes," *IEEE Transactions on Medical Imaging*, vol. 37, no. 12, pp. 2663–2674, Dec 2018.

25. X Feng, K Qing, NJ Tustison, CH Meyer, and Q Chen, "Deep convolutional neural network for segmentation of thoracic organs-at-risk using cropped 3D images," *Medical Physics*, vol. 46, no. 5, pp. 2169–2180, May 2019.

26. B Ibragimov and L Xing, "Segmentation of organs-at-risks in head and neck CT images using convolutional neural networks," *Medical Physics*, vol. 44, no. 2, pp. 547–557, 2017.

27. X Fang and P Yan, "Multi-organ segmentation over partially labeled datasets with multi-scale feature abstraction," *IEEE Transactions on Medical Imaging*, vol. 39, no. 11, pp. 3619–3629, 2020. doi: 10.1109/TMI.2020.3001036

28. S Wong, A Gatt, V Stamatescu, and MD Mcdonnell, "Understanding data augmentation for classification: when to warp?," in *Digital Image Computing Techniques and Applications*, 2016, pp. 1–6.

29. O Ronneberger, P Fischer, and T Brox, "U-net: convolutional networks for biomedical image segmentation," in *International Conference on Medical Image Computing and Computer-Assisted Intervention*, 2015: Springer, pp. 234–241.

30. HR Roth et al., "DeepOrgan: multi-level deep convolutional networks for automated pancreas segmentation," in *Medical Image Computing and Computer Assisted Intervention*, 2015, pp. 556–564.

31. N Nguyen, and S Lee, "Robust boundary segmentation in medical images using a consecutive deep encoder-decoder network," *IEEE Access*, vol. 7, pp. 33795–33808, 2019.

32. WT Zhu et al., "AnatomyNet: deep learning for fast and fully automated whole-volume segmentation of head and neck anatomy," (in English), *Medical Physics*, vol. 46, no. 2, pp. 576–589, Feb 2019.

33. D Lachinov, E Vasiliev, and V Turlapov, "Glioma segmentation with cascaded UNet," in *International MICCAI Brainlesion Workshop*, 2018: Springer, pp. 189–198.

34. J Nalepa, M Marcinkiewicz, and M Kawulok, "Data augmentation for brain-tumor segmentation: a review," *Frontiers in Computer Neuroscience*, vol. 13, p. 83, 2019.

35. A Galdran et al., "Data-driven color augmentation techniques for deep skin image analysis," *arXiv: Computer Vision and Pattern Recognition*, 2017.

36. PJ Hu, F Wu, JL Peng, YY Bao, F Chen, and DX Kong, "Automatic abdominal multi-organ segmentation using deep convolutional neural network and time-implicit level sets," (in English), *International Journal of Computer Assisted Radiology*, vol. 12, no. 3, pp. 399–411, Mar 2017.

37. NV Chawla, KW Bowyer, LO Hall, and WP Kegelmeyer, "SMOTE: synthetic minority over-sampling technique," *Journal of Artificial Intelligence Research*, vol. 16, pp. 321–357, 2002.

38. H Inoue, "Data augmentation by pairing samples for images classification," *arXiv preprint arXiv:1801.02929*, 2018.

39. H Zhang, M Cisse, YN Dauphin, and D Lopez-Paz, "mixup: beyond empirical risk minimization," *arXiv preprint arXiv:1710.09412*, 2017.

40. I Goodfellow et al., "Generative adversarial nets," in *Advances in Neural Information Processing Systems*, 2014, pp. 2672–2680.

41. A Radford, L Metz, and S Chintala, "Unsupervised representation learning with deep convolutional generative adversarial networks," *arXiv preprint arXiv:1511.06434*, 2015.

42. M Arjovsky, S Chintala, and L Bottou, "Wasserstein gan," *arXiv preprint arXiv:1701.07875*, 2017.

43. J-Y Zhu, T Park, P Isola, and AA Efros, "Unpaired image-to-image translation using cycle-consistent adversarial networks," in *Proceedings of the IEEE International Conference on Computer Vision*, 2017, pp. 2223–2232.

44. V Sandfort, K Yan, PJ Pickhardt, and RM Summers, "Data augmentation using generative adversarial networks (CycleGAN) to improve generalizability in CT segmentation tasks," *Scientific Reports -Uk*, vol. 9, no. 1, p. 16884, 2019.

45. M Frid-Adar, E Klang, M Amitai, J Goldberger, and H Greenspan, "Synthetic data augmentation using GAN for improved liver lesion classification," in *2018 IEEE 15th International Symposium on Biomedical Imaging (ISBI 2018)*, pp. 289–293, April 4, 2018.

46. Y-B Tang, S Oh, Y-X Tang, J Xiao, and RM Summers, "CT-realistic data augmentation using generative adversarial network for robust lymph node segmentation," in *Journal of Medical Imaging 2019: Computer-Aided Diagnosis*, 2019, vol. 10950: International Society for Optics and Photonics, p. 109503V.

47. D Zou, Q Zhu, and P Yan, "Unsupervised domain adaptation with dual-scheme fusion network for medical image segmentation," presented at the International Joint Conference on Artificial Intelligence (IJCAI), Yokohama, Japan, July 11–17, 2020.

48. J Yang et al., "Data from lung CT segmentation challenge," The Cancer Imaging Archive, 2017. http://doi.org/10.7937/K9/TCIA.2017.3r3fvz08

49. K Clark et al., "The Cancer Imaging Archive (TCIA): maintaining and operating a public information repository," *Journal of Digital Imaging*, vol. 26, no. 6, pp. 1045–1057, 2013.

50. J Yang et al., "Autosegmentation for thoracic radiation treatment planning: a grand challenge at AAPM 2017," *Medical Physics*, vol. 45, no. 10, pp. 4568–4581, 2018.

51. HR Roth, A Farag, E Turkbey, L Lu, J Liu, and RM Summers, "Data from Pancreas-CT," *The Cancer Imaging Archive*, 2016. https://doi.org/10.7937/K9/TCIA.2016.tNB1kqBU

52. Ö Çiçek, A Abdulkadir, SS Lienkamp, T Brox, and O Ronneberger, "3D U-Net: learning dense volumetric segmentation from sparse annotation," in *International Conference on Medical Image Computing and Computer-Assisted Intervention*, 2016.

53. Z Peng et al., "A method of rapid quantification of patient-specific organ doses for CT using deep-learning-based multi-organ segmentation and GPU-accelerated Monte Carlo dose computing," *Medical Physics*, vol. 47, no. 6, pp. 2526–2536, Mar 10 2020.
54. V Badrinarayanan, A Kendall, and R Cipolla, "SegNet: a deep convolutional encoder-decoder architecture for image segmentation," (in English), *IEEE Transactions on Pattern Analysis*, vol. 39, no. 12, pp. 2481–2495, Dec 2017.
55. CM Bishop, *Pattern Recognition and Machine Learning*. Springer, 2006.
56. S Arlot, and A Celisse, "A survey of cross validation procedures for model selection," *Statistics Surveys*, vol. 4, pp. 40–79, 2010.
57. DP Kingma, and J Ba, "Adam: a method for stochastic optimization," *arXiv preprint arXiv:1412.6980*, 2014.
58. M Abadi et al., "Tensorflow: a system for large-scale machine learning," in *12th {USENIX} Symposium on Operating Systems Design and Implementation ({OSDI} 16)*, 2016, pp. 265–283.
59. LR Dice, "Measures of the amount of ecologic association between species," *Ecology*, vol. 26, no. 3, pp. 297–302, 1945.
60. H Lee, SJ Hwang, and J Shin, "Rethinking data augmentation: self-supervision and self-distillation," *arXiv preprint arXiv:1910.05872*, 2019.

12 Identifying Possible Scenarios Where a Deep Learning Auto-Segmentation Model Could Fail

Carlos E. Cardenas

CONTENTS

12.1 BACKGROUND

12.1.1 SITE-SPECIFIC MODELS

A majority of radiation oncology auto-segmentation studies focus on developing algorithms for single anatomical sites. This is driven by several reasons, some of which include limited scan field-of-view (FOV), availability of expert contours, and an increase in complexity to account for in multi-site solutions.

In radiotherapy, image acquisition FOV of treatment simulation scans is driven by preset imaging protocols for each disease site (defined based on disease site, the tumor location, and the extent of the disease as imaging the patient outside of these regions would deliver an unnecessary radiation dose to the patient). Limiting the FOV potentially reduces an unnecessary radiation dose in a computed tomography (CT) scan or could drastically reduce the scanning time in magnetic resonance imaging (MRI).

Generating manual contours for normal tissues outside of the treatment region offers little benefit in the treatment planning process. For example, most commercially available treatment planning systems do not calculate doses outside of the treatment region, so contouring the lungs for a prostate cancer patient's plan offers no benefit as a dose to the lungs (if visible in the simulation scan) would not be calculated and, therefore, not reflect accurate dose estimates for that organ. Additionally, it is well established that manual contouring is time-consuming, and it would be cost prohibitive to contour all normal tissues outside of the treatment region in routine clinical practice.

The location and size of a tumor play a critical role in determining patient setup during simulation and the course of treatment. Patient setup could be standard for certain sites (i.e. head and neck cancers are supine with thermoplastic mask) whereas many options can be available for others. For example, rectal cancers can be treated either in the supine or prone positions, depending on the treatment delivery technique 3D conformal radiotherapy (3DRT) or Volumetric Modulated Arc Therapy (VMAT) and individual treatment centers preferences for clinical practice. Furthermore, a wide range of immobilization devices are used in radiotherapy, playing a role in treatment simulation setup, with some devices designed to pull/compress organs in or out of the treatment field. The wide-variability in patient setup and the effects of some immobilization devices (i.e. use of a belly board on prone setup for treatment of rectal cancers) on patients' anatomy increase the complexity in developing a multi-site auto-segmentation solution.

12.1.2 Limitations of Training Data

Despite the success and superior performance of deep learning-based auto-segmentation algorithms [1], individual model performance and generalizability on unseen data is often overlooked and unreported. As detailed in the previous subsection, several factors such as a scan's craniocaudal FOV, patient setup, and anatomical variabilities due to immobilization devices could have a significant impact on a model's ability to produce clinically acceptable segmentations. Other factors such as image quality, presence of medical devices, implants or hardware, and anatomical changes from prior patient history (i.e. collapsed lung), or due to surgical procedures have not been widely investigated.

At the time this chapter is written, a single study has reported results highlighting the limitations caused by the lack of diversity in publicly available data when used for thoracic organ segmentation [2]. Feng et al. used the 2017 AAPM Thoracic Auto-segmentation Challenge data to train a two-stage 3D convolutional network model [3] using a variant of the 3D U-net [4]. This model (trained on the challenge data) was then evaluated on institutional data where some discrepancies between clinical and auto-segmented volumes were noticed, with significant differences reported for heart contours [2]. In their study, the authors found that differences in motion in the management technique between the challenge data and institutional patient scans resulted in unacceptable auto-segmentations for the heart. Figure 12.1 illustrates differences between challenge and institutional data from the study by Feng et al. The authors explain that most of their thoracic cancer patients are treated using an abdominal compression immobilization technique resulting in the heart (shown as manually contoured in the figure) being pushed cranially into the thoracic cavity. Utilization of this immobilization technique leads to over-contouring of the heart, as shown in Figure 12.2a.

It remains unknown how deep learning auto-segmentation models trained using limited datasets (n < 50) behave when used for a wide variety of clinical applications. In this chapter, the 2017 AAPM Thoracic Auto-segmentation Challenge data was used as a case study to train a commonly used deep learning auto-segmentation method and evaluate its performance on a variety of treatment sites (i.e. non-thorax) and clinical presentations (i.e. atelectasis).

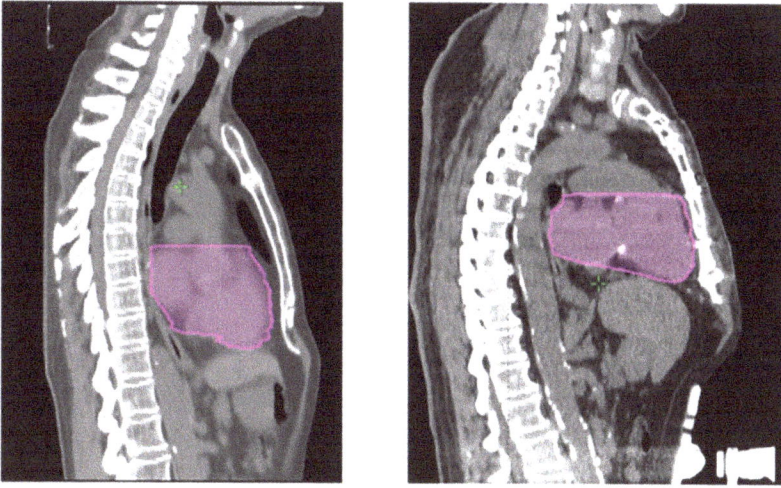

FIGURE 12.1 Figure from Feng et al. [2] illustrating differences in motion management techniques between challenge data (left) and institutional data (right) using sagittal views from thoracic CT scans. Most patients at their institution were treated using an abdominal compression technique which the hardware can be seen on the bottom right corner of the right panel. To highlight these differences, the heart contour is provided (purple) for both patient images.

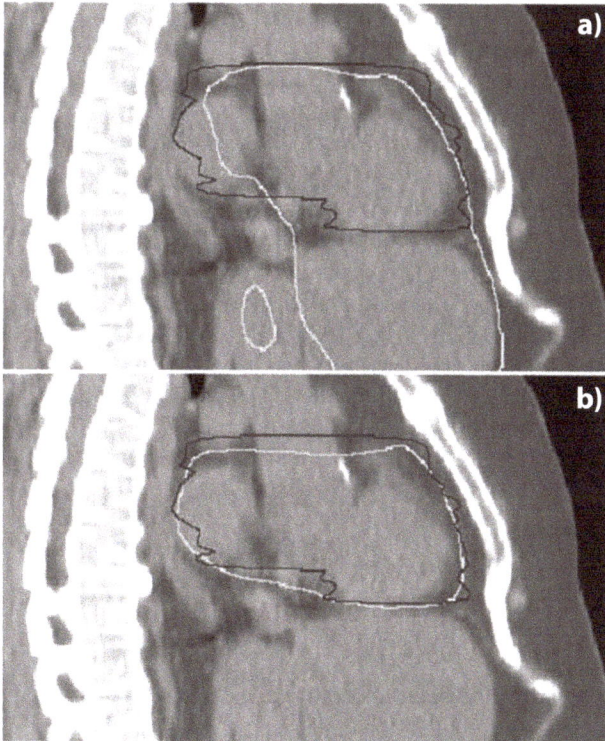

FIGURE 12.2 Adapted figure from a study by Feng et al. [2] Both panels show clinical (dark) and auto-segmented (light) heart contours, (a) highlights disagreement between the clinical and auto-segmentation when the model is only trained using the challenge dataset, whereas (b) shows an improvement in agreement between clinical and auto-segmented heart contours after the model has been re-trained using both challenge and an additional 30 thoracic cancer patient scans with abdominal compression immobilization techniques.

12.2 DEEP LEARNING ARCHITECTURE

12.2.1 Two-Stage U-Net Model

In this chapter, a two-stage U-net model similar to the approach previously introduced by Feng et al. [3] is used. The two stages each employ 3D U-net architecture to first localize the normal tissues using a multi-class (background and five normal tissues) model, and then to segment individual normal tissues about the regions identified in the localization stage (Figure 12.3). This approach of combining neural networks that focus on localization first and then perform the segmentation is widely used in the literature [5–9]. The following subsections describe in more detail the selected image preprocessing steps and the two stages (localization and fine-detail segmentation neural networks) used for this case study.

12.2.1.1 Image Preprocessing

Image preprocessing is a critical step in any image-related task. To account for the variability in pixel spacing and slice thickness, the voxel size is standardized to be 1.25 mm × 1.25 mm × 2.75 mm (in the x-, y-, and, z-directions) for both CT images and their corresponding region of interest (ROI) masks which are shaped to have size $n_z \times n_y \times n_x$; this convention of ordering images in z-, y-, and x-directions is used for all inputs in the model. These values were chosen as it ensures that every input image goes through this standardization process as none of the provided images used these pixel spacing or slice thickness values. To reduce the Housfield unit (HU) range of the CT images, all voxels outside of −500 HU and +500 HU were set to have values of −500 and +500, respectively. The threshold pixel values were then linearly transformed to be within the range [0, 1].

For the first stage in this approach, the standardized voxel image and ROI masks were resized to 64 × 128 × 128 using a tri-linear interpolator. Prior to this resizing step, a binary 3D dilation was applied to the esophagus and spinal cord masks to ensure these ROIs are preserved and not averaged out during downsampling. This 3D dilation is only applied when generating input data for the first stage of this model as this model is only intended for localizing these normal tissues rather than getting an anatomically correct segmentation.

In the second stage of this approach, the standardized voxel image and ROI masks were used as "full resolution" inputs for the fine-detail segmentation models. Here, the individual ROIs were cropped by identifying their left-right (LR), superior-inferior (SI), and anterior-posterior (AP) borders and then a margin was applied to ensure that there was a buffer space large enough to encompass the random translations applied during training. Individual ROI input sizes are determined based on the volume sizes on the standardized voxel masks. For all cases in the training set, the displacement in the RL, CC, and SI directions was calculated for a specific ROI. Then, Equations 12.1–12.3 were used to determine the optimal value for each individual direction.

$$a = percentile(A, 80th) \tag{12.1}$$

$$quotient_a = floor(a * 2 / 3) \tag{12.2}$$

$$size_a = (quotient_a + 1) * multiple \tag{12.3}$$

Here A is a vector containing all displacements for a specific direction (either LR, SI, or AP) and *multiple* is a constant used to ensure that the input size values have a base 2. This constant was chosen to be 32 (2^5). The resulting input sizes for each ROI are listed in Table 12.1.

12.2.1.2 Stage 1: Localization through Coarse Segmentations

Several studies have demonstrated the effectiveness of segmentation networks to initially localize a region of interest prior to generating a final segmentation. Detection networks require large amounts

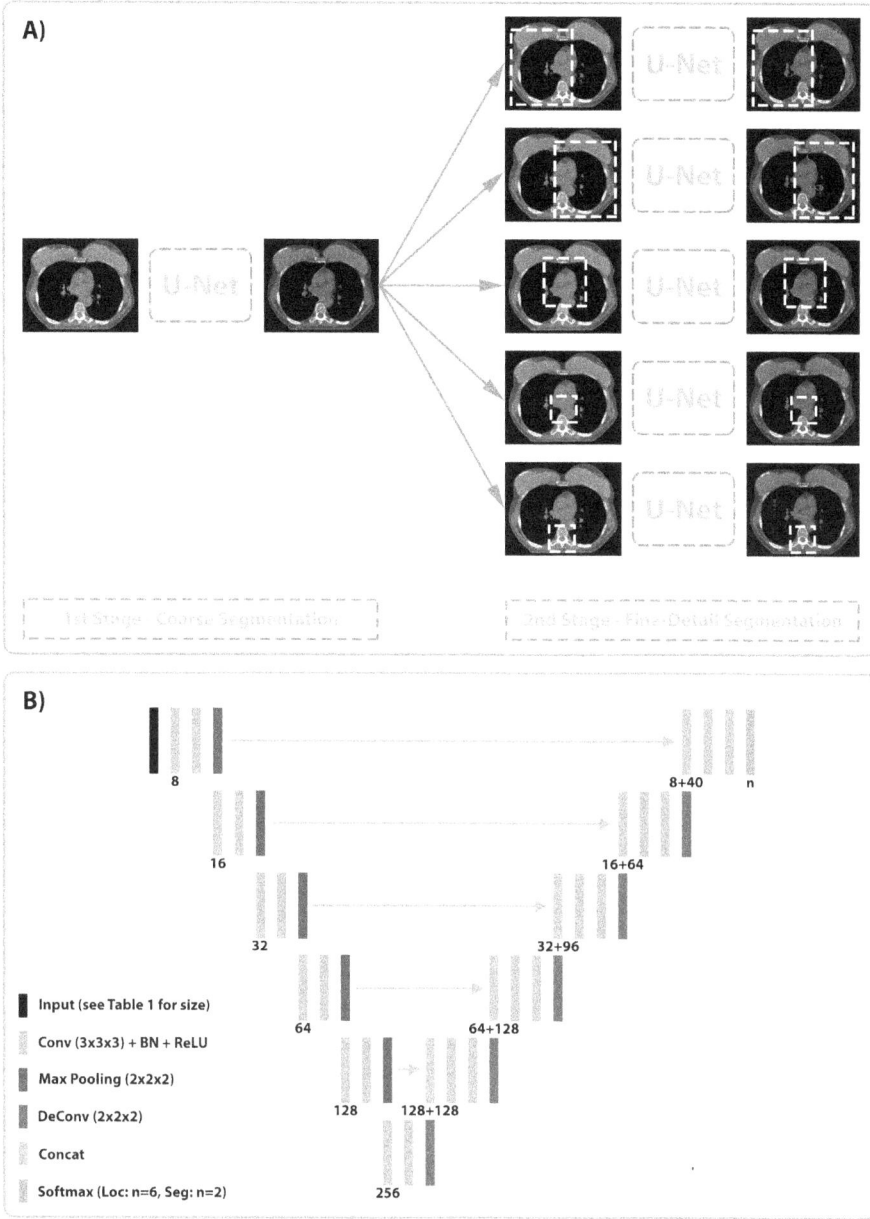

FIGURE 12.3 (A) Illustration of two-stage approach in this chapter. The first stage generates coarse segmentations for all OARs, whereas the second stage uses these coarse segmentations to focus individual segmentation networks about the desired OARs to be automatically segmented. (B) The U-net network used in this work. Input sizes are detailed in Table 12.1. The numbers at each resolution stage represent the number of filters used for the stage; concatenation layers combine features from encoding path with decoding path features. The number of features is doubled at each max pooling layer and halved at each de-convolutional layer. The final number of features in the softmax layer depends on the stage (i.e. six features/classes each representing an OAR + background in first stage, and two features/classes representing background and foreground in second stage).

TABLE 12.1

Input Size for 3D U-Net Networks Used in this Chapter. For Individual Segmentation Networks, Input Sizes Were Determined Individually for Each Normal Tissue by Sampling Volume Sizes for Each Structure on the Training Dataset

Network	SI Direction (z)	AP D(y)	LR Direction (x)
Localization	64	128	128
Esophagus	96	96	64
Heart	64	128	128
Left Lung	96	160	128
Right Lung	96	160	128
Spinal Cord	128	96	64

SI: superior-inferior (i.e. cranio-caudal), AP: anterior-posterior, LR: left-right

of training data to identify useful patterns to accurately localize objects within an image. In a different manner, segmentation networks can take advantage of the label map information to train a network with higher accuracy than detection networks, especially when training data is limited, to localize an ROI within an image. It is for this reason that several works have highlighted the success of using this approach in medical imaging segmentation.

There are several advantages to using a segmentation network to localize an ROI as a first step in a segmentation model. First, the localization network is being used to find a specific ROI within the image space, which can be compared to "finding a needle in a haystack" for some volumes. In medical imaging, the image space can be large, and reducing this FOV through a localization (and then cropping to this ROI) can lead to faster and more efficient training of a neural network. Secondly, since the localization network is used to find a region to focus on during the segmentation stage, there is no requirement in maintaining the original resolution of the image space (i.e. pixel or slice spacing). This allows for the resizing of the medical image to a smaller input size which can then better accommodate graphics processor unit (GPU) memory limitations when training a network. There are a few things to consider when resizing an image to a smaller size. One unintended consequence of reducing the image size in medical imaging is that small or thin volume masks can often lose useful information or be completely averaged out by the interpolation method chosen during the resizing process. Another disadvantage of using a localization stage in a segmentation task is that it can be computationally expensive during training and increase the time to auto-segment a new patient. Lastly, using a multi-class localization stage assumes that all cases in the training data have ground truth volumes available for all normal tissues or structures to be auto-segmented. While this is true for well curated datasets such as those often found in public segmentation challenges, this may not be necessarily true for clinical data where often only critical organs at risk within or near the treatment region are contoured. A way to address this could be to use multiple localization networks for organs that are less frequently contoured or contoured under a specific protocol for a limited number of cases.

In this chapter, a generic implementation of the 3D U-net was used for the localization stage (Figure 12.3). This network uses an input size of 64 × 128 × 128 (see Table 12.1), a kernel size of 3 × 3 × 3, and has two convolutional layers at each resolution level (all convolutional layers apply padding), which are followed by a max pooling operation. The network has a depth of six resolution levels; the first level uses eight convolutional filters for each convolutional layer, which then are doubled after each max pooling operation. Each convolutional layer (Conv3D) is followed by batch normalization (BN), which is then followed by the rectified linear unit (ReLU) activation function

TABLE 12.2

Limit Values Selected for Augmentations Used During Training

Augmentation Type	Limits
Roll (rotation)	$\pm 15°$
Pitch (rotation)	$\pm 5°$
Translation[a]	$[-24, 24]$[b]
Zoom	$[90\%, 110\%]$
Gaussian (sigma)	$[0, 0.5]$

[a] x, y, and z translations were independently defined.
[b] In some cases, this could have been smaller to prevent cropping outside of the image space.

(Conv3D + BN + ReLU) prior to the next convolutional layer or max pooling layer. The localization network outputs coarse segmentations for six classes which include the five organs-at-risk (OARs) (left and right lung, heart, esophagus, and spinal cord) and the background.

During training, resized volumes (Table 12.1) were used and traditional augmentations such as translation, pitch and roll rotations, and zoom were applied. In addition, a Gaussian filter was applied with randomly selected sigma values to make the network robust to variations in contrast and sharpness. Values used for augmentations were randomly generated using a uniform distribution within the predefined limits shown in Table 12.2. Image padding was applied when needed using the reflection of the edge of the image requiring padding. Here, a batch size of four per iteration was used. To train the model, the commonly used Dice loss function was used as it generally converges faster than categorical cross-entropy. Chapter 10 gives more details regarding the use of loss function in training deep learning for auto-segmentation. The Adam optimizer was used with a learning rate of 0.001, with β_1 and β_2 values of 0.9 and 0.999, respectively. Early stopping was used to prevent overfitting by randomly selecting six training cases (out of 36) prior to the start of training and these were used to independently assess the progress of the model during training.

When running an inference on a new patient, the resulting segmentations of the localization stage are used to identify which regions of the "full resolution" image contain individual OARs to focus individual segmentation models for the resulting fine-detail segmentations using a cropping approach.

12.2.1.3 Stage 2: OAR Segmentation through Fine-Detail Segmentation

The second stage of this approach focuses on fine-detail segmentations by training individual ROI 3D U-net models using "full resolution" inputs. There are a few advantages to using individual models per ROI. First, is that each model can be trained to focus on learning intensity features that are characteristic of each ROI. Second, using individual models per ROI allows for additional flexibility in model design and input size. Third, individual ROI models can be updated independently without making changes to other ROI models. This is advantageous when a segmentation model produces high quality segmentations for most ROIs but produces inaccurate segmentations for a few, more challenging, ROIs. Lastly, using individual models per ROI allows for the use of segmentations from sparse datasets where most cases do not contain the full list of ROIs to be auto-segmented. Using individual ROI models has its disadvantages though; training individual ROI models can be computationally expensive and less efficient when computational resources are limited. Also, using multiple models can result in an increase in the time required to auto-segment a large list of ROIs; here, individual models need to be loaded to the GPU, individual ROI input data have to be generated, with this input data then fed through the neural network to predict the resulting auto-segmentations.

During training, the CT images were cropped to regions around the ground truth segmentations for each individual ROI provided in the training data. Individual ROI models use different input sizes, which were determined based on training data volumes (see Section 12.2.1.1), except for the left and right lung where the image size is the same (Table 12.1). The architecture used to train each ROI model has identical parameters to the 3D U-net model trained for the localization stage; here, the training details (loss, optimizer, etc.) and image augmentations remain the same as those described in Section 12.2.1.2.

In the testing phase, a clustering approach was used to identify cropped CT image inputs for each model by using the segmentations generated from the localization stage network. This approach is described in more detail in Section 12.2.2.

12.2.2 Test-Time Cluster Cropping Technique

Instead of using a traditional tile and stitch approach, a cluster cropping technique was introduced in order to focus the cropped CT image volumes within the localized region identified in the first stage of the model. Here, the resulting coarse segmentation masks (size of $64 \times 128 \times 128$) from the first stage model were resized back to the standardized voxel size for a specific patient. Then, K-means clustering was used on the coarse segmentations to identify cluster centroids which will serve as the center of the region of the standardized voxel CT image to be cropped and used as inputs for testing using the weights trained for the second stage models (Figure 12.4). The K-means clustering assigns individual voxels within an ROI to the nearest cluster centroid by distance resulting in an even distribution of clusters (and therefore centroids) throughout a volume mask. Using many clusters and/or a large input size will ensure large overlap between patches which will then increase the confidence in the probability of belonging to the foreground assigned to individual voxels. To increase the confidence in the predictions, the number of clusters (K) was set to 24 for all ROIs.

The predictions of the cluster patch inputs are then stacked so that individual voxel averages can be calculated using the overlapping prediction values for each voxel. The resulting probability prediction map is resized from the "full resolution" image back to the original image size for that patient. Here probabilities are converted to a mask (using $p \geq 0.5$) where post-processing takes place prior to converting the auto-segmentations to Digital Imaging and Communications in Medicine (DICOM) format.

12.3 QUANTITATIVE AND QUALITATIVE REVIEW OF AUTO-SEGMENTATIONS

12.3.1 Performance on Challenge Test Set

The model's performance was evaluated on the live test challenge data by comparing the ground truth and auto-segmented volumes using the DSC, MSD, and 95HD. These results are summarized in Table 12.3 and are comparable to other published results using this dataset [3, 10]. Overall, the auto-segmented volumes agreed well with the ground truth volumes. The most challenging volume to auto-segment was the esophagus. This was expected as the esophagus is a difficult volume to segment manually due to the elasticity of this volume and the potential lack of tissue contrast which prevents a clear visualization of this organ's boundaries.

For the lungs, most disagreement between the ground truth and auto-segmented volumes was found to be within the most caudal and cranial extent of the lungs. These differences were minor for most cases (Figure 12.5); however, some disagreement was found for cases (LCTSC-S2-203 and LCTSC-S3-203, Figure 12.6) where lung volumes were affected by patients' disease. When considering the heart, the worst performing auto-segmentation (based on DSC, DSC = 0.88) was on case LCTSC-S3-204 (Figure 12.7). Here, the auto-segmentation lacks anterior coverage of the pericardium (top left and bottom left panels) which is properly covered by the ground truth heart

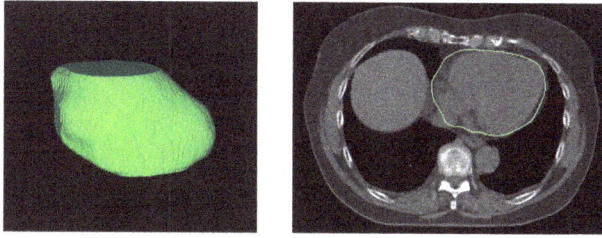

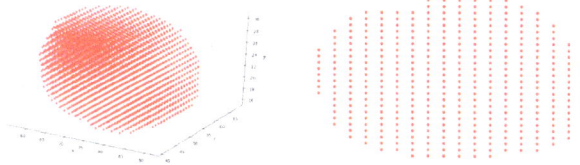

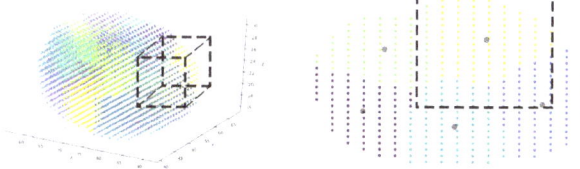

FIGURE 12.4 Illustration of K-means clustering approach to generate cropped CT image patches for testing. The left and right panels represent the approach in 3D and 2D (single slice for illustration purposes), respectively. Using the coarse segmentation masks, clusters were generated, and the cluster centers (dark circles in 2D illustration) were used to center the image crop input for the fine-detail segmentation network.

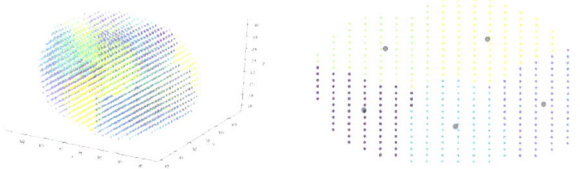

TABLE 12.3

Summary of Overlap and Distance Metrics Calculated between Auto-Segmentations and Ground Truth Volumes from the Live Test Challenge Data

	DSC			MSD (mm)			95HD (mm)		
Lung_L	0.97	+/−	0.03	0.6	+/−	0.4	2.5	+/−	1.4
Lung_R	0.97	+/−	0.02	0.8	+/−	0.6	3.0	+/−	2.5
Heart	0.93	+/−	0.02	2.0	+/−	0.6	5.8	+/−	1.2
Esophagus	0.67	+/−	0.11	2.4	+/−	1.8	9.5	+/−	8.5
Spinal Cord	0.84	+/−	0.05	0.8	+/−	0.3	2.4	+/−	0.7

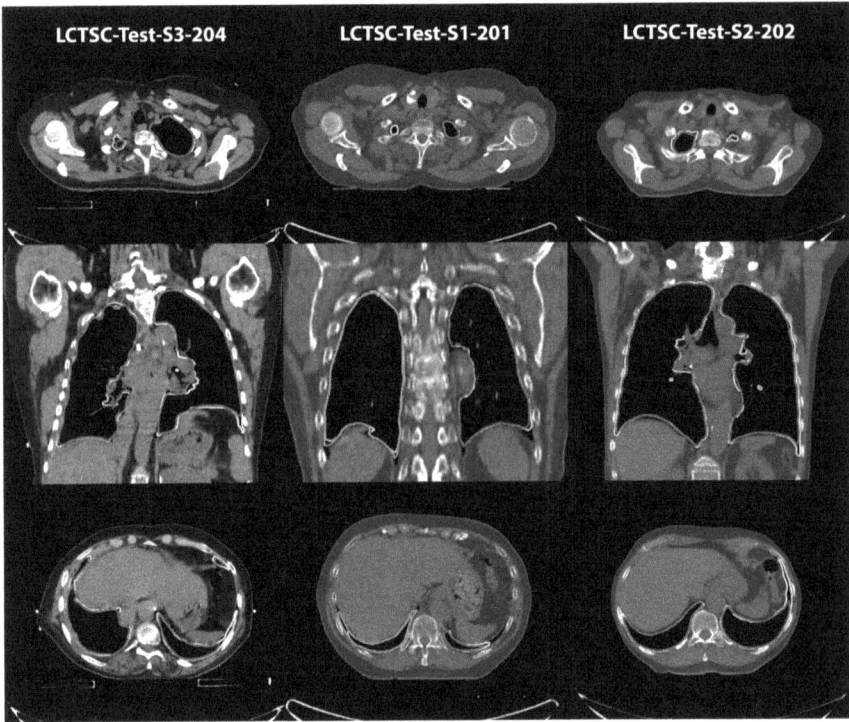

FIGURE 12.5 Comparison between the ground truth (darker) and auto-segmented (lighter) lungs. The top row shows minor differences between these volumes towards the cranial extent of the lungs, whereas the bottom row highlights differences in the most caudal extent of the lungs.

volume. In some cases, there were slight differences between the ground truth and auto-segmented volumes around the apex of the heart.

When considering the spinal cord and esophagus, poorer performance (in terms of the overlap and distance metrics) was noticed compared to that found for the heart and lung segmentations. The Radiation Therapy Oncology Group (RTOG) contouring guideline, RTOG 1106, defines the spinal cord by the bony limits of the spinal canal starting at the level just below the cricoid. The auto-segmentation model produced contours that followed the spinal cord and, in some slices, failed to fully encompass the spinal canal (Figure 12.8). Interestingly, the cranial extent of auto-segmented spinal cord volumes varied widely sometimes reaching the base of the skull as shown on Figure 12.8a. After review of esophageal auto-segmentations, it was noticed that auto-segmentations often fail on the CT slices where the heart is present. As the esophagus runs behind the heart, it tends to vary widely in appearance as it is compressed between the heart vessels, vertebral bodies, and the trachea. Interestingly, the esophagus auto-segmentations were more consistent with ground truth volumes when air bubbles are present throughout the esophageal cavity (Figure 12.9).

12.3.2 DIFFERENT ANATOMICAL SITES

In this section, the model's ability to auto-segment all five OARs on CT scans from non-thoracic primary tumor sites was qualitatively evaluated. A subjective evaluation of the auto-segmentation model could highlight potential problems when translating the trained thoracic OAR auto-segmentation model to auto-segment these normal tissues on simulation CT scans from other treatment sites.

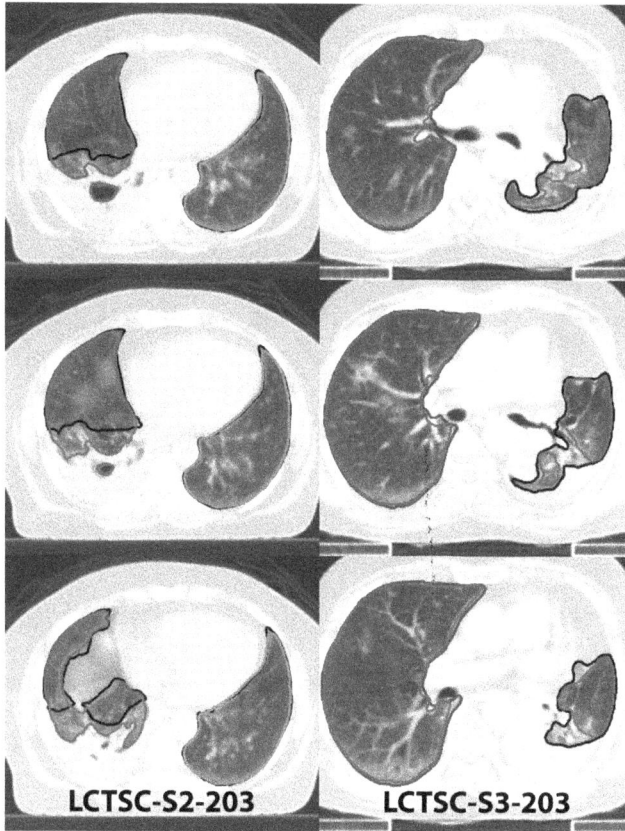

FIGURE 12.6 Highlight of disagreement between ground truth (darker) and auto-segmented (lighter) lung volumes due to the presence of different pathologies.

12.3.2.1 Head and Neck Scans

Head and neck radiotherapy treatment planning requires manual contours for several organs at risk and target volumes. For the past 20 years, several approaches have been proposed for the auto-segmentation of these volumes [5, 11–16]. More recently, several publications using deep learning approaches have shown that head and neck normal tissue auto-segmentations can be of high quality and may not require manual edits leading to an improvement over previously proposed techniques.

From the five organs at risk available in the challenge dataset, the heart is usually not included in the FOV of head and neck simulation scans, but the spinal cord, esophagus, and both lungs are typically (at least partially) present within the CT scan. For this reason, the qualitative evaluation focuses on the auto-segmentations of the spinal cord, esophagus, and both lungs. Here, five head and neck cancer patients previously treated at MD Anderson Cancer Center with radiotherapy were randomly selected. DICOM files for the simulation CT scans were exported and auto-segmentations were produced using the two-stage deep learning model described in Section 12.2.

Visual inspection of the auto-segmentations showed similar results for the spinal cord as those observed in the challenge test data. For example, the cranial extent of the spinal cord auto-segmentations varied across the cases inspected. For some cases it was found that the spinal cord auto-segmentation reached the brainstem while for others the spinal cord was only auto-segmented partially through the head and neck region. Esophagus auto-segmentations covered the organ at risk appropriately for all cases; this was consistent with the findings on the challenge test that auto-segmentations overlapped the ground truth volumes well in the cervical and upper thoracic sections of the esophagus.

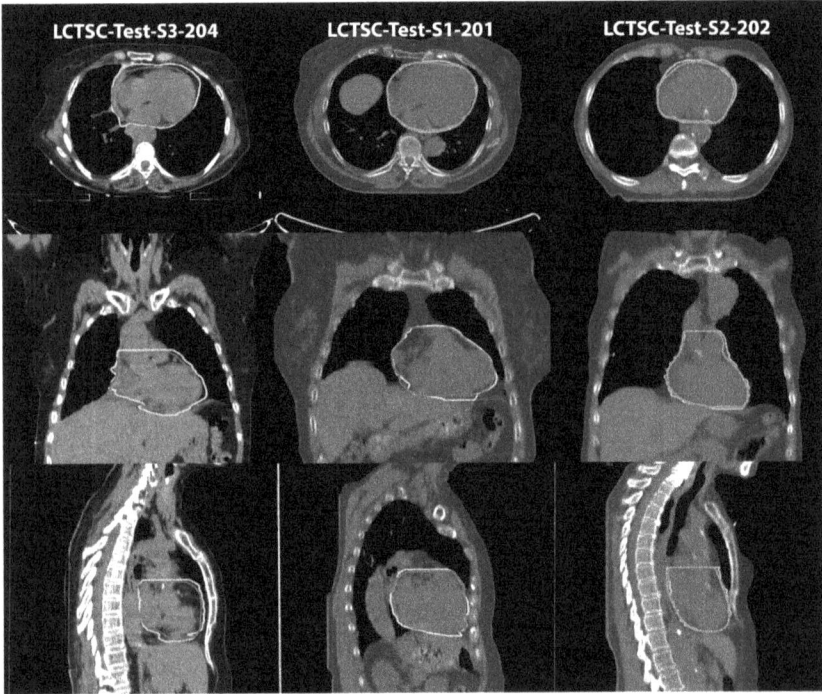

FIGURE 12.7 Axial, coronal, and sagittal views of ground truth (darker) and auto-segmented (lighter) heart volumes.

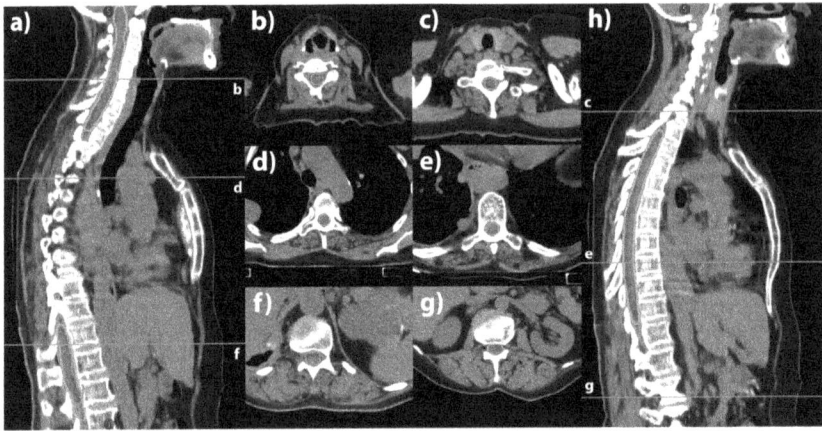

FIGURE 12.8 Illustration highlighting the differences between ground truth (darker) and auto-segmented (lighter) spinal cord volumes. All panels (a–h) are from challenge data case LCTSC-Test-S3-204. Panels (a) and (h) show sagittal views across different planes to encompass the curvature of the spine. Panels (b–g) show axial views of the spinal cord contours. Panels (b) and (f) show typical discrepancies between ground truth and auto-segmented volumes.

Surprisingly, the lung auto-segmentations (Figure 12.10) failed to properly cover the lung volumes at the caudal edge of the CT scan. Upon inspection of the prediction probability maps, it was found that both U-net architectures used in the first stage and second stage of the model under-contoured the lung volumes in these cases. Interestingly, the most medial portions of the lungs were contoured properly in most cases, but the segmentations failed to cover the most lateral sides of the lungs even when these regions were sampled in the second stage of the model.

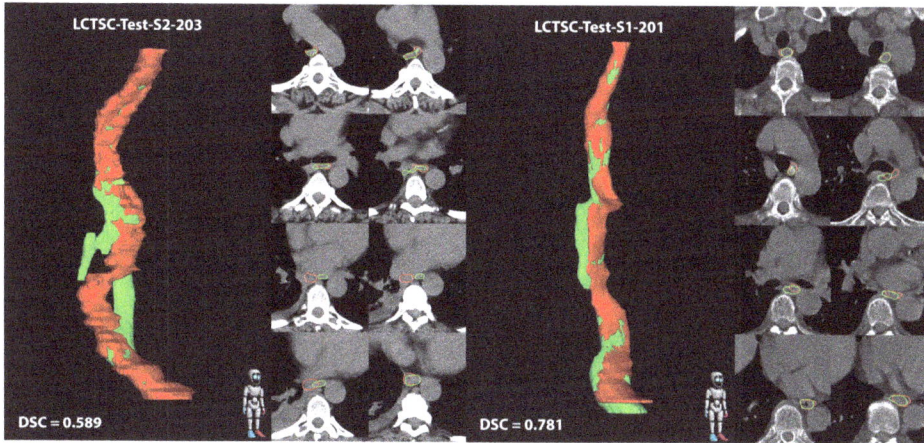

FIGURE 12.9 Comparison of ground truth (red) and auto-segmented (green) esophagus volumes. The two cases shown illustrate the difference between the 15th and 85th percentile DSC scores volumes. It is easy to visualize the disagreement for the case on the left half (LCTSC-Test-S2-203), the auto-segmentation fails to follow the curvature of the esophagus as it runs posteriorly to the heart. The auto-segmentation in the case displayed on the right half (LCTSC-Test-S1-201) follows the trajectory of the esophagus with higher agreement.

12.3.2.2 Abdominal Scans

Simulation CT scans of the abdomen are like head and neck CT scans in that they generally only provide a partial view of the thoracic organs at risk. Here, abdominal scans include the most caudal extent of these organs, whereas head and neck scans include the most cranial extent of these volumes, as seen in Section 12.3.2.1. As the heart sits above the diaphragm, it is usually included in the scan's FOV.

For this subsection, liver cancer patients who were previously treated with radiotherapy at MDAnderson Cancer Center were randomly selected. The thoracic auto-segmentation model was applied to these patients' simulation CT scans. Upon visual inspection of the auto-segmentations (Figure 12.11), it was found that most auto-segmented volumes were appropriately defined on these cases. In contrast to the findings on head and neck scans (Section 12.3.2.1), the partial lung volumes were properly auto-segmented in the abdominal CT scans. The spinal cord auto-segmentations spanned from the most cranial to the most caudal extent of the CT scan for all cases. The esophagus auto-segmentations were of reasonable quality with some variability in coverage on the caudal edge near the gastroesophageal junction. In some cases, the full volume of the heart was present in the CT scan's FOV; however, this was not true for all scans. When the heart was close to the edge of the FOV of the scan, slight inaccuracies in the heart auto-segmentations were noted, which can be appreciated in the top row (center) panel of Figure 12.11. This panel shows how the posterior portion of the heart is under-contoured as the heart volume gets closer to the most superior edge of the scan.

12.3.2.3 Breast Cancer Simulation Scans

Breast cancer patients' simulation CT scans generally have a similar FOV to those found in thoracic cancer patients. A way thoracic and breast cancer patients' radiotherapy simulation CT scans can differ is in how patients are positioned during simulation and treatment. For example, thoracic cancer patients are typically set up in the supine position with both arms above the head. Breast cancer patients' setup can vary depending on the extent of the disease. At a minimum the ipsilateral arm is up, but sometimes both arms are placed above the head. An adjustable breast board is generally used to set up the patient in an incline (typically 5–15 degrees) to isolate the breast tissue below the clavicle. Specialized breast treatment immobilization devices such as breast compression paddles are often used throughout treatment.

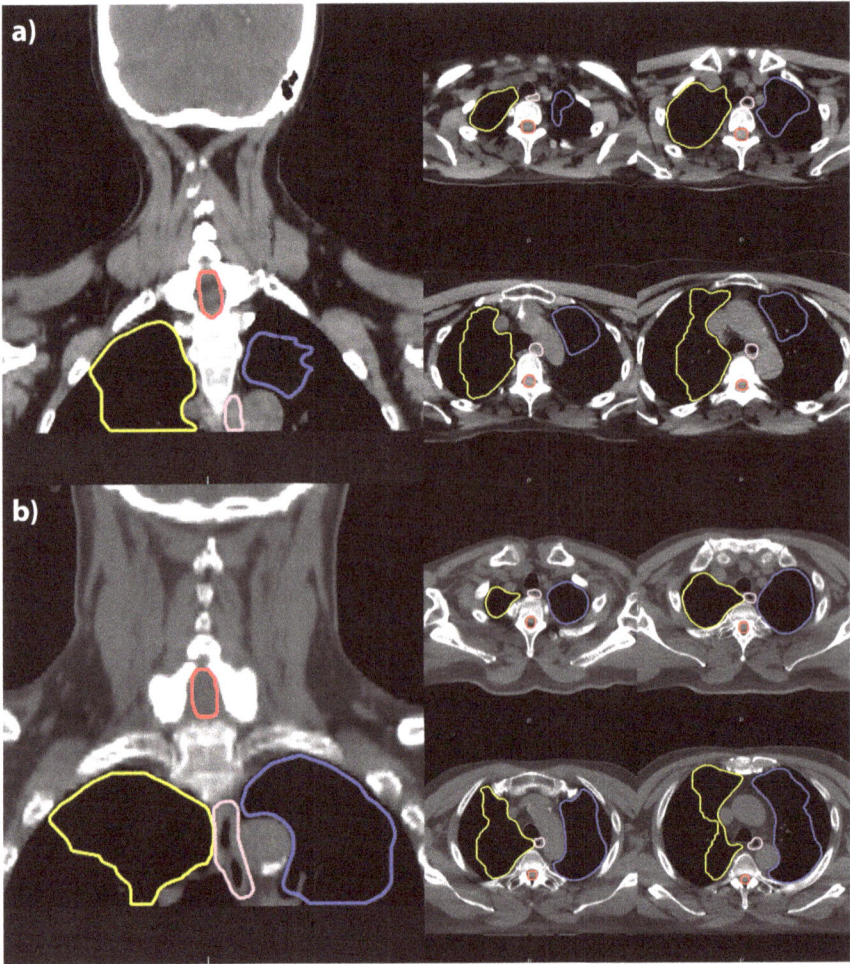

FIGURE 12.10 Illustration of auto-segmented thoracic organs at risk on two (a and b) head and neck cancer CT simulation scans. The auto-segmented spinal cord (red), esophagus (pink), and right (yellow) and left (blue) lungs are displayed. For each case (a and b) a coronal and four axial views are displayed illustrating how the lung auto-segmentations result in under-contouring of the lung volumes.

Auto-segmentations on breast cancer patient simulation CT scans were evaluated to determine the effect of the inclusion of breast cancer-specific immobilization devices and patient set up on organs at risk auto-segmentations. The inclusion of breast compression paddles is evaluated. Figure 12.12 shows a patient simulated for treatment of the left breast (a) who was later simulated for a boost treatment using a compression paddle (b). The addition of the breast paddle did not show any noticeable effects on the auto-segmentations across multiple cases.

Figure 12.13 shows two breast cancer patients who were treated using a similar setup during radiotherapy simulation. For these cases, a metal rod is used to mark and identify on the CT scan the superior border of the non-divergent fields used to treat these patients. Interestingly, it was noticed, for the case on the left column, that the presence of the rod influenced the left lung auto-segmentation (blue) as can be appreciated in the axial and coronal views. Here, the lateral border of the lung is under-contoured towards on adjacent slices where the rod is present. This effect is not observed for all cases where this rod marker is used as can be seen for the case on the right column of Figure 12.13.

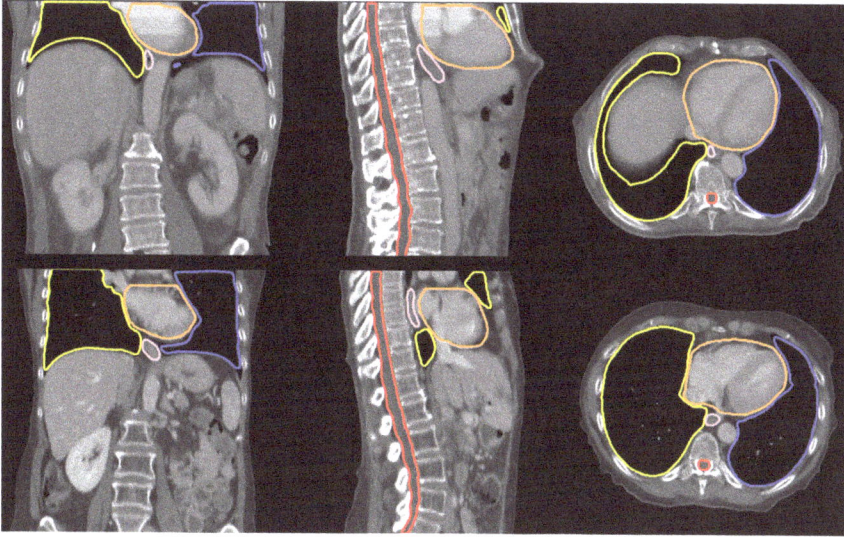

FIGURE 12.11 Auto-segmented organs at risk on abdominal CT scans from two liver cancer patients (different patients separated by row). The auto-segmented spinal cord (red), esophagus (pink), heart (orange), and right (yellow) and left (blue) lungs are displayed on the coronal, sagittal, and axial views.

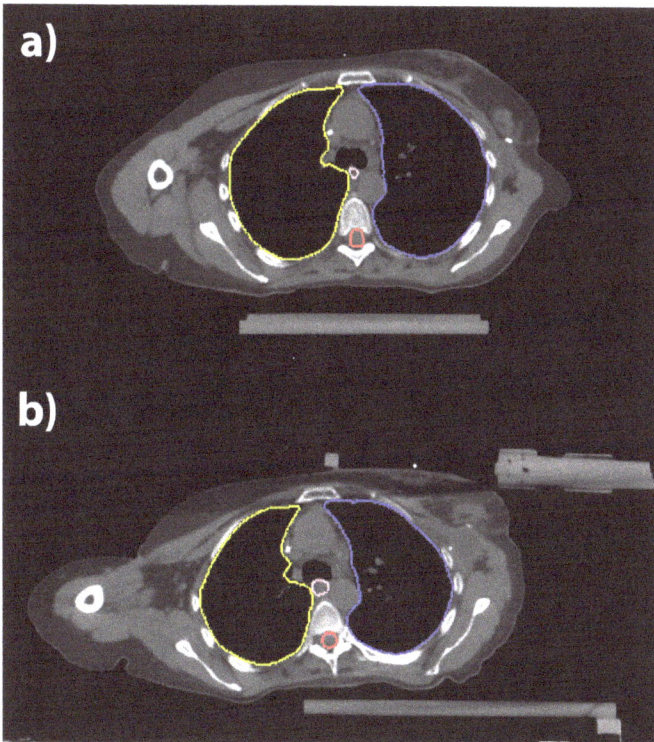

FIGURE 12.12 Left breast cancer patient with two simulation scans for (a) whole breast irradiation and (b) boost radiotherapy plans.

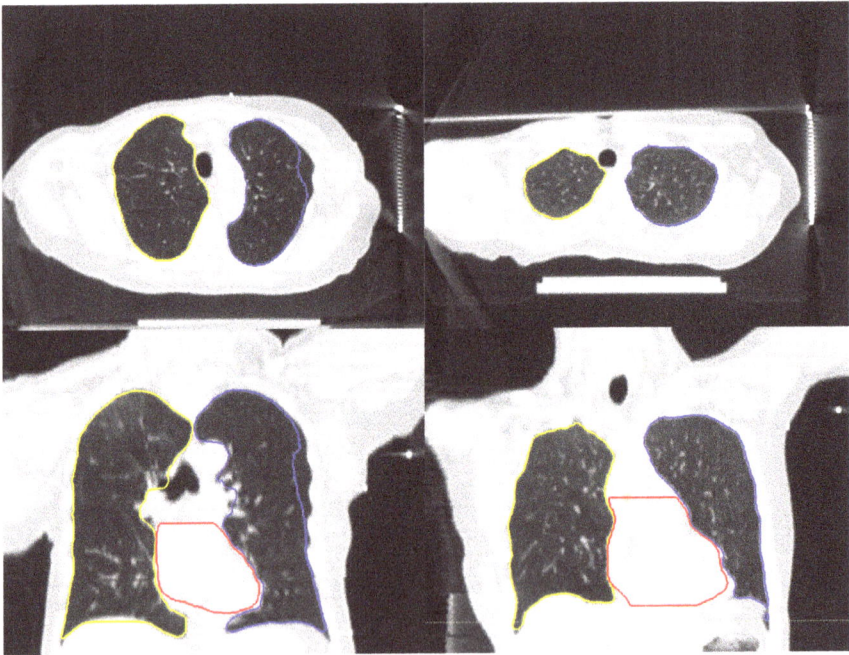

FIGURE 12.13 Two breast cancer patients (different patients separated by column) using the same immobilization device which includes a metal rod to identify the non-divergent border of the tangent fields on the CT scan. The heart (red) and left (blue) and right (yellow) lung auto-segmentations are displayed on the axial and coronal views.

12.3.3 DIFFERENT CLINICAL PRESENTATIONS

In this section, the effect of anatomical changes to the thorax produced by pre-existing conditions (atelectasis and pleural effusions) and patient immobilization techniques are investigated, as well as the effect of the presence of contrast and implanted devices on the auto-segmentation of OARs from a deep learning model trained using the challenge dataset.

12.3.3.1 Atelectasis and Pleural Effusion

Atelectasis and pleural effusions are not uncommon in cancer patients, with some studies reporting 7–22% of lung cancer patients presenting with these conditions prior to the start of treatment [visual inspection of challenge training data showed that 5/36 (14%) of cases showed signs of atelectasis (LCTSC-S1-008, LCTSC-S1-011, LCTSC-S3-005, LCTSC-S3-008, and LCTSC-S3-012)]. Atelectasis is a condition in which the airways and air sacs in the lung collapse or do not expand properly. This condition can happen in the presence of a tumor or because of a pleural effusion which is a condition that affects the tissue that covers the outside of the lungs and lines the inside of the chest cavity. These occur when an infection, medical condition, or chest injury causes fluid, pus, blood, air, or other gases to build up in the pleural space. Both atelectasis and pleural effusions can greatly alter the anatomy within the thorax; therefore, the auto-segmented organs at risk in cases presenting with these conditions were qualitatively evaluated.

Pleural effusions resulting in total collapse of the lungs showed to have an impact on the quality of the auto-segmentations. The left column on Figure 12.14 shows axial and coronal views from a patient who presented with pleural effusion and atelectasis of the left lung. Auto-segmentations on this patient's CT scan did not produce accurate results; the left lung auto-segmentation fails to fully cover the remaining healthy lung whereas the heart auto-segmentation extends outside of the pericardium and into the fluid in the pleura. The patient shown in the axial and coronal views in

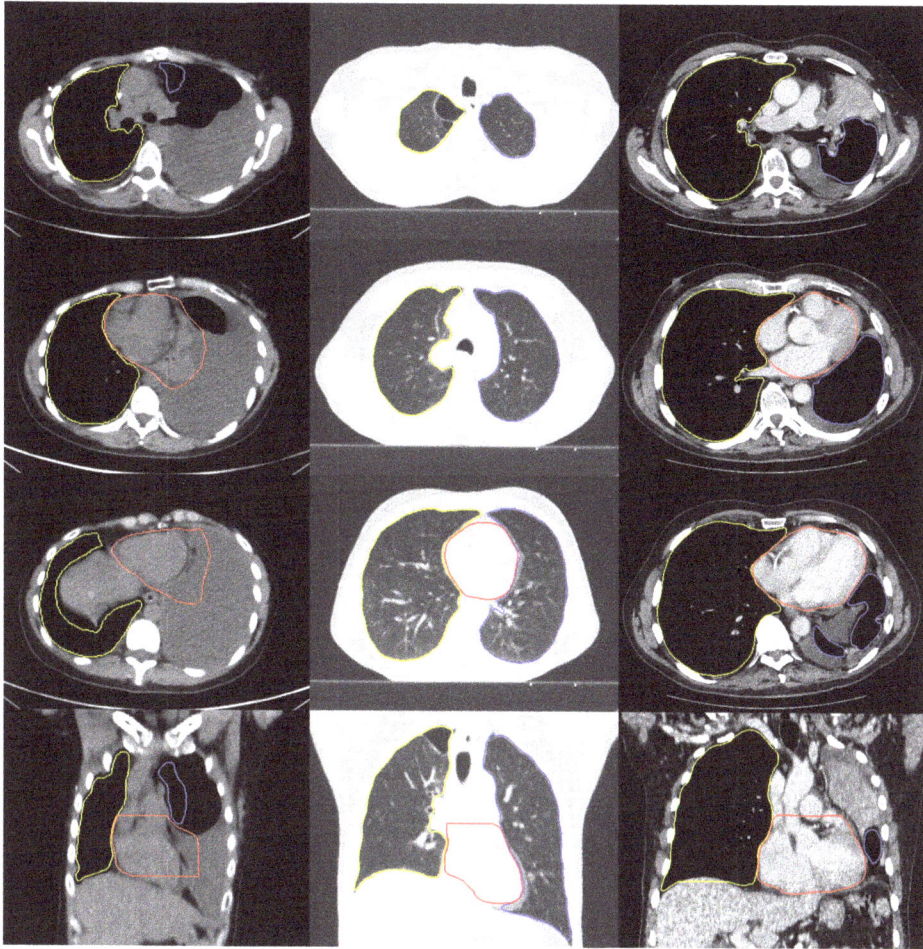

FIGURE 12.14 Auto-segmentation of organs at risk in the presence of atelectasis and/or pleural effusions. The heart (red) and left (blue) and right (yellow) lungs are displayed in the axial and coronal views for three lung cancer patients (columns).

the center panel of Figure 12.14 presented with atelectasis of the right upper lobe of the lung (top center panel). For this case, the auto-segmentation model was able to accurately define both lungs and the heart. The third case shown in Figure 12.14 (right panel) presented with atelectasis of the left lung due to the presence of a tumor. Upon visual inspection of this case, it was found that the auto-segmentations accurately contoured all organs at risk.

12.3.3.2 Presence of Motion Management Devices

Previously, Feng et al. [2] showed that the use of abdominal compressions at their institution led to an over-contouring of the heart in the caudal direction (Figure 12.2a). Here, replication of this failure mode is attempted by predicting on a case previously treated with abdominal compressions. Unfortunately, this immobilization technique is rarely used at MD Anderson Cancer Center and only a single case was found on the database to replicate their findings. For this single case, good agreement was found between the auto-segmented (green) and clinically contoured (red) heart volumes as shown in Figure 12.15. A difference between the model used in this chapter and the one used by Feng et al. is that the heart input size is larger in the SI direction ($64 \times 128 \times 128$ vs $32 \times 160 \times 192$). It is possible that the greater field of view in the SI direction used in this chapter's model

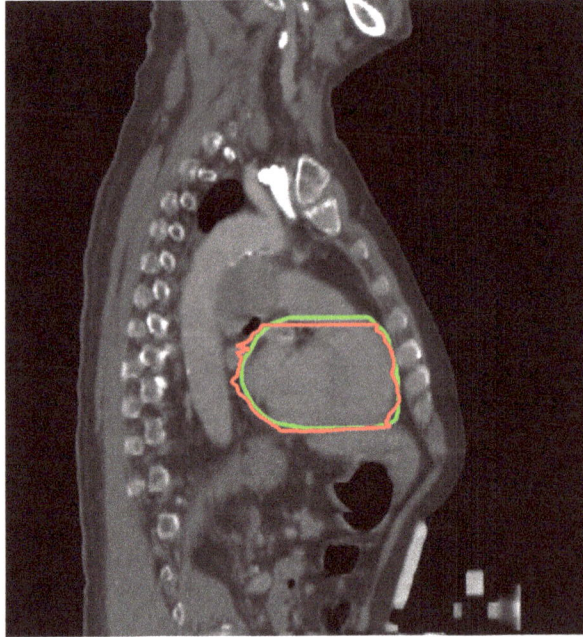

FIGURE 12.15   Comparison of ground truth (red) and auto-segmented (green) heart volumes on a lung cancer patient where abdominal compression immobilization was used.

allows for a broader view of the SI borders of the heart; however, this remains to be investigated and is outside of the scope of this analysis.

12.3.3.3   Use of Contrast and the Presence of Implanted Devices

Contrast-enhanced CT scans are commonly used in cancer diagnosis, staging, and post-treatment follow-up. In radiotherapy, the use of intravenous (IV) contrast during treatment simulation CT scan acquisition can allow for enhanced visualization of target volumes and adjacent organs at risk facilitating more accurate delineations of these volumes [17]. For non-small cell and small cell lung cancers, the National Comprehensive Cancer Network recommends the use of IV contrast with or without oral contrast for better target and organ delineation in patients with central tumors or nodal disease [18]. The presence of contrast in CT imaging changes the local distribution of HU values which can have an impact on auto-segmentations. Previous works have addressed this challenge by using training sets that contain contrast-enhanced and non-contrast CT images [19]. In this subsection, the model trained on the challenge data was used to evaluate the quality of auto-segmentations in cases where there are high density regions such as those in contrast-enhanced CT images and implanted devices.

Figure 12.16 shows three patients' CT scans (axial and coronal or sagittal views) that illustrate the possible failure modes observed when auto-segmenting OARs in cases where contrast or hardware are present. The patient shown on the left panel of Figure 12.16 had contrast in their bowels. Here, the left lung (blue) is not auto-segmented when it is near the high-density regions in the scan (as seen on coronal and axial views). In contrast, the right lung (yellow) was accurately auto-segmented for this case. For this particular case, it was found that most CT image patches used to predict the left lung included regions where contrast was present; this was not the case for the right lung where the liver was found in the most caudal regions of a large number of image patches. The patient shown in the center panels of Figure 12.16 had a previously implanted pacemaker (as shown on the top two center panels).

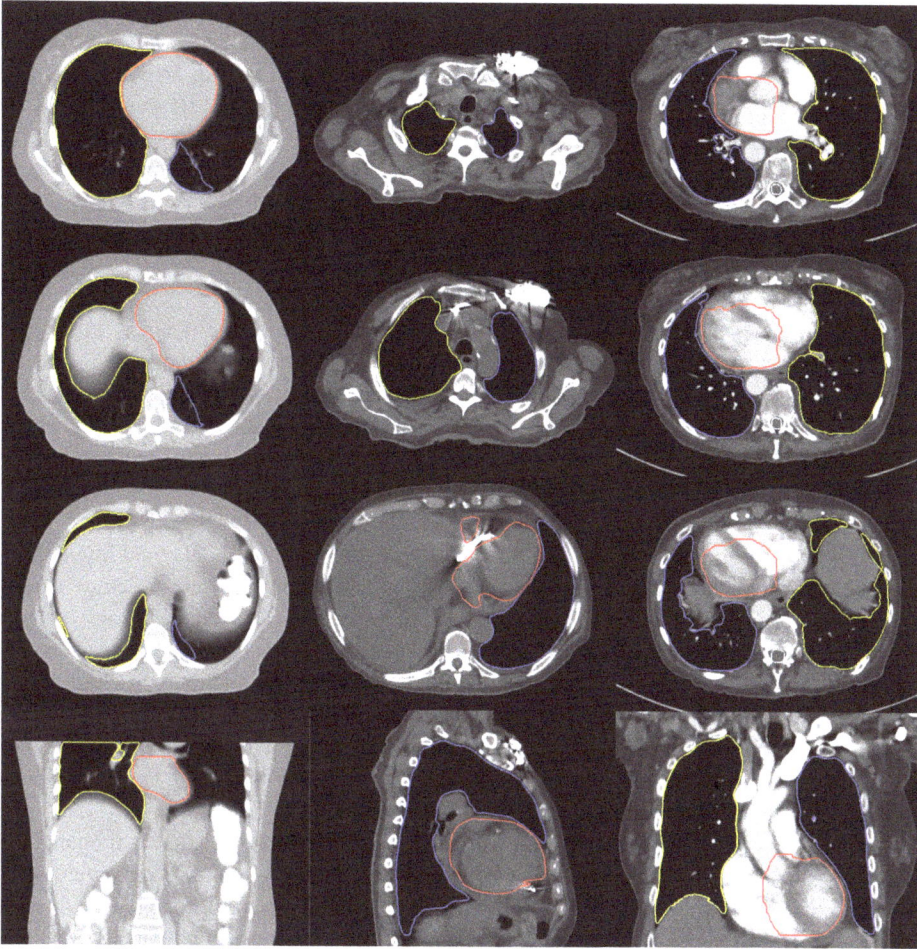

FIGURE 12.16 CT images from a patient with contrast in the abdomen (left), pacemaker (center), and contrast in the heart (right). The heart (red) and left (blue) and right (yellow) lungs are displayed to illustrate failures in auto-segmentations.

The pacemaker leads appear as bright wires that transverse from the collar bone to the chambers of the heart. For this patient, the presence of the high-density leads traveling through the heart affected the quality of the heart auto-segmentation (as seen on the bottom axial and sagittal views). Lastly, the right panel of Figure 12.16 shows a patient who had a chest CT with contrast for radiotherapy simulation. Here the presence of contrast through the heart caused the deep learning-based auto-segmentation algorithm to significantly under-contour the heart volume. Interestingly, the presence of contrast did not affect the quality of the auto-segmentations of the lungs. These findings merit additional investigations to better understand the mechanisms for why the presence of high-density regions affects some auto-segmentations.

12.3.3.4 Adapting to the Unseen

When a deep learning auto-segmentation model fails to produce high-quality segmentations due to limitations on the training data (low number of training cases or limited anatomical diversity), there are several options to overcome these challenges. Feng et al. [2] showed how they adapted their previously trained convolutional neural network by retraining their model with both challenge and additional institutional data. In their study, the authors illustrate how both transfer learning and end-to-end training can improve the quality of the segmentations (Figure 12.2b) by learning patterns from the additional training data. Other plausible options include changing the architecture (see Chapters 7, 8, and 9), changing model parameters such as input size and loss function (see Chapter 10), or to change how training data is input through a model during training (see Chapter 11). All of these could have a significant impact on the resulting auto-segmentations. Lastly, curation of additional training data could help address many of the limitations observed due to diversity in anatomical presentations and/or model applications.

12.4 DISCUSSION AND CONCLUSIONS

In this chapter, a two-stage deep learning auto-segmentation model was trained using the 2017 AAPM Thoracic Auto-segmentation Challenge data and the ability of this model to produce accurate auto-segmentations on a variety of clinical uses and scenarios was evaluated. The model used a generic 3D U-net architecture that first localizes (stage 1) the organs at risk and then auto-segments (stage 2) individual organs using cropped CT scan volumes about each organ. The resulting model was validated using the challenge test dataset producing similar results to those reported in the literature.

To identify possible scenarios where the trained auto-segmentation model could fail, the model's performance was evaluated on two types of scenarios. First, it was investigated whether using simulation CT scans from other treatment sites would have an influence on the accuracy of the auto-segmentations. Potential issues were identified with auto-segmentations that are near the edge of the scan's FOV. Several reasons could be attributed to this observation and there is a high likelihood that these issues could be resolved by using a simple 2D architecture, yet this remains to be evaluated. Another failure mode in the auto-segmentations was noticed when streaking artifacts from a breast cancer radiotherapy immobilization device were located near the lungs. Under-contouring in these regions near the artifact varied from patient to patient (Figure 12.13).

The second type of scenario investigated was the potential effect of anatomical changes to the thorax produced by pre-existing conditions (atelectasis and pleural effusions) and patient immobilization technique, as well as the effect of the presence of contrast and implanted devices on the auto-segmentation. When considering atelectasis and pleural effusions, it was found that the accuracy of the auto-segmentations was greatly reduced when there was significant buildup of fluid in the pleura. Only a single case was located at MD Anderson Cancer Center where the abdominal compression immobilization technique was used during radiotherapy simulation; all organs at risk were accurately auto-segmented for this case but additional cases are needed to confirm that slight parameter differences in approach resolved the auto-segmentation errors previously reported by Feng et al. [2]. Lastly, the presence of contrast and high-density materials from medical implants was shown to have a negative impact on the quality of the auto-segmentations.

The case study presented has a few limitations. First, a single architecture is considered highlighting issues that might not be observed with other similar 3D approaches. Second, only a limited range of possible scenarios were considered in the qualitative evaluation of this case study. Potential scenarios such as anatomical changes from surgical procedures (i.e. pulmonary lobectomy) could have significant impact on the accuracy of auto-segmentations. Lastly, the analysis presented is limited as it only uses the 2017 AAPM Thoracic Auto-segmentation Challenge data for the case study. Expanding this analysis to other publicly available datasets could confirm the shortcomings observed in this study.

In conclusion, while the model presented produced high quality auto-segmentations when compared to the challenge test data, some shortcomings were observed when applying such a model to a diverse set of scenarios and clinical uses, highlighting the limitations of the proposed model and the use of limited training datasets. Identifying failure modes of a deep learning auto-segmentation model can be a critical step in the clinical deployment of such models and could improve patient safety throughout the radiotherapy treatment planning process.

ACKNOWLEDGMENTS

The author would like to acknowledge the support of the High Performance Computing facility at the University of Texas MD Anderson Cancer Center for providing computational resources (including consulting services) that have contributed towards the training of models and generation of data used for this book chapter.

REFERENCES

1. Cardenas CE, Yang J, Anderson BM, Court LE, Brock KB. Advances in auto-segmentation. *Semin Radiat Oncol* 2019;29(3):185–197. doi:10.1016/j.semradonc.2019.02.001
2. Feng X, Bernard ME, Hunter T, Chen Q. Improving accuracy and robustness of deep convolutional neural network based thoracic OAR segmentation. *Phys Med Biol* 2020;65(7). doi:10.1088/1361-6560/ab7877
3. Feng X, Qing K, Tustison NJ, Meyer CH, Chen Q. Deep convolutional neural network for segmentation of thoracic organs-at-risk using cropped 3D images. *Med Phys* 2019;46(5):2169–2180. doi:10.1002/mp.13466
4. Çiçek Ö, Abdulkadir A, Lienkamp SS, Brox T, Ronneberger O. 3D U-net: learning dense volumetric segmentation from sparse annotation. *Lect Notes Comput Sci* 2016;9901 LNCS:424–432. doi:10.1007/978-3-319-46723-8_49
5. Rhee DJ, Cardenas CE, Elhalawani H, et al. Automatic detection of contouring errors using convolutional neural networks. *Med Phys* September 2019:mp.13814. doi:10.1002/mp.13814
6. Balagopal A, Kazemifar S, Nguyen D, et al. Fully automated organ segmentation in male pelvic CT images. *Phys Med Biol* 2018;63(24). doi:10.1088/1361-6560/aaf11c
7. Wang Y, Zhao L, Wang M, Song Z. Organ at risk segmentation in head and neck CT images using a two-stage segmentation framework based on 3D U-Net. *IEEE Access.* 2019;XX:1–1. doi:10.1109/access.2019.2944958
8. Karimi D, Samei G, Kesch C, Nir G, Salcudean SE. Prostate segmentation in MRI using a convolutional neural network architecture and training strategy based on statistical shape models. *Int J Comput Assist Radiol Surg* 2018;13(8):1211–1219. doi:10.1007/s11548-018-1785-8
9. Yang Y, Jiang H, Sun Q. A multiorgan segmentation model for CT volumes via full convolution-deconvolution network. *Biomed Res Int* 2017;2017:1–9. doi:10.1155/2017/6941306
10. Um H, Jiang J, Thor M, et al. Multiple resolution residual network for automatic thoracic organs-at-risk segmentation from CT. 2020:1–5. http://arxiv.org/abs/2005.13690.
11. Yang J, Zhang Y, Zhang L, Dong L. Automatic segmentation of parotids from CT scans using multiple atlases. *Med Image Anal Clin A Gd Chall.* CreateSpace Independent Publishing Platform, 2010:323–330. www.amazon.com/Medical-Image-Analysis-Clinic-Challenge/dp/1453759395
12. Stapleford LJ, Lawson JD, Perkins C, et al. Evaluation of automatic atlas-based lymph node segmentation for head-and-neck cancer. *Int J Radiat Oncol Biol Phys* 2010;77(3):959–966. doi:10.1016/j.ijrobp.2009.09.023
13. Cardenas CE, McCarroll RE, Court LE, et al. Deep learning algorithm for auto-delineation of high-risk oropharyngeal clinical target volumes with built-in dice similarity coefficient parameter optimization function. *Int J Radiat Oncol Biol Phys* 2018;101(2):468–478. doi:10.1016/j.ijrobp.2018.01.114
14. McCarroll R, Yang J, Cardenas CE, et al. Machine learning for the prediction of physician edits to clinical auto-contours in the head-and-neck. *Med Phys* 2017;44(6):3160.
15. Cardenas CE, Anderson BM, Aristophanous M, et al. Auto-delineation of oropharyngeal clinical target volumes using 3D convolutional neural networks. *Phys Med Biol* 2018;63(21). doi:10.1088/1361-6560/aae8a9
16. Men K, Chen X, Zhang Y, et al. Deep deconvolutional neural network for target segmentation of nasopharyngeal cancer in planning computed tomography images. *Front Oncol* 2017;7(December):1–9. doi:10.3389/fonc.2017.00315
17. Bae KT. Intravenous contrast medium administration and scan timing at CT: considerations and approaches. *Radiology* 2010;256(1):32–61. doi:10.1148/radiol.10090908
18. National Comprehensive Cancer Network. *Clinical Practice Guidelines in Oncology for Non-Small Cell Lung Cancer.* 2020. www.nccn.org/professionals/physician_gls/pdf/nscl.pdf.
19. Anderson BM, Lin EY, Cardenas C, et al. Automated contouring of variable contrast CT liver images. *Adv Radiat Oncol* 2020. doi:10.1016/j.adro.2020.04.023

Part III

Clinical Implementation Concerns

13 Clinical Commissioning Guidelines

Harini Veeraraghavan

CONTENTS

13.1 INTRODUCTION

Rapid advances in image-guided radiation therapy have brought forth a range of treatments including high dose radiation therapy, whereby high dose radiation is delivered over a few fractions with high conformity to the target tumors. Clinical studies have shown excellent local control in multiple cancers and with it, the potential to improve progression-free and overall survival [1–4] of some patients.

An important requirement for achieving highly conformal doses is very precise and accurate segmentation of the target and the nearby normal organ at risk (OAR) structures [5, 6]. In current clinical practice, targets and OARs are manually delineated by clinicians on CT images. Manual delineation is difficult because of the low soft tissue contrast especially on non-contrast CT images [7, 8]. Manual delineation is also subject to inter-rater variabilities [9] that can adversely impact tumor control probability [10]. The advent of new image-guided treatments including magnetic resonance imaging (MRI)-guided radiotherapy have made manual delineation more accurate due to improved soft-tissue contrast on MRI [11]. Nevertheless, the problem of variable segmentations persists, and segmentation remains the most time-consuming step in radiotherapy [12].

The commonly used atlas-based methods have been shown to both reduce inter-rater variability [13, 14] and decrease the manual editing times [12, 14]. Clinical validation studies have shown that atlas-based methods produce excellent agreement with manual delineations and reduce user effort [15, 16]. However, these methods can have systematic biases in the segmentations, which may necessitate extensive user editing of some organs [15, 17]. For example, small organs may need consistent manual editing [15], or certain organs like parotid glands may require re-segmentation [17].

More importantly, the large geometric uncertainties resulting from manual segmentation constrain the dose that can be safely delivered to target tumors. This is because of the large field treatment margins [18] necessary to account for the geometric uncertainties, which invariably lead to higher doses delivered to the nearby normal OARs. In-treatment-room X-ray-based cone beam computed tomography (CBCT) imaging is currently available as part of standard equipment. CBCT has been used for positional and setup corrections during treatments. Incorporating geometric corrections has been shown to lead to improved accuracy of the conformal treatment [4, 19], and with it the potential for improving outcomes. However, a key obstacle for these treatments is a lack of robust, fast, and accurate segmentation methods. Atlas-based methods are computationally expensive and are impacted by changing anatomy; anatomical changes are common during radiation therapy and imaging appearance may change, reducing the accuracy of atlas-based deformation techniques.

The more recent deep learning methods are computationally fast (usually in the order of seconds or minutes, compared to minutes or hours for atlas-based methods), and are robust of inter-rater variability [20]. Importantly, deep learning methods have been shown to reduce the manual editing times for multiple OARs [21] more than atlas-based methods. As a result, deep learning has been applied very extensively in numerous image segmentation problems in radiation oncology [20, 22–28].

However, the use of either deep learning or atlas-based methods in routine clinical care is highly limited. This is due in part to the difficulty in establishing the reliability of these methods for commissioning [29], as well as a lack of tools to identify when and where the algorithm fails, leading to manual override of the delineations. More importantly, while some of these methods may show phenomenal performance on limited testing sets, they fail to generalize in clinical datasets, as discussed in Chapter 12. Discrepancies in the performance may stem from large differences in the training/testing sets and the actual clinical use. For example, it is not uncommon to remove difficult conditions like images with large artifacts, large tumors, or those with abnormal anatomy like collapsed lungs from limited training/testing cohorts to assess basic performance of the developed methods. However, methods developed under these conditions fail to scale to actual clinical scenarios when images may have large artifacts (like dental artifacts on head and neck CT images) or abnormal anatomy (e.g. collapsed lungs, missing structures due to surgery, mass effect due to the presence of large tumors).

All this motivates the need for clear guidelines and metrics for evaluating auto-segmentation methods prior to clinical commissioning. In the rest of the chapter, the approaches used for clinical validation as outlined in some prior studies, and the challenges involved for clinical commissioning are briefly discussed, and some solutions towards mitigating these issues through better data curation and evaluation metrics are presented.

13.2   STAGES IN CLINICAL COMMISSIONING

A phased approach consisting of technique identification and verification, testing before clinical deployment, and ongoing clinical quality assurance (QA) is generally recommended when introducing a new technology into a clinic [30]. Clinical commissioning of an AI method at the minimum should include testing of the AI technique on the institution dataset prior to clinical implementation followed by routine QA of selected clinical cases with a group of multi-disciplinary experts. Prior to assessment of the auto-segmentation system, it is also important to obtain some details of the technique regarding the evaluations and quantitative performance of the metrics to be used. Evaluation of the system should include, where possible, testing with multi-institutional datasets, comparison against multiple methods, and evaluation on established metrics. In order to ensure that the most suitable methods are used in the clinic, there is a need for both robust and reliable metrics to evaluate these methods as well as well-curated datasets arising from multi-institutional and internal institution datasets. Next, after introduction into clinical use a frequent QA of selected cases should be done to determine that the system continues to perform at the desired level. Problems in performance should be logged and AI techniques may need to be retrained if the imaging technology and imaging protocols change.

13.3 NEED FOR ROBUST AND CLINICALLY USEFUL METRICS

The most commonly used metrics for evaluating the accuracy of segmentation methods are based on those used in computer vision, whereby the spatial and geometric overlap accuracy of the algorithm is measured against manual delineation by an expert. One such metric is the Dice similarity coefficient that measures the overlap in the number of voxels that match between the algorithm and manual delineation [15, 16, 31–33]. Because of the need to guarantee spatial accuracy in terms of metric distances, Hausdorff distances have also been commonly used to assess medical image segmentation applied to radiation therapy. While these metrics are reasonable when analyzing objects from real world images that have well defined boundaries and can be easily identified and delineated by people, medical image analysis and delineations requires significant domain expertise. Also, even delineations by experts are subject to inter-rater variability, which inevitably results in a lack of an absolute gold standard segmentation [34, 35]. In medical images, there is no "gold standard" ground truth. More importantly, these metrics are neither clear indicators of clinical efficiency, such as the amount of reduction in editing times [36], nor are they indicative of target coverage improvement [37]. This motivates the need for practical metrics that can be used for clinical commissioning of auto-segmentation methods. Chapter 15 further considers quantitative methods for evaluation of auto-segmentation in clinical commissioning.

13.4 NEED FOR CURATED DATASETS FOR CLINICAL COMMISSIONING

A related problem is the lack of benchmark datasets to assess and compare different algorithms. Using a common reference dataset to establish performance allows direct comparison of the various methods. Benchmarking datasets that encompass the expected variabilities to be seen in the clinic, including imaging variations, imaging artifacts, and large deformations in patient anatomy are necessary to evaluate the utility of the auto-segmentation methods for clinical use. Such datasets, if available, are also useful to evaluate upgrades to auto-segmentation software already used in the clinic to ensure safe use for treatment planning. The recent push towards increased reproducibility in research initiated by multiple top-tier machine learning conferences like Neural Information Processing Systems and some medical journals has accelerated the implementation of more advanced methods by biomedical scientists and the evaluation of the new methods against established state-of-the-art. However, there is a need for a community-wide testing framework with common datasets and common metrics to evaluate the various methods. Grand challenges [29, 38–40] that provide limited size datasets with clearly defined metrics for evaluation and delineations done using clearly defined criteria represent a successful first step in this direction.

More importantly, lack of well-curated and large datasets is a fundamental problem in the successful application of data-hungry methods like deep learning. Variabilities in the delineation guidelines used across the institutions may make these methods less portable across institutions; the accuracy and robustness heavily depend on the size and quality of the training datasets [41]. Hence, well-curated datasets delineated according to published contouring guidelines, with reasonable size, and arising from different institutions to capture the variability in the imaging is crucial to ensure the development of clinically useful deep learning methods. Chapter 14 addresses the challenges of data curation for auto-segmentation in more detail.

On the flip side, internal commissioning within an institution would require a well-curated dataset, where the contouring adheres to the clinical preferences and the needs of treated patients within that institution. More details on some of the considerations needed for clinical commissioning data curation are presented in the subsequent subsection on data curation guidelines for radiation oncology.

13.5 AUTO-SEGMENTATION CLINICAL VALIDATION STUDIES – CURRENT STATE-OF-THE-ART

Although there are several studies for clinical validation, explored in depth in Chapter 15, a few representative studies are outlined here to describe the range of assessment metrics used to evaluate the auto-segmentation (atlas or deep learning methods). Most of the evaluated methods reported here are atlas-based, because deep learning methods are relatively new in radiation oncology. In general, the most common metric used for evaluating the auto-segmentation methods is the Dice similarity coefficient (DSC) [15, 16, 31–33]. DSC compares the segmentations against a clinical "gold standard" manual delineation using a form of overlap ratio. DSC is simple to compute, and is used pervasively in the literature, which allows for easy comparison to prior state-of-the-art.

In addition, the clinical usefulness has commonly been measured by evaluating the time savings in the contour editing [12, 14, 16, 21, 31]. This is a slightly more useful metric as it takes in to account the efficiency of using an auto-segmentation in clinical practice. Another approach has been to employ visual grading of the segmentations to assess how useful they are likely to be for clinical use [16].

But these measurements do not consider the endpoint and do not provide any information on the impact of the auto-segmentation on improving treatment accuracy or outcomes. A different approach has been to compute whether the auto-segmentations improve the robustness with respect to multiple raters. For instance, editing variations between radiation oncologists were reported to be reduced when using auto-segmented contours as a starting point for segmentation [42, 43]. This is important because large field margins are often the result of increased uncertainties. Finally, clinical evaluation studies which studied the relation between the geometric metrics used for evaluating auto-segmentation, such as the DSC and Hausdorff distance, and the clinically relevant dosimetric metrics [13, 44] found that there is little correlation between these two types of assessment. For example, imperfect DSC scores did not necessarily translate to inferior normal organ dosimetry [44], especially if the OAR was in the low dose region, while even small deviations in the geometric accuracy could have a large impact if the organ in question was in the high dose region [13]. Novel metrics of geometric accuracy such as the surface DSC metric [45] and quantitative measures of time savings like the added path length [21] have been shown to be more clinically representative surrogates for the time-clock measures required for editing incorrect segmentations in the clinic. Table 15.1 in Chapter 15 summarizes representative studies with the metrics used for algorithm validation.

13.6 DATA CURATION GUIDELINES FOR RADIATION ONCOLOGY

A basic requirement for the application of auto-segmentation tools is that the definition of volumes produced using these tools adheres to clinical contouring guidelines according to the anatomical region. However, adherence to such guidelines requires that the training set delineations were done following well-defined guidelines either defined institutionally, or better through community-wide agreed-upon guidelines. Methods developed on one set of training data from one institution may be difficult to apply to a dataset from a different institution. Difficulties arise due to variabilities in the imaging acquisition protocols applied across institutions, but also due to differences in the clinician preferences for accepting contours for treatment. Finally, patient privacy concerns make it difficult to share datasets across institutions. As a result, the algorithm evaluations and comparisons are done using methods developed and tested on widely different datasets, which makes the performances of various methods hard to compare. Even if the source code for performing evaluations are available, as the number of models grows, comparative evaluations themselves become quite laborious.

"Grand challenges" offer an unbiased approach to reduce some of the aforementioned issues. Multiple grand challenges in the area of segmentation are becoming increasingly common, through which limited datasets with common scoring and editing guidelines are available to compare the various algorithms. Some grand challenges publish the delineation guidelines, which could be used in creating the institutional and larger training datasets to allow for improved training of the deep learning methods as well as the comparison against multiple methods. Some issues to consider when developing datasets both for training and testing of auto-segmentation methods include:

a. *Adherence to well-defined clinical guidelines of target and normal OAR structures*: There is inherent variability between users when performing segmentations. One approach to reduce variabilities is to use clearly defined guidelines such as use of the Radiation Therapy Oncology Group (RTOG) guidelines, RTOG 1106, in the 2017 AAPM Thoracic Auto-segmentation Challenge [29], RTOG 0920 in the Medical Image Computer and Computer-Assisted Intervention (MICCAI) Head and Neck Segmentation Challenge [39], as well as the AAPM RT MRI auto-contouring (RTMAC) 2019 grand challenge [40]. Another approach is to combine these descriptive guidelines with visual examples as done in web-based tools like the e-contour [46].

b. *Use of multiple expert delineations*: While it is difficult to obtain even a single expert delineation on a reasonably sized dataset, one clinically useful measure to assess auto-segmentation is whether or not it reduces the variability with respect to multiple experts [45]. Another approach is to use multiple algorithm delineations with manual editing to be used as benchmark for comparison against new methods [47]. While not a perfect substitute to model inter-rater variabilities, such datasets are useful to estimate how well a new algorithm compares to others.

c. *Multi-institutional datasets for algorithm selection*: While obtaining multi-institutional datasets is very challenging for individual researchers, grand challenges offer an easier option for a group of researchers to put together multi-institutional datasets. Multi-institutional datasets are invaluable to evaluate the algorithms under a wider range of realistic clinical conditions used in these institutions as opposed to a specific imaging protocol used for training and testing the algorithms. The algorithms, especially those using deep learning methods, have matured to a point that it is no longer useful to restrict their training and testing to homogenized imaging protocols without any corner cases. Particularly, if these methods are to be used in the clinic, it is imperative that the datasets consider the wider range of imaging variations.

d. *Creating an internal benchmarking dataset for internal acceptance*: Multi-institutional datasets are useful for selecting the appropriate method among other methods. However, once selected, it is necessary that this method satisfies the requirements of clinicians and clinical preferences in segmentation. For example, the delineation requirements in an institution may require that the organs are never under-segmented such that the radiographer or dosimetrist do not have to spend additional time including the parts left out by the algorithm. In the author's own institutional experience in commissioning a head and neck auto-segmentation method, it was found that the method trained on the well-known external institution Public domain database for computation anatomy (PDDCA) dataset produced segmentations closer to the inside boundary of the organs but the institution preference was to include the external boundary of the organs (like mandible, for instance). This is a small difference but can lead to excess additional work for individual cases. A more practical solution for this problem was to retrain the algorithm on the internally curated datasets.

Other considerations for internal commissioning should include the range of cases seen in the institution. For instance, in a tertiary cancer center, the number of patients with abnormal anatomies are often higher than is encountered in multi-institutional challenge datasets. Ignoring abnormal anatomies can create problems as the algorithms fail to segment or even lead to false positives. Thus, it is necessary to include difficult and abnormal conditions that are more commonly seen in the clinic as detailed below.

e. *Inclusion of difficult clinical conditions commonly seen in-clinic*: As the goal of training deep learning methods is typically to develop an algorithm that achieves reasonable performance, datasets often ignore or even remove difficult conditions like images with artifacts, images with large tumors, or anatomical conditions like collapsed lungs, absence of structures like submandibular glands due to surgery, and large tumors. However, for clinical use, the algorithms need to be robust to these clinical conditions. A common reason why the methods are not used in the clinic is because they fail to generalize even slightly outside the most homogenous conditions under which they were trained. Data curation, either institutional or for a grand challenge would benefit by combining datasets that incorporate some more challenging situations to "stress test" the algorithms.

13.7   EVALUATION METRICS GUIDELINES FOR CLINICAL COMMISSIONING

Auto-segmentation methods are usually evaluated with respect to contour similarity metrics and the potential for time savings in contouring when using these methods [35]. The commonly used metrics for evaluating contour similarity are the Dice similarity coefficient, Jaccard index (which is related to the Dice similarity coefficient), and spatial distance metrics like the Hausdorff distances, mean surface distance, and average distances. However, there is no clear consensus on the best method for assessing performance. Importantly, these metrics may not be effective in distinguishing random from systematic errors or to clearly separate false positives from false negative segmentations [35, 48]. Also, these metrics operate "out of context", in the sense that these metrics are independent of what impact their errors mean for treatment accuracies and the outcomes.

Importantly, volume overlap-based metrics like DSC are not well suited for measuring geometric deviations as they weight all misplaced segmentations, both internal and external, equally [45]. Similarly, volume comparison measures like volume ratio only give a rough idea of whether the segmentations are over and under the expected volumes as produced by an expert. However, these are not indicative of errors on the surface of the organs and target, which are more relevant for ensuring appropriate shaping of the treatment beams. Surface distance-based measures, such as the surface DSC that measures the extent of deviations of the algorithm contour from the manual delineation at the surface may be more suitable to measure the absolute deviations (in mm) from the organ surface by the algorithm [45]. These measures have also been shown to be more correlated to clinical usability measured using the time to perform manual editing on the auto-segmentations [21, 36].

The ultimate goal for clinical commissioning is whether or not the auto-segmentation method improves clinical efficiency and whether it improves treatment accuracy. In this regard, the time savings-based metrics as used by several works on clinical validation [12, 13, 16, 21, 31] are more useful measures of clinical efficiency improvements. However, measuring wall clock times in editing is also difficult and may be problematic to require clinicians to time themselves. A non-invasive (in terms of additional work for clinicians), and more robust metric was recently introduced [36], which measured the added path length of the edited contours. These metrics can be directly measured after the clinician edits by comparing against the algorithm contours. Also, this measure together with the surface DSC metric was shown to be highly correlated with the time savings measurements.

Ultimately, the real impact of the segmentations is in the dose calculations. In this regard, measuring the dose impact using dose-volume histograms (DVH) computed from new treatment plans computed from the auto-segmentations when compared to the "gold standard" segmentation plans would be most useful. However, such an evaluation is difficult to do as it requires access to fast treatment optimization plans that must be computed twice (once for the "gold standard" and second for the algorithm delineations). Also, average comparisons of the dosimetric measures derived from the DVH may not be sufficiently informative as the real impact of these metrics is for organs that lie in the high dose regions.

Importantly, the metrics used to evaluate the algorithms are often uncorrelated with each other. For instance, a high DSC score achieved by an algorithm does not necessarily imply a low Hausdorff distance [34]. Similarly, high DSC accuracy does not necessarily correspond to the improved dosimetric accuracies, because dosimetric accuracies are impacted by the location of the organ in the high-dose region [13]. In this regard, a better strategy might be to come up with new and comprehensive measures that combine the various metrics. Grand challenges themselves typically use DSC and Hausdorff, and additional metrics based on one or more expert delineations [49]. These metrics could be treated separately or as a combination to create a combined score, as was used in the 2017 AAPM Thoracic Auto-segmentation Challenge, also incorporating the inter-rater differences as a baseline in the accuracy calculation [29]. An advantage of using such combined metrics is that it provides one comprehensive score to compare multiple algorithms and provides an easier way to rank the performance of these methods.

However, the combination of metrics is still only useful to assess how good the algorithms are with respect to each other. In the case of clinical commissioning and for clinical use, this may not necessarily be useful as a metric. This is because in day-to-day clinical use a more useful measure is the degree of confidence in different parts of the segmentation. Such a measure could be, for example, computed using a multi-atlas method comparing multi-atlas segmentations to the clinical contour (if available) to indicate areas that potentially need a second inspection to reduce inter-rater variability. Alternatively, segmentation uncertainties on the voxel-by-voxel basis could also inform users where corrections are necessary. These uncertainty metrics could be tied to the algorithm alone or could also consider how the algorithm contouring variability (due to different levels of uncertainty) impacts treatment plans (assuming fast treatment planning methods are available to automatically compute multiple plans). Figure 13.1 shows an example of a multi-atlas registration-based segmentation uncertainty map [50] visualized for a selected organ (the left parotid gland). As seen, the voxel-level visualization of uncertainties, which can also easily be obtained using deep

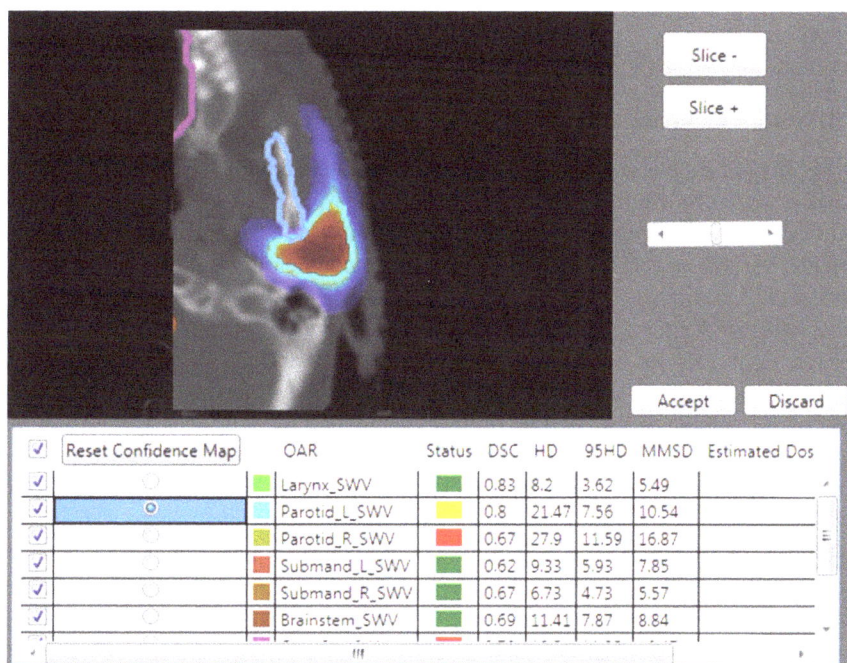

FIGURE 13.1   Voxel-wise segmentation confidences (red indicates higher certainty) visualized for a representative case computed using multi-atlas registration-based segmentation. The geometric accuracy metrics are also shown by evaluating against clinical delineations.

learning methods, can potentially help ascertain where segmentation corrections may need to be done for QA in clinical workflows. The geometric uncertainties could also be combined with the dose volume histogram calculations to evaluate the extent of deviations between the various contours as a measure of contour stability.

In summary, the best metric or set of metrics needed for clinical commissioning is still an open research question. However, depending on the clinical requirements (accuracy vs time efficiency), a combination of the aforementioned metrics could be used in a hierarchical manner. For instance, geometric metrics could be used on a common benchmark dataset to ascertain the best method. Next, clinical commissioning and ongoing evaluations during various algorithm upgrades or in the event of imaging protocol changes could be evaluated using time-based, dosimetric, or a combination of these metrics.

13.8 COMMISSIONING AND SAFE USE IN THE CLINIC

Finally, commissioning any auto-segmentation method should go through multiple phases starting from identifying the appropriate method to ongoing evaluation and periodic testing. A typical commissioning procedure could use a two-phase approach, where the first phase involves training/ validation and identification of a suitable method and the second phase involves testing and verification for use in the clinic [30]. However, the algorithm development and training phase itself does not need to be done in the commissioning institution as long as the vendor providing the software can provide relevant details for evaluating the algorithm with the internal datasets.

13.8.1 TRAINING AND VALIDATION PHASE

The training/validation phase should consider evaluating a reasonable number of different methods, employ multiple metrics, including dosimetric metrics [35], and if possible evaluate the methods with open-source datasets to identify the best method and the general conditions in which the method works.

13.8.2 TESTING AND VERIFICATION PHASE

In the testing/verification phase, the method should be evaluated with an internally curated dataset. In this phase, prior to clinical implementation, it is useful to have a multidisciplinary team such as the algorithm developers, physicists, radiation oncologists, etc. who can review the method's performance under variable scenarios that are reasonably encountered in the clinic, and understand its performance and limitations.

In Memorial Sloan Kettering Cancer Center, a similar training/testing schedule as proposed in Vandewinckele et al. [30] for clinical commissioning was employed. In addition to the training/ validation and the pre-commissioning testing, a feedback-oriented testing and developmental cycle for clinical commissioning of the deep learning segmentation methods was used. A team of multidisciplinary experts including computer scientists and developers involved in the algorithm development, physicists, a radiation oncologist (for the specific disease site), and an anatomy contourist meets on a weekly basis to visually review cases from the internal clinical datasets. Cases for review are selected from the internally curated dataset, as well as new or abnormal cases identified by the radiation oncologist. Any problems encountered with the algorithm are discussed, and if necessary the algorithms are improved or retrained and reviewed the following week. This feedback-oriented development cycle continues until the team is satisfied with the algorithm's performance and the algorithm is applied to a set of new incoming clinical cases for the following month and verified in a blinded experiment by the anatomist, physicists, and radiation oncologist. If no issues are encountered in the new cases, the auto-segmentation method is commissioned for clinical use. The auto-segmentation methods are periodically reviewed (every three months) and improved as necessary.

Any improvements are further reviewed before recommissioning of the improved method. This method has worked for the institution for two different deep learning auto-segmentation methods currently in use in the clinic for prostate and head and neck cancer treatments.

13.9 SUMMARY

Some of the issues related to clinical commissioning and data curation of auto-segmentation methods for radiation therapy treatments were reviewed. These issues are explored further in the next two chapters. Some guidelines were presented, and suggestions were proposed that could be implemented for these aforementioned two issues and help with better and more efficient commissioning mechanisms in different institutions.

REFERENCES

1. Bradley JD, Hu C, Komaki RR, Masters GA, Blumenschein GR, Schild SE, Bogart JA, Forster KM, Magliocco AM, Kavadi VS, Narayan S, Iyengar P, Robinson CG, Wynn RB, Koprowski CD, Olson MR, Meng J, Paulus RWJC, Jr, Choy H: Long-term results of NRG oncology RTOG 0617: standard- versus high-dose chemoradiotherapy with or without cetuximab for unresectable stage III non–small-cell lung cancer. *J Clin Oncol* 38:706–714, PMID:31841363, 2020
2. Bradley JD, Paulus R, Komaki R, Masters G, Blumenschein G, Schild S, Bogart J, Hu C, Forster K, Magliocco A, Kavadi V, Garces YI, Narayan S, Iyengar P, Robinson C, Wynn RB, Koprowski C, Meng J, Beitler J, Gaur R, Curran W, Jr., Choy H: Standard-dose versus high-dose conformal radiotherapy with concurrent and consolidation carboplatin plus paclitaxel with or without cetuximab for patients with stage IIIA or IIIB non-small-cell lung cancer (RTOG 0617): a randomised, two-by-two factorial phase 3 study. *Lancet Oncol* 16:187–199, PMC4419359, PMID:25601342, 2015
3. Kavanaugh J, Hugo G, Robinson CG, Roach MC: Anatomical adaptation—early clinical evidence of benefit and future needs in lung cancer. *Semin Radiat Oncol* 29:274–283, 2019
4. Weiss E, Fatyga M, Wu Y, Dogan N, Balik S, Sleeman WT, Hugo G: Dose escalation for locally advanced lung cancer using adaptive radiation therapy with simultaneous integrated volume-adapted boost. *Int J Radiat Oncol Biol Phys* 86:414–419, PMC3665644, PMID:23523321, 2013
5. Mackie TR, Kapatoes J, Ruchala K, Lu W, Wu C, Olivera G, Forrest L, Tome W, Welsh J, Jeraj R, Harari P, Reckwerdt P, Paliwal B, Ritter M, Keller H, Fowler J, Mehta M: Image guidance for precise conformal radiotherapy. *Int J Radiat Oncol Biol Phys* 56:89–105, PMID:12694827, 2003
6. Sonke JJ, Belderbos J: Adaptive radiotherapy for lung cancer. *Semin Radiat Oncol* 20:94–106, PMID:20219547, 2010
7. Vinod SK, Jameson MG, Min M, Holloway LC: Uncertainties in volume delineation in radiation oncology: a systematic review and recommendations for future studies. *Radiother Oncol* 121:169–179, PMID:27729166, 2016
8. Yin LJ, Yu XB, Ren YG, Gu GH, Ding TG, Lu Z: Utilization of PET-CT in target volume delineation for three-dimensional conformal radiotherapy in patients with non-small cell lung cancer and atelectasis. *Multidiscip Respir Med* 8:21, PMC3608960, PMID:23506629, 2013
9. Caldwell CB, Mah K, Ung YC, Danjoux CE, Balogh JM, Ganguli SN, Ehrlich LE: Observer variation in contouring gross tumor volume in patients with poorly defined non-small-cell lung tumors on CT: the impact of 18FDG-hybrid PET fusion. *Int J Radiat Oncol Biol Phys* 51:923–931, PMID:11704312, 2001
10. Jameson MG, Kumar S, Vinod SK, Metcalfe PE, Holloway LC: Correlation of contouring variation with modeled outcome for conformal non-small cell lung cancer radiotherapy. *Radiother Oncol* 112:332–336, PMID:24853367, 2014
11. Kupelian P, Sonke JJ: Magnetic resonance-guided adaptive radiotherapy: a solution to the future. *Semin Radiat Oncol* 24:227–232, PMID:24931098, 2014
12. Gambacorta MA, Valentini C, Dinapoli N, Boldrini L, Caria N, Barba MC, Mattiucci GC, Pasini D, Minsky B, Valentini V: Clinical validation of atlas-based auto-segmentation of pelvic volumes and normal tissue in rectal tumors using auto-segmentation computed system. *Acta Oncol* 52:1676–1681, PMID:23336255, 2013
13. Kaderka R, Gillespie EF, Mundt RC, Bryant AK, Sanudo-Thomas CB, Harrison AL, Wouters EL, Moiseenko V, Moore KL, Atwood TF, Murphy JD: Geometric and dosimetric evaluation of atlas based auto-segmentation of cardiac structures in breast cancer patients. *Radiother Oncol* 131:215–220, PMID:30107948, 2019

14. Teguh DN, Levendag PC, Voet PW, Al-Mamgani A, Han X, Wolf TK, Hibbard LS, Nowak P, Akhiat H, Dirkx ML, Heijmen BJ, Hoogeman MS: Clinical validation of atlas-based auto-segmentation of multiple target volumes and normal tissue (swallowing/mastication) structures in the head and neck. *Int J Radiat Oncol Biol Phys* 81:950–957, PMID:20932664, 2011

15. Isambert A, Dhermain F, Bidault F, Commowick O, Bondiau PY, Malandain G, Lefkopoulos D: Evaluation of an atlas-based automatic segmentation software for the delineation of brain organs at risk in a radiation therapy clinical context. *Radiother Oncol* 87:93–99, PMID:18155791, 2008

16. Hoang Duc AK, Eminowicz G, Mendes R, Wong SL, McClelland J, Modat M, Cardoso MJ, Mendelson AF, Veiga C, Kadir T, D'Souza D, Ourselin S: Validation of clinical acceptability of an atlas-based segmentation algorithm for the delineation of organs at risk in head and neck cancer. *Med Phys* 42:5027–5034, PMID:26328953, 2015

17. Sims R, Isambert A, Grégoire V, Bidault F, Fresco L, Sage J, Mills J, Bourhis J, Lefkopoulos D, Commowick O, Benkebil M, Malandain G: A pre-clinical assessment of an atlas-based automatic segmentation tool for the head and neck. *Radiother Oncol* 93:474–478, PMID:19758720, 2009

18. Sonke JJ, Aznar M, Rasch C: Adaptive radiotherapy for anatomical changes. *Semin Radiat Oncol* 29:245–257, PMID:31027642, 2019

19. Purdie TG, Moseley DJ, Bissonnette JP, Sharpe MB, Franks K, Bezjak A, Jaffray DA: Respiration correlated cone-beam computed tomography and 4DCT for evaluating target motion in stereotactic lung radiation therapy. *Acta Oncol* 45:915–922, PMID:16982558, 2006

20. Balagopal A, Kazemifar S, Nguyen D, Lin MH, Hannan R, Owrangi A, Jiang S: Fully automated organ segmentation in male pelvic CT images. *Phys Med Biol* 63:245015, PMID:30523973, 2018

21. Lustberg T, van Soest J, Gooding M, Peressutti D, Aljabar P, van der Stoep J, van Elmpt W, Dekker A: Clinical evaluation of atlas and deep learning based automatic contouring for lung cancer. *Radiother Oncol* 126:312–317, PMID:29208513, 2018

22. Ibragimov B, Xing L: Segmentation of organs-at-risks in head and neck CT images using convolutional neural networks. *Med Phys* 44:547–557, PMC5383420, PMID:28205307, 2017

23. Dolz J, Kirişli HA, Fechter T, Karnitzki S, Oehlke O, Nestle U, Vermandel M, Massoptier L: Interactive contour delineation of organs at risk in radiotherapy: clinical evaluation on NSCLC patients. *Med Phys* 43:2569, PMID:27147367, 2016

24. Trullo R, Petitjean C, Ruan S, Dubray B, Nie D, Shen D: Segmentation of organs at risk in thoracic CT images using a sharpmask architecture and conditional random fields. *Proc IEEE Int Symp Biomed Imaging* 2017:1003–1006, PMC5649634, PMID:29062466, 2017

25. Feng X, Qing K, Tustison NJ, Meyer CH, Chen Q: Deep convolutional neural network for segmentation of thoracic organs-at-risk using cropped 3D images. *Med Phys* 46:2169–2180, PMID:30830685, 2019

26. Wu X, Zhong Z, Buatti J, Bai J: Multi-scale segmentation using deep graph cuts: robust lung tumor delineation in MVCBCT. *Proc IEEE Int Symp Biomed Imaging* 2018:514–518, PMC6878112, PMID:31772718, 2018

27. Dong X, Lei Y, Wang T, Thomas M, Tang L, Curran WJ, Liu T, Yang X: Automatic multiorgan segmentation in thorax CT images using U-net-GAN. *Med Phys* 46:2157–2168, PMC6510589, PMID:30810231, 2019

28. Jiang J, Hu YC, Tyagi N, Zhang P, Rimner A, Deasy JO, Veeraraghavan H: Cross-modality (CT-MRI) prior augmented deep learning for robust lung tumor segmentation from small MR datasets. *Med Phys* 46:4392–4404, PMC6800584, PMID:31274206, 2019

29. Yang J, Veeraraghavan H, Armato SG, 3rd, Farahani K, Kirby JS, Kalpathy-Kramer J, van Elmpt W, Dekker A, Han X, Feng X, Aljabar P, Oliveira B, van der Heyden B, Zamdborg L, Lam D, Gooding M, Sharp GC: Autosegmentation for thoracic radiation treatment planning: a grand challenge at AAPM 2017. *Med Phys* 45:4568–4581, PMC6714977, PMID:30144101, 2018

30. Vandewinckele L, Claessens M, Dinkla A, Brouwer C, Crijns W, Verellen D, van Elmpt W: Overview of artificial intelligence-based applications in radiotherapy: recommendations for implementation and quality assurance. *Radiother Oncol* 153:55–66, 2020.

31. Sjöberg C, Lundmark M, Granberg C, Johansson S, Ahnesjö A, Montelius A: Clinical evaluation of multi-atlas based segmentation of lymph node regions in head and neck and prostate cancer patients. *Radiat Oncol* 8:229, PMC3842681, PMID:24090107, 2013

32. Ahn SH, Yeo AU, Kim KH, Kim C, Goh Y, Cho S, Lee SB, Lim YK, Kim H, Shin D, Kim T, Kim TH, Youn SH, Oh ES, Jeong JH: Comparative clinical evaluation of atlas and deep-learning-based auto-segmentation of organ structures in liver cancer. *Radiat Oncol* 14:213, PMC6880380, PMID:31775825, 2019

33. Walker GV, Awan M, Tao R, Koay EJ, Boehling NS, Grant JD, Sittig DF, Gunn GB, Garden AS, Phan J, Morrison WH, Rosenthal DI, Mohamed AS, Fuller CD: Prospective randomized double-blind study of atlas-based organ-at-risk autosegmentation-assisted radiation planning in head and neck cancer. *Radiother Oncol* 112:321–325, PMC4252740, PMID:25216572, 2014

34. Sharp G, Fritscher KD, Pekar V, Peroni M, Shusharina N, Veeraraghavan H, Yang J: Vision 20/20: perspectives on automated image segmentation for radiotherapy. *Med Phys* 41:050902, PMC4000389, PMID:24784366, 2014

35. Valentini V, Boldrini L, Damiani A, Muren LP: Recommendations on how to establish evidence from auto-segmentation software in radiotherapy. *Radiother Oncol* 112:317–320, PMID:25315862, 2014

36. Vaassen F, Hazelaar C, Vaniqui A, Gooding M, van der Heyden B, Canters R, van Elmpt W: Evaluation of measures for assessing time-saving of automatic organ-at-risk segmentation in radiotherapy. *Phys Imaging Radiat Oncol* 13:1–6, 2020

37. Voet PW, Dirkx ML, Teguh DN, Hoogeman MS, Levendag PC, Heijmen BJ: Does atlas-based autoseg-mentation of neck levels require subsequent manual contour editing to avoid risk of severe target under-dosage? A dosimetric analysis. *Radiother Oncol* 98:373–377, PMID:21269714, 2011

38. Emre Kavur A, Sinem Gezer N, Barış M, Aslan S, Conze P-H, Groza V, Duy Pham D, Chatterjee S, Ernst P, Özkan S, Baydar B, Lachinov D, Han S, Pauli J, Isensee F, Perkonigg M, Sathish R, Rajan R, Sheet D, Dovletov G, Speck O, Nürnberger A, Maier-Hein KH, Bozdağı Akar G, Ünal G, Dicle O, Selver MA: CHAOS challenge – combined (CT-MR) healthy abdominal organ segmentation. arXiv e-prints, 2020, pp arXiv:2001.06535

39. Raudaschl PF, Zaffino P, Sharp GC, Spadea MF, Chen A, Dawant BM, Albrecht T, Gass T, Langguth C, Luthi M, Jung F, Knapp O, Wesarg S, Mannion-Haworth R, Bowes M, Ashman A, Guillard G, Brett A, Vincent G, Orbes-Arteaga M, Cardenas-Pena D, Castellanos-Dominguez G, Aghdasi N, Li Y, Berens A, Moe K, Hannaford B, Schubert R, Fritscher KD: Evaluation of segmentation methods on head and neck CT: auto-segmentation challenge 2015. *Med Phys* 44:2020–2036, PMID:28273355, 2017

40. Cardenas C, Mohamed, A., Sharp, G., Gooding, M., Veeraraghavan, H., Yang J.: Data from AAPM RT-MAC grand challenge 2019, in archive TCI (ed). *The Cancer Imaging Archive*, NIH, 2019

41. Shen C, Nguyen D, Zhou Z, Jiang SB, Dong B, Jia X: An introduction to deep learning in medical phys-ics: advantages, potential, and challenges. *Phys Med Biol* 65:05tr01, PMC7101509, PMID:31972556, 2020

42. Chao KS, Bhide S, Chen H, Asper J, Bush S, Franklin G, Kavadi V, Liengswangwong V, Gordon W, Raben A, Strasser J, Koprowski C, Frank S, Chronowski G, Ahamad A, Malyapa R, Zhang L, Dong L: Reduce in variation and improve efficiency of target volume delineation by a computer-assisted system using a deformable image registration approach. *Int J Radiat Oncol Biol Phys* 68:1512–1521, PMID:17674982, 2007

43. Mattiucci GC, Boldrini L, Chiloiro G, D'Agostino GR, Chiesa S, De Rose F, Azario L, Pasini D, Gambacorta MA, Balducci M, Valentini V: Automatic delineation for replanning in nasopharynx radio-therapy: what is the agreement among experts to be considered as benchmark? *Acta Oncol* 52:1417–1422, PMID:23957565, 2013

44. van Rooij W, Dahele M, Ribeiro Brandao H, Delaney AR, Slotman BJ, Verbakel WF: Deep learning-based delineation of head and neck organs at risk: geometric and dosimetric evaluation. *Int J Radiat Oncol Biol Phys* 104:677–684, PMID:30836167, 2019

45. Nikolov S, Blackwell S, Mendes R, Fauw JD, Meyer C, Hughes C, Askham H, Romera-Paredes B, Karthikesalingam A, Chu C, Carnell D, Boon C, D'Souza D, Moinuddin SA, Sullivan K, Consortium DR, Montgomery H, Rees G, Sharma R, Suleyman M, Back T, Ledsam JR, Ronneberger O: Deep learning to achieve clinically applicable segmentation of head and neck anatomy for radiotherapy. *Clin Orthop Relat Res* abs/1809.04430, 2018

46. Sherer MV, Lin D, Puri K, Panjwani N, Zhang Z, Murphy JD, Gillespie EF: Development and usage of econtour, a novel, three-dimensional, image-based web site to facilitate access to contouring guidelines at the point of care. *JCO Clin Cancer Inform* 3:1–9, PMID:31756136, 2019

47. Kalpathy-Cramer J, Zhao B, Goldgof D, Gu Y, Wang X, Yang H, Tan Y, Gillies R, Napel S: A compari-son of lung nodule segmentation algorithms: methods and results from a multi-institutional study. *J Digit Imaging* 29:476–487, PMC4942386, PMID:26847203, 2016

48. Hanna GG, Hounsell AR, O'Sullivan JM: Geometrical analysis of radiotherapy target volume delinea-tion: a systematic review of reported comparison methods. *Clin Oncol (R Coll Radiol)* 22:515–525, PMID:20554168, 2010

49. Menze BH, Jakab A, Bauer S, Kalpathy-Cramer J, Farahani K, Kirby J, Burren Y, Porz N, Slotboom J, Wiest R, Lanczi L, Gerstner E, Weber MA, Arbel T, Avants BB, Ayache N, Buendia P, Collins DL, Cordier N, Corso JJ, Criminisi A, Das T, Delingette H, Demiralp C, Durst CR, Dojat M, Doyle S, Festa J, Forbes F, Geremia E, Glocker B, Golland P, Guo X, Hamamci A, Iftekharuddin KM, Jena R, John NM, Konukoglu E, Lashkari D, Mariz JA, Meier R, Pereira S, Precup D, Price SJ, Raviv TR, Reza SM, Ryan M, Sarikaya D, Schwartz L, Shin HC, Shotton J, Silva CA, Sousa N, Subbanna NK, Szekely G, Taylor TJ, Thomas OM, Tustison NJ, Unal G, Vasseur F, Wintermark M, Ye DH, Zhao L, Zhao B, Zikic D, Prastawa M, Reyes M, Van Leemput K: The multimodal Brain Tumor Image Segmentation Benchmark (BRATS). *IEEE Trans Med Imaging* 34:1993–2024, PMC4833122, PMID:25494501, 2015

50. Haq R, Berry SL, Deasy JO, Hunt M, Veeraraghavan H: Dynamic multi-atlas selection based consensus segmentation of head and neck structures from CT images. *Med Phys*, PMID:31587300, 2019

14 Data Curation Challenges for Artificial Intelligence

Ken Chang, Mishka Gidwani, Jay B. Patel,
Matthew D. Li, and Jayashree Kalpathy-Cramer

CONTENTS

14.1 INTRODUCTION

The last decade marks a significant leap in the capabilities of artificial intelligence (AI) algorithms, with the advent of powerful processing units, open-source frameworks for deep learning, the availability of large-scale datasets [1–3]. These algorithms are capable of learning the output(s) of interest from raw/preprocessed data for a variety of tasks such as image classification, speech recognition, and natural language processing [4–6]. More specifically, this state-of-the-art performance is accomplished through the chaining together of many layers of high-dimensional, non-linear transforms (i.e. convolutions, activations, pooling, etc.) which together form a network that allows for the learning of complex patterns with a high degree of abstraction. Within the medical domain, a logical application of this technique is to medical imaging, where clinicians have long noticed the relationship between imaging patterns and diagnosis, prognosis, genomics, and treatment response. As such, AI has brought on a paradigm shift in automated methods for medical image processing, with recent studies showing its potential utility for clinical applications within dermatology, ophthalmology, pathology, oncology, cardiology, infectious disease, radiation oncology, and radiology [7–12].

Although there are numerous possible tasks for AI algorithms within medical imaging, they generally fall into three main categories: classification, detection, and segmentation (Figure 14.1). Perhaps the most common task is segmentation, which may be considered the voxel or pixel-wise classification of anatomical areas into categories. In medicine, this is especially useful when the regions of abnormalities, tissues, and organs need to be precisely delineated, such as in the case of disease burden quantification, treatment response assessment, and radiation therapy planning [13–16].

Segmentation, referred to as contouring, is a critical step in development of a radiation therapy treatment plan. Members of the radiology and radiation oncology team including technicians,

Classification

Does this patient have pneumothorax?

Detection

Which region is the pneumothorax in?

Segmentation

What are the boundaries of the pneumothorax?

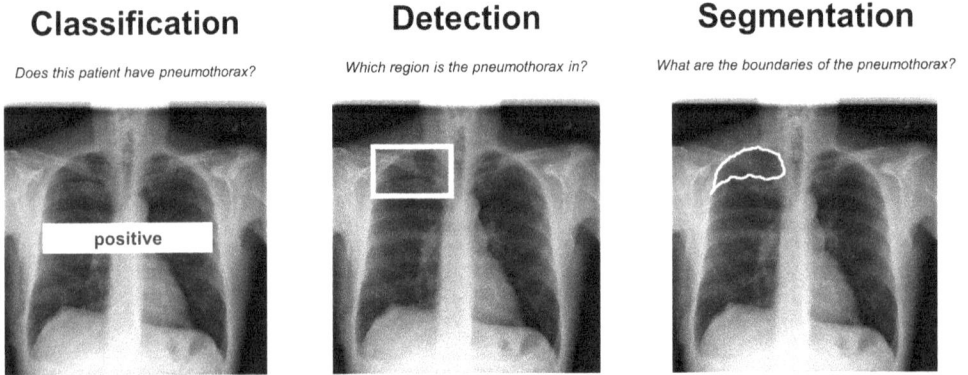

FIGURE 14.1 Within medical imaging, there are three general categories of tasks: classification, detection, and segmentation.

dosimetrists, and radiation oncologists frequently delineate normal structures and lesions in a manual or semi-automatic manner. These structures in turn inform the creation of a dose map, which is intended to deliver the maximal dose to target sites while sparing normal tissue. This highlights the importance of accurate segmentation of these structures. Additionally, radiation therapy is frequently informed by characteristics of tumors which can be derived from the gross tumor volume (GTV) contour. These include the total volume (in voxels or milliliters), the maximal diameter (in pixels or centimeters), and the number of discontinuous lesions [17].

Advances in deep learning techniques for segmentation have been catalyzed in part by large-scale competitions [18–21], which provide a common framework for comparison and evaluation. Specific advances, such as the U-net architecture [22], ensembling [23], and adaptable input sizes [24] have shown to be particularly effective, elevating the performance of segmentation to the levels of inter-rater variability expected of human experts (Figure 14.2). Progress in methods development has also been expedited by availability of open-source code and software [25–27].

14.2 THE COMPLEXITY OF MEDICAL IMAGING

It is an understatement to say that medical imaging is complex. Although medical imaging is commonly stored in the Digital Imaging and Communications in Medicine (DICOM) format which harmonizes meta-information and images, there are still many sources of variation. Firstly, imaging can come in various resolutions and bit-depths. While regulatory agencies and device manufacturers recommend imaging parameters for stability and safety, many imaging settings are determined by the anatomy being imaged and the operator acquiring the image, to say nothing of environmental factors. Secondly, imaging can come in multiple views with the possibility of supplemental views in scenarios where the imaging contains artifacts, ambiguity, or suspicious pathology [28]. This can complicate the harmonization of segmentation labels, which may be superimposed on a single view. Imaging can also be acquired in a 2D format, such as x-ray or histopathology, a 3D format such as CT, magnetic resonance imaging (MRI) or ultrasound, or even a 4D format such as free-breathing CT (4DCT) and cine cardiac MRI. Furthermore, images can be stored and displayed as either grayscale (such as radiographs and computed tomography, CT) or color scale (positron emission tomography and doppler ultrasound), depending on modality [28]. Finally, the native dimension of the images directly influences the construction of a deep-learning algorithm. High resolution multimodal 3D imaging must be treated differently to low resolution unimodal 2D data, primarily due to graphics processing unit (GPU) memory concerns. Even when considering only 3D imaging formats, one must specifically design networks to handle differences between isotropic and non-isotropic data. Moreover, other factors such as convolutional filter size and filter numbers must

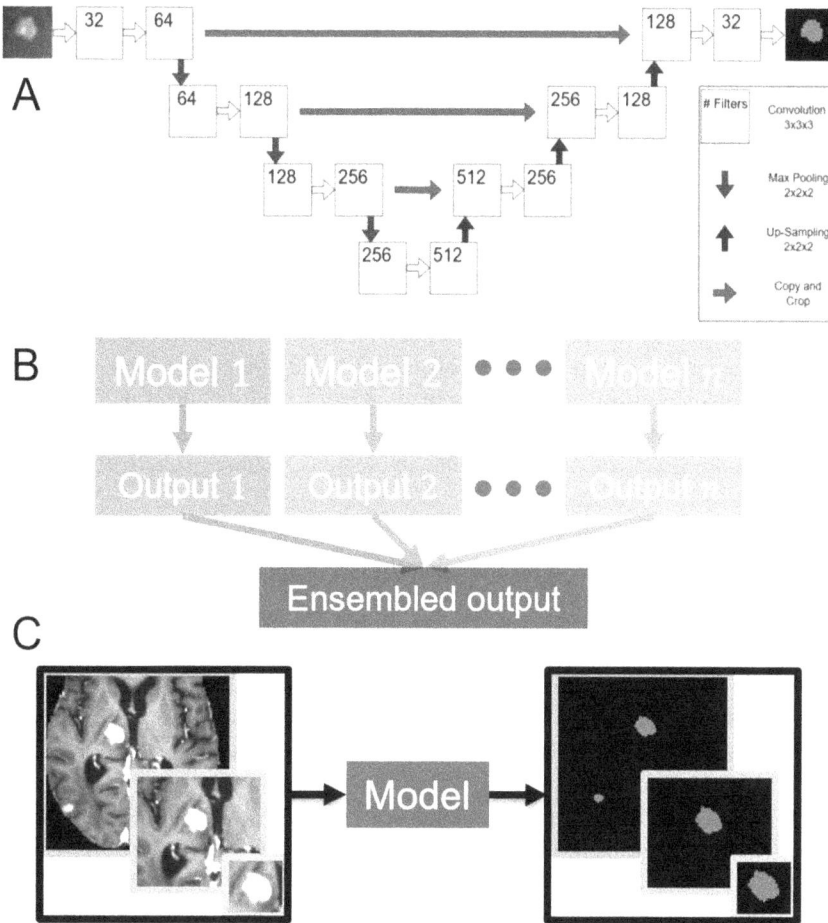

FIGURE 14.2 (A) U-net architecture for image segmentation. The network is composed of an encoding arm, which repeatedly downsamples the input into a lower dimensional space, and a decoding arm, which recovers the output of interest. The skip connections between the encoder and decoder allow for feature re-use, which helps the network combine low-level with high-level information. (B) A schematic showing model ensembling across *n* models. The outputs of the individual models are averaged together to create a final, more refined prediction. (C) An example of a segmentation network that can accept variable input sizes, which can be useful for learning features at different scales.

change accordingly with image dimension. Operations that decrease the dimension of the image, such as pooling layers, need to be carefully tracked to ensure that spatial information is preserved when expected.

Image intensity values can have units with physical meaning (such as the Hounsfield unit in CT) or be unit-less such as with conventional MRI. Studies can also have multiple sequences that need to be interpreted in combination such as diffusion weighted imaging (DWI) and apparent diffusion coefficient (ADC) MRI sequences. Because of this, intensity values cannot be directly compared across sequences. For example, while an anatomical feature may be contrast-enhancing under certain imaging conditions (e.g. T1 post-contrast MRI), it may be more absorptive under different conditions (e.g. T2 MRI). When using a contrast-enhancing agent, or a molecular imaging probe or fluorophore, the challenge of comparing images is compounded. Standardizing concentrations of reagents and imaging parameters such as field depth and exposure, as well as normalizing by the size of the targets of the imaging (e.g. number of cells for molecular imaging) can mitigate some of these differences. The diverse distributions of image intensity values call for standardization during preprocessing, as deep-learning models are more accepting of uniform values as input.

The research setting often involves imaging across scales and model systems. The necessary technical differences when imaging cell culture, organoids, xenograft models, and humans, can limit the ability of the researcher to compare images across scales. In the clinical setting, images from different modalities often need to be considered when making a diagnosis. Additionally, disease can present across multiple anatomical regions, such as primary and metastatic disease. Lastly, the longitudinal tracking of disease is often critical for evaluation and current imaging must be compared with prior imaging [29]. When multiple scans are required for the development of a machine learning model or for a granular view of the pathology of interest, the images acquired are not independent and the effect of repeated imaging on the subject should be ascertained. This may involve movement of the subject, loss of contrast agent or fluorophore, or the molecular effects from ionizing radiation, among others. Unifying imaging information across scales, modalities, physiological sites, and time are some of the many challenges facing the data collector.

14.3　THE CHALLENGE OF GENERALIZABILITY AND DATA HETEROGENEITY

One critical hurdle that prevents the widespread utilization of deep learning models in clinical workflows is the lack of generalizability (or transferability) of trained models across datasets and institutions [30–33]. Indeed, this fact is masked within the medical literature, as very few published studies have external validation [34]. In general, the lack of generalizability is attributed to data heterogeneity, that is divergence between the distribution of data that was used for training and the distribution that was used for evaluation.

Data heterogeneity can stem from several causes. Firstly, there may be differences in patient demographics (such as age, sex, and ethnicity) and disease prevalence between different institutions. The imaging acquisition can also vary. For example, different mammography systems can have different X-ray tube targets, filters, digital detector technology, and control of automatic exposure [35]. Along the same line, MRI at different institutions may utilize different field strength, resolution, and scanning protocols. In fact, MRI acquisitions can differ even within an institution if there are machines from multiple different manufacturers. Lastly, there can be variability in the labeling by human annotators, a phenomenon that has been documented across many medical disciplines [36–38] (Figure 14.3).

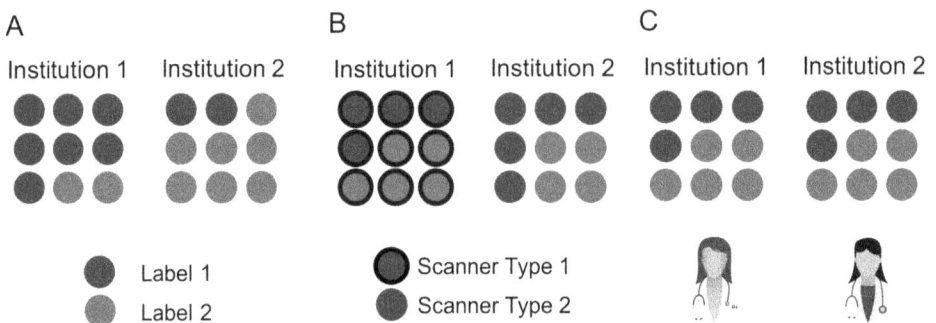

FIGURE 14.3 Various types of heterogeneity can exist in real patient data, such as (A) imbalanced labels or patient characteristics, (B) different image acquisition setting or scanner systems, and (C) inter-rater variability in labeling.

14.4   DATA SELECTION

The choice of a well-defined sample population is important when beginning data curation. In order to obtain primary medical data of a pathology of interest, an authorized user must access the electronic medical record (EMR) with the approval of an institutional review board (IRB). This may not be the researcher themselves. Therefore, this requires clear communication and definitions of the cohort of interest. The authorized user further needs to be able to search and harvest data from the EMR, which is complicated by heterogeneously tagged data and differing formats. When medical data is downloaded and saved, often in an anonymized fashion, relevant information may be lost, such as prior treatments, comorbidities, and genetic characteristics, all of which may have influence data analysis. Importantly, the DICOM header may contain useful information such as the resolution, orientation, and acquisition settings that may be stripped away during anonymization. One other aspect of data selection that should be mentioned is concept drift, that is how the classification of disease (and thus, relevant annotations) changes over time as knowledge of medicine evolves over time. In addition, there can be technology shift, where newer imaging systems replace older ones. It is important that data selection captures the most recent clinical standards in order to ensure trained algorithms can be used prospectively. As such, data selection, curation, and annotation must be a continuous process. Broadening this statement, the training and testing data should be similar. If the training data significantly predates (or differs in other ways) from the testing data, the performance of the algorithm may degrade. Additionally, it is critical that the positive and negative cases are acquired under the same conditions. If not, the algorithm may learn differences in image acquisition instead of the disease of interest, achieving deceptively high performance [39].

14.5   THE NEED FOR LARGE QUANTITIES OF DATA

The large quantities of data needed to train effective, robust models is mainly driven by this issue of data heterogeneity. This data requirement is further increased in use-cases where there is a high amount of anatomical and disease phenotypic variability. For general computer vision tasks, public datasets can be quite large (on the order of $10^5 - 10^9$ samples) [2, 40, 41]. Conversely, publicly available medical imaging datasets are considerably smaller (on the order of $10^1 - 10^5$ samples), and thus it becomes much more challenging to meet the data requirement [20, 21, 42, 43]. Furthermore, using publicly available data can introduce additional complexity. For example, the data may come with incomplete annotations or may have already been processed via some unknown transformations (which is effectively equivalent to adding unwanted noise to the input data distribution). Indeed, various studies have shown that the performance of algorithms improves substantially with the incorporation of more training data [44, 45]. Importantly, the size of the dataset includes both the absolute quantity of data but also the number of images from patients that have the pathology of interest. For example, a very large dataset that only has a very small percentage of patients with disease may not be effective in training an algorithm as it may not be exposed to adequate phenotypic diversity. That said, curating matched "normal" data to serve as algorithm controls can likewise be limited due to the infrequency of healthy patient visits as well as the high prevalence of abnormalities that are not clinically significant [46]. Moreover, enough data to not only train the network, but also perform both internal and external validation to assess for model generalizability needs to be acquired.

In cases where a minimum acceptable threshold of data is unable to be curated, pre-training can be a powerful tool. Pre-training is when a model is initially trained on a large, diverse dataset on a tangentially related task before being fine-tuned on the smaller dataset of the task of interest. In general, the upper layers of a neural network learn only generic, non-task-specific information (i.e. edge filters, shape detectors, color filters, etc.), and thus can be transferred to other tasks without modification. Indeed, it is common to only fine-tune the final layers of the network on the new task of interest, freezing the rest of the network to preserve the pre-trained weights. This process of

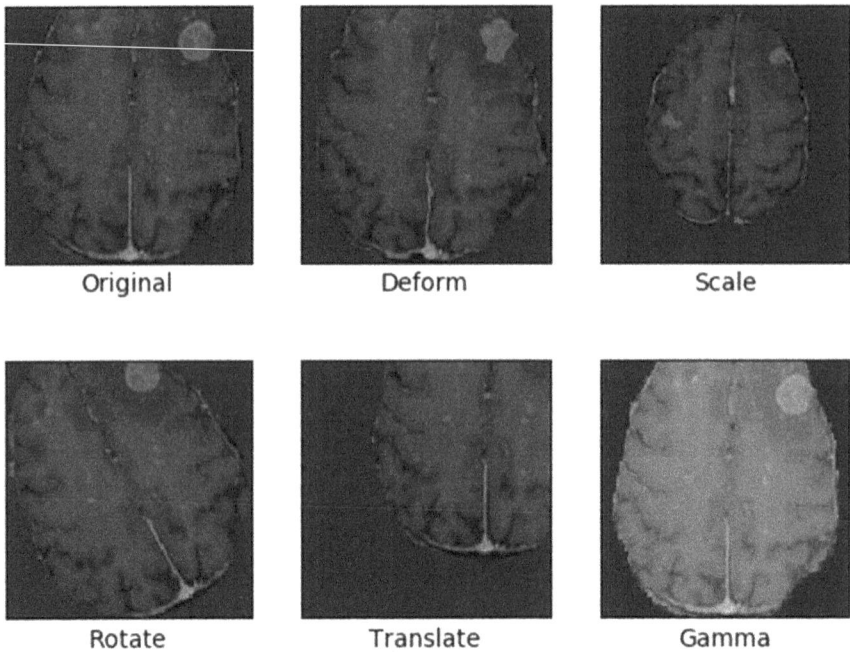

FIGURE 14.4 Examples of common spatial and intensity data augmentation transforms that can be applied to imaging data. It is common to compose these transformations together to generate large amounts of variation from a potentially limited training dataset.

utilizing large quantities of related data (also known as transfer learning) is an effective paradigm within medical imaging and has been shown to improve performance by allowing the network to learn domain-specific imaging features without needing to learn generic filters [47, 48]. Pre-training can also be important with the occurrence of concept or technology shift, whereby historical data can be used for transfer learning.

Regardless of access to the absolute quantity of data, training can still be improved with careful application of methods that artificially augment the dataset. These methods rely on manipulations to the existing data to increase its diversity, such as random spatial transforms (translations, mirroring, scaling, rotating, elastic deformations) and intensity transforms (gamma corrections, saturation, noise) [24, 49] (Figure 14.4). Other modulations such as random jittering, kernel filters, and erasing can be applied as well [24, 50–52]. One counter-intuitive approach that has been shown to be effective is to mix, or average, two images together, which allows the generation of N^2 more training images from N training images [53, 54]. More advanced techniques include using neural networks to learn optimal augmentation policies, making alterations to the image via style transfer, or generating new training images [55–58]. While it is expected that augmentation will generally improve the generalizability of the model, this is not necessarily the case for medical imaging [49, 50]. Specifically, one needs to ensure that the chosen augmentations are physiologically possible. For example, the heart contracts and relaxes in a very regular pattern, and applying random elastic deformations that are not carefully constrained can produce augmented images that lie outside the true data distribution, which may in fact lower performance. Data augmentation is explored further in Chapter 11.

14.6   BARRIERS TO SHARING PATIENT DATA AND DISTRIBUTED DEEP LEARNING

Another major hurdle in data curation is the difficulty in sharing patient data, specifically in dealing with the concerns of patient privacy, data anonymization, patient consent, intellectual property, and data storage requirements. Firstly, protection of patient privacy and confidentiality is of critical importance both within medical care as well as in research. Indeed, studies have shown that the leakage of just a few clinical variables or a single image can allow for reidentification of patients [59–61]. One approach to prevent leakage of information is to convert DICOM images into other formats, such as Joint Photographic Expert Group (JPEG), Neuroimaging Informatics Technology Initiative (NIfTI), and Portable Network Graphics (PNG), which removes identifiable information that is present in the DICOM header [28]. In addition, other potentially identifiable information can be removed through various approaches including defacing of imaging data, anonymization of clinical reports, and removal of patient health information imprinted into the image itself [62–64]. Despite the efforts in automation, there is still the possibility of information leakage and there is still the need for laborious manual audits. For example, there can be identity leakage from accessories such as necklaces and wristbands [28]. In cases where there are no automated methods for patient de-identification, manual auditing and anonymization can be prohibitively expensive especially for large-scale datasets. Additionally, depending on the institution and their IRB, patient consent may be needed to share the data which adds an additional barrier [65]. Furthermore, data may be regarded as a valuable commodity and institutions may simply prefer not to share patient data with external groups due to the interests of the organization. Lastly, there may be a high cost of data storage, especially given the increasing utilization of high-resolution imaging with multiple modalities [66].

One alternative approach to sharing data is to train deep learning models via a distributed learning approach. Under this paradigm, the deep learning model weights, updates, or intermediate outputs are shared instead of the patient data. This would alleviate the need for full data anonymization and would eliminate the need for a secure central database. With this approach, each institution installs software that allows them to connect to other institutions for collaborative training. Techniques such as cyclical weight transfer, federated learning, and split learning have shown the potential to achieve performance comparable to sharing patient data [67–72]. Recently, proof-of-concept studies have demonstrated the utility of federated learning for brain tumor segmentation. [73, 74].

14.7 DATA QUALITY ISSUES

Data quality issues are detrimental to all medical imaging applications. These issues can include motion artifact, reconstruction artifact, low signal, noise, out-of-focus imaging, low-resolution, operator error, technique limitations, and physical artifacts [75–78]. Additionally, conventional imaging formats that are not or historically have not been rendered digitally, such as pathology slides or X-rays, can result in batch differences that the researcher must address [79]. The use of an imaging standard, commonly called a "phantom", can help differentiate issues in image quality stemming from the equipment and protocol from those arising from the subject matter. Registration to a template of fixed anatomy after an image has been taken can help identify deformations such as rotation, blurring, or shearing.

As data is curated at scale, there will inevitably be low-quality images within the dataset. Just as a radiation oncologist cannot perform treatment planning on a severely motion-corrupted image, effective deep learning algorithms cannot be trained using low-quality images. Thus, it is of vital importance to be able to identify, and if possible correct, issues with low quality samples in the dataset. Identification of such samples can be done via out-of-distribution detection methods, which aim at selecting samples that by some metric are classified as outliers [80]. Another solution is to utilize novel deep learning uncertainty approaches, which aim at flagging samples that the algorithm is not confident in [81–85].

After identification, correction should be applied when possible. For example, low resolution imaging/annotations should be appropriately resampled via spline interpolation (ensuring that annotations are resampled only using a spline of order 0). More modern approaches to this problem involve domain-specific content-aware resampling (e.g. neural network trained specifically to upsample low-resolution brain images to a higher resolution) [86]. Such methods come with the same caveats of model brittleness and lack of generalizability that have been mentioned previously, and so should be used with caution. Other algorithms are capable of removing burned-in text, such as that from an ultrasound [64].

Overall, numerous algorithms have been developed to correct for artifacts or improve the resolution of images and it is the duty of the researcher to identify which of these algorithms or methods will work best on their data [87, 88]. If all else fails, then manual inspection of the imaging may be needed to remove them from the dataset.

14.8 DATA ANNOTATIONS

One important part of training a supervised deep learning algorithm is the need for ground truth labels. While the preferred ground truth would be manual expert annotations, these can be difficult to acquire at scale. Specifically, labeling should be performed by trained clinicians with domain expertise and experience (which can include sub-specialty training and years of specialty practice). This labeling can be time-consuming and expensive to utilize. Unsurprisingly, studies have shown that annotators with more domain experience label more accurately than those with less experience [89–91]. A major challenge to manual expert annotation is the inherent human variability [36–38]. Even under highly controlled settings with well-defined annotation criteria, there will still be variable distributions of class frequencies across users, with some experts being more conservative and others being more liberal in their annotations (an observation known as inter-rater variability) [28]. Furthermore, certain annotators may even exhibit poor self-consistency (an observation known as intra-rater variability). Both inter-rater and intra-rater variability can weaken the ground truth labels, negatively affecting training due to the added noise. Additionally, these forms of variability can affect the generalizability of the model, since the "ground truth" in the external validation set may be produced differently than the "ground truth" for training.

Another approach is to use natural language processing algorithms to extract labels from clinical reports. Studies have shown that natural language processing approaches allow for accurate and high throughput extraction of labels from unstructured narrative reports [92–95]. Another alternative for high-throughput annotation is through citizen science and crowd-sourcing [96–99]. By decreasing the annotation burden on any individual, this approach is scalable. To ensure high quality annotations, consensus and verification approaches can be utilized [100].

As an alternative to requiring high quality annotations for the entire dataset, using methods under the umbrella of weakly supervised learning may also be considered [101]. Weakly supervised learning reduces the annotation burden by combining information learning from gold-standard labels with that of unlabeled or weakly labeled information during the training process [101]. There are three major types of weakly supervised learning: incomplete, inexact, and inaccurate. Incomplete weakly supervised learning is a scenario in which only a small subset of the entire dataset is labeled [101]. An example of this is active learning, where the algorithm suggests which images, if labeled, would be most informative to training the neural network. Ideally, this would identify highly difficult or representative cases, either of which would help the algorithm learn [102]. Another example is semi-supervised learning, where the algorithm exploits unlabeled data by learning from the small subset of labeled data with the assumption that both the labeled and unlabeled data come from the same distribution [101, 103]. The second type of weakly supervised learning is inexact supervision, in which a coarse label is used for a more granular task. For example, image-level annotations can be utilized in a weakly supervised algorithm to produce pixel-level predictions [9, 104]. The third type of weakly supervised learning is inaccurate supervision in which there are non-negligible errors in the labels [105]. For example, labels that are automatically extracted from clinical reports using imperfect algorithms can still be used to train high performing algorithms [106, 107].

14.9   DATA CURATION VIA COMPETITIONS

Competitions can also be an effective means of compiling multi-institutional and multi-national datasets. In this approach, multiple institutions work together to collect, prepare, anonymize, pool, and annotate the dataset [96, 108–114]. This is part of a broader effort to incorporate citizen science for annotation of experimental datasets [99, 115, 116]. There are several advantages to creating a dataset from a competition approach. First, the resulting dataset is substantially larger than a dataset that can be curated at any single institution. Additionally, it is efficient because similar data curation workflows can be used across all institutions. Also, the final dataset is diverse in terms of patient populations and acquisition settings, which allows for robust training of algorithms and evaluation of the generalizability of algorithms across these dataset differences. Furthermore, a competition framework allows for fair and direct comparison of the performance of different algorithms. Lastly, competitions facilitate collaboration via open datasets and open-source code with the shared goal of integrating tools into the clinical workflow and improving patient care. This is catalyzed by social media, online forums, blogs, and preprint servers, all with the culture of sharing insight and experience. Notably, there has been cross-pollination of participants from a variety of backgrounds, including clinicians, trainees, computer science, engineering, and data science [28]. Some recent examples of datasets curated using the competition framework include the RSNA Brain CT Hemorrhage [108], Pediatric Bone Age [109], and Pneumonia Detection Challenges [96] , and the 2017 AAPM Thoracic Auto-segmentation Challenge used throughout this book.

14.10   BIAS AND CURATION OF FAIR DATA

While it is well documented that minorities are disproportionately underrepresented in clinical trials and population health profiling studies [117, 118], these inequities also extend to medical AI. Since AI models learn the characteristics of training data, if the data provided are not equitable, they will not generalize well to unseen or minority classes. A study of published AI systems built on publicly available X-ray data found worse performance on minority genders when the model was not trained with a minimum gender balance [119]. When designed fairly, AI systems can mitigate inherent bias for race, age, gender, sexual orientation, and socioeconomic status. Proposed methods of bias mitigation include minimum quota of minority populations, subsampling of these data, careful consideration of label definitions of "normal" and "abnormal", and inclusion of minority populations in AI systems design [120]. Another source of bias mitigation is the use of an external validation cohort, as data gathered from a single institution represent a sample of the patient population served and are therefore subject to geographic bias which can span demographic stratifications. Finally, thorough ethical review during problem formulation, model development, and system deployment can increase awareness of bias even when it cannot be corrected [121] (Figure 14.5).

14.11   OVERVIEW OF DATA CURATION PROCESS

As an overview, there are several key steps in the data curation process. The first step is selection of pertinent data that is both diverse and captures the most recent clinical standards. Importantly, effort should be made to avoid biases from factors such as race, age, gender, sexual orientation, and socioeconomic status. The second step is obtaining appropriate approval to share patient data and ensuring patient privacy via anonymization. Alternatively, a distributed learning approach may be considered in scenarios where sharing patient data is difficult. The third step is handling data quality issues via correction of artifacts and removal of low-quality images. The last step is data annotation, either delineated manually from experts, extracted from clinical reports, or crowd-sourced from competitions. Alternatively, weakly supervised approaches can be considered in scenarios where labels are incomplete, inexact, or inaccurate.

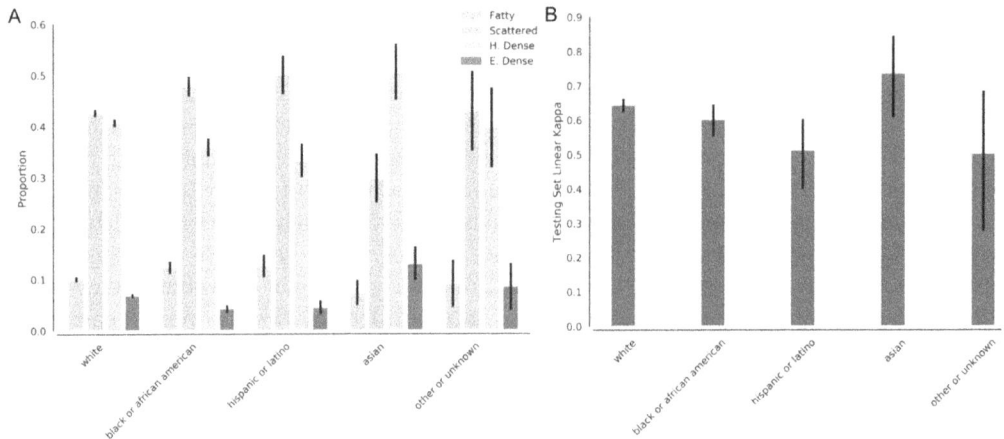

FIGURE 14.5 (A) Distribution of mammographic breast density varies by race showing that different races have different attributes for the digital mammograms in the Digital Mammographic Imaging Screening Trial (DMIST) [122]. (B) Testing set performance of a deep learning model for mammographic breast density trained on DMIST, demonstrating that performance can be different across different races. Bias in deep learning models can occur when populations are underrepresented in the training set.

14.12 CONCLUSION

In summary, deep learning has changed the way researchers approach medical imaging analysis. Deep learning has brought on remarkable performance, on the level of human experts, but also requires large, high quality, and diverse datasets. The curation of such datasets is challenging, and factors such as the complexity of medical imaging data, patient privacy, data quality issues, and annotation must be considered. Automated anonymization methods and distributed learning can serve as two different approaches to protect patient privacy. Automated methods can also be used to flag and correct for data quality issues. Collection of high-quality annotations can be time-consuming and expensive but has been partially addressed by the advent of natural language processing and weakly supervised learning algorithms. Lastly, competitions are a promising framework to facilitate collaboration in the construction of large, multi-institutional datasets. When beginning a medical imaging project, the clinician or researcher should bear in mind that the data corpus collected determines the quality and confidence of the results. In radiation oncology, which is an imaging-driven discipline, amassing the data may not be rate-limiting, but care should still be given to data curation techniques in order to train fair, highly performant, and generalizable deep-learning algorithms.

REFERENCES

1. Abadi M, Barham P, Chen J, et al. TensorFlow: a system for large-scale machine learning this paper is included in the proceedings of the TensorFlow: a system for large-scale machine learning. *Proc 12th USENIX Conf Oper Syst Des Implement.* 2016.
2. Russakovsky O, Deng J, Su H, et al. ImageNet large scale visual recognition challenge. *Int J Comput Vis.* 2015. https://doi.org/10.1007/s11263-015-0816-y
3. Paszke A, Gross S, Massa F, et al. *PyTorch: An Imperative Style, High-Performance Deep Learning Library.* 2019.
4. Krizhevsky A, Sutskever I, Hinton GE. ImageNet classification with deep convolutional neural networks. *Adv Neural Inf Process Syst.* 2012:1–9. http://dx.doi.org/10.1016/j.protcy.2014.09.007
5. Hinton G, Deng L, Yu D, et al. Deep neural networks for acoustic modeling in speech recognition: the shared views of four research groups. *IEEE Signal Process Mag.* 2012;29(6):82–97. https://doi.org/10.1109/MSP.2012.2205597

6. Collobert R, Weston J. A unified architecture for natural language processing: deep neural networks with multitask learning. *Proc 25th Int Conf Mach Learn*. 2008:160–167. https://doi.org/10.1145/1390156.1390177

7. Esteva A, Kuprel B, Novoa RA, et al. Dermatologist-level classification of skin cancer with deep neural networks. *Nature*. 2017;542(7639):115–118. https://doi.org/10.1038/nature21056

8. Brown JM, Campbell JP, Beers A, et al. Automated diagnosis of plus disease in retinopathy of prematurity using deep convolutional neural networks. *JAMA Ophthalmol*. May 2018. https://doi.org/10.1001/jamaophthalmol.2018.1934

9. Campanella G, Hanna MG, Geneslaw L, et al. Clinical-grade computational pathology using weakly supervised deep learning on whole slide images. *Nat Med*. 2019. https://doi.org/10.1038/s41591-019-0508-1

10. Ghorbani A, Ouyang D, Abid A, et al. Deep learning interpretation of echocardiograms. *NPJ Digit Med*. 2020. https://doi.org/10.1038/s41746-019-0216-8

11. Chang K, Bai HX, Zhou H, et al. Residual convolutional neural network for the determination of IDH status in low- and high-grade gliomas from mr imaging. *Clin Cancer Res*. 2018;24(5):1073–1081. https://doi.org/10.1158/1078-0432.CCR-17-2236

12. Li MD, Arun NT, Gidwani M, et al. Automated assessment of COVID-19 pulmonary disease severity on chest radiographs using convolutional siamese neural networks. *medRxiv*. May 2020. https://doi.org/10.1101/2020.05.20.20108159

13. Chen L, Bentley P, Rueckert D. Fully automatic acute ischemic lesion segmentation in DWI using convolutional neural networks. *NeuroImage Clin*. 2017;15:633–643. https://doi.org/10.1016/j.nicl.2017.06.016

14. Chang K, Beers AL, Bai HX, et al. Automatic assessment of glioma burden: a deep learning algorithm for fully automated volumetric and bidimensional measurement. *Neuro Oncol*. 2019. https://doi.org/10.1093/neuonc/noz106

15. Beers A, Chang K, Brown J, Gerstner E, Rosen B, Kalpathy-Cramer J. Sequential neural networks for biologically-informed glioma segmentation. In: Angelini ED, Landman BA, eds. *Medical Imaging 2018: Image Processing*. Vol 10574. SPIE; 2018:108. https://doi.org/10.1117/12.2293941

16. Elguindi S, Zelefsky MJ, Jiang J, et al. Deep learning-based auto-segmentation of targets and organs-at-risk for magnetic resonance imaging only planning of prostate radiotherapy. *Phys Imaging Radiat Oncol*. 2019. https://doi.org/10.1016/j.phro.2019.11.006

17. Ashamalla H, Rafla S, Parikh K, et al. The contribution of integrated PET/CT to the evolving definition of treatment volumes in radiation treatment planning in lung cancer. *Int J Radiat Oncol Biol Phys*. 2005;63(4):1016–1023. https://doi.org/10.1016/j.ijrobp.2005.04.021

18. Bakas S, Reyes M, Jakab A, et al. Identifying the best machine learning algorithms for brain tumor segmentation, progression assessment, and overall survival prediction in the BRATS challenge. November 2018. http://arxiv.org/abs/1811.02629. Accessed December 15, 2019.

19. Winzeck S, Hakim A, McKinley R, et al. ISLES 2016 and 2017-benchmarking ischemic stroke lesion outcome prediction based on multispectral MRI. *Front Neurol*. 2018. https://doi.org/10.3389/fneur.2018.00679

20. Simpson AL, Antonelli M, Bakas S, et al. A large annotated medical image dataset for the development and evaluation of segmentation algorithms. February 2019. http://arxiv.org/abs/1902.09063. Accessed May 20, 2020.

21. Menze BH, Jakab A, Bauer S, et al. The multimodal Brain Tumor Image Segmentation Benchmark (BRATS). *IEEE Trans Med Imaging*. 2015;34(10):1993–2024. https://doi.org/10.1109/TMI.2014.2377694

22. Ronneberger O, Fischer P, Brox T. U-Net: convolutional networks for biomedical image segmentation. *Med Image Comput Comput Interv – MICCAI 2015*. May 2015:234–241. https://doi.org/10.1007/978-3-319-24574-4_28

23. Kamnitsas K, Bai W, Ferrante E, et al. Ensembles of multiple models and architectures for robust brain tumour segmentation. In: *Lecture Notes in Computer Science (Including Subseries Lecture Notes in Artificial Intelligence and Lecture Notes in Bioinformatics)*. 2018. https://doi.org/10.1007/978-3-319-75238-9_38

24. Isensee F, Petersen J, Klein A, et al. nnU-Net: self-adapting framework for U-Net-based medical image segmentation. In: *Informatik Aktuell*. 2019. https://doi.org/10.1007/978-3-658-25326-4_7

25. Kamnitsas K, Ledig C, Newcombe VFJ, et al. Efficient multi-scale 3D CNN with fully connected CRF for accurate brain lesion segmentation. *Med Image Anal*. 2017;36:61–78. https://doi.org/10.1016/j.media.2016.10.004

26. Beers A, Brown J, Chang K, et al. DeepNeuro: an open-source deep learning toolbox for neuroimaging. August 2018. http://arxiv.org/abs/1808.04589. Accessed August 30, 2018.

27. Gibson E, Li W, Sudre C, et al. NiftyNet: a deep-learning platform for medical imaging. *Comput Methods Programs Biomed.* 2018;158:113–122. https://doi.org/10.1016/j.cmpb.2018.01.025

28. Prevedello LM, Halabi SS, Shih G, et al. Challenges related to artificial intelligence research in medical imaging and the importance of image analysis competitions. *Radiol Artif Intell.* 2019. https://doi.org/10.1148/ryai.2019180031

29. Li MD, Chang K, Bearce B, et al. Siamese neural networks for continuous disease severity evaluation and change detection in medical imaging. *NPJ Digit Med.* 2020;3(1):48. https://doi.org/10.1038/s41746-020-0255-1

30. Zech JR, Badgeley MA, Liu M, Costa AB, Titano JJ, Oermann EK. Variable generalization performance of a deep learning model to detect pneumonia in chest radiographs: a cross-sectional study. *PLoS Med.* 2018. https://doi.org/10.1371/journal.pmed.1002683

31. Albadawy EA, Saha A, Mazurowski MA. Deep learning for segmentation of brain tumors: impact of cross-institutional training and testing: impact. *Med Phys.* 2018. https://doi.org/10.1002/mp.12752

32. Mårtensson G, Ferreira D, Granberg T, et al. The reliability of a deep learning model in clinical out-of-distribution MRI data: a multicohort study. *Med Image Anal.* May 2020:101714. https://doi.org/10.1016/j.media.2020.101714

33. Chang K, Beers AL, Brink L, et al. Multi-institutional assessment and crowdsourcing evaluation of deep learning for automated classification of breast density. *J Am Coll Radiol.* 2020;17(12):1653–1662.

34. Kim DW, Jang HY, Kim KW, Shin Y, Park SH. Design characteristics of studies reporting the performance of artificial intelligence algorithms for diagnostic analysis of medical images: results from recently published papers. *Korean J Radiol.* 2019;20(3):405–410. https://doi.org/10.3348/kjr.2019.0025

35. Keavey E, Phelan N, O'Connell AM, et al. Comparison of the clinical performance of three digital mammography systems in a breast cancer screening programme. *Br J Radiol.* 2012. https://doi.org/10.1259/bjr/29747759

36. Sprague BL, Conant EF, Onega T, et al. Variation in mammographic breast density assessments among radiologists in clinical practice: a multicenter observational study. *Ann Intern Med.* 2016. https://doi.org/10.7326/M15-2934

37. Campbell JP, Kalpathy-Cramer J, Erdogmus D, et al. Plus disease in retinopathy of prematurity: a continuous spectrum of vascular abnormality as a basis of diagnostic variability. *Ophthalmology.* 2016;123(11):2338–2344. https://doi.org/10.1016/j.ophtha.2016.07.026

38. Komuta K, Batts K, Jessurun J, et al. Interobserver variability in the pathological assessment of malignant colorectal polyps. *Br J Surg.* 2004;91(11):1479–1484. https://doi.org/10.1002/bjs.4588

39. Maguolo G, Nanni L. A critic evaluation of methods for COVID-19 automatic detection from X-ray images. April 2020. http://arxiv.org/abs/2004.12823. Accessed May 31, 2020.

40. Lin TY, Maire M, Belongie S, et al. Microsoft COCO: common objects in context. In: *Lecture Notes in Computer Science (Including Subseries Lecture Notes in Artificial Intelligence and Lecture Notes in Bioinformatics).* 2014. https://doi.org/10.1007/978-3-319-10602-1_48

41. Mahajan D, Girshick R, Ramanathan V, et al. Exploring the limits of weakly supervised pretraining. In: *Lecture Notes in Computer Science (Including Subseries Lecture Notes in Artificial Intelligence and Lecture Notes in Bioinformatics).* 2018. https://doi.org/10.1007/978-3-030-01216-8_12

42. Irvin J, Rajpurkar P, Ko M, et al. CheXpert: a large chest radiograph dataset with uncertainty labels and expert comparison. January 2019. http://arxiv.org/abs/1901.07031. Accessed October 30, 2019.

43. Johnson AEW, Pollard TJ, Berkowitz SJ, et al. MIMIC-CXR, a de-identified publicly available database of chest radiographs with free-text reports. *Sci Data.* 2019. https://doi.org/10.1038/s41597-019-0322-0

44. Gulshan V, Peng L, Coram M, et al. Development and validation of a deep learning algorithm for detection of diabetic retinopathy in retinal fundus photographs. *JAMA.* 2016;304(6):649–656. https://doi.org/10.1001/jama.2016.17216

45. Dunnmon JA, Yi D, Langlotz CP, Ré C, Rubin DL, Lungren MP. Assessment of convolutional neural networks for automated classification of chest radiographs. *Radiology.* 2019. https://doi.org/10.1148/radiol.2018181422

46. Swensen SJ. CT screening for lung cancer. *AJR Am J Roentgenol.* 2002;179(4):833–836. https://doi.org/10.2214/ajr.179.4.1790833

47. Parakh A, Lee H, Lee JH, Eisner BH, Sahani D V., Do S. Urinary stone detection on CT images using deep convolutional neural networks: evaluation of model performance and generalization. *Radiol Artif Intell.* 2019. https://doi.org/10.1148/ryai.2019180066

48. Mukherjee P, Zhou M, Lee E, et al. A shallow convolutional neural network predicts prognosis of lung cancer patients in multi-institutional computed tomography image datasets. *Nat Mach Intell.* 2020;2(5):274–282. https://doi.org/10.1038/s42256-020-0173-6

49. Safdar M, Kobaisi S, Zahra F. A comparative analysis of data augmentation approaches for Magnetic Resonance Imaging (MRI) scan images of brain tumor. *Acta Informatica Medica.* 2020;28(1):29. https://doi.org/10.5455/aim.2020.28.29-36

50. Hussain Z, Gimenez F, Yi D, Rubin D. Differential data augmentation techniques for medical imaging classification tasks. In: *AMIA Annu Symp Proceedings AMIA Symp.* 2017.

51. Kang G, Dong X, Zheng L, Yang Y. PatchShuffle regularization. July 2017. http://arxiv.org/abs/1707.07103. Accessed May 24, 2020.

52. Zhong Z, Zheng L, Kang G, Li S, Yang Y. Random erasing data augmentation. August 2017. http://arxiv.org/abs/1708.04896. Accessed May 24, 2020.

53. Inoue H. Data augmentation by pairing samples for images classification. January 2018. http://arxiv.org/abs/1801.02929. Accessed May 24, 2020.

54. Zhang H, Cisse M, Dauphin YN, Lopez-Paz D. MixUp: beyond empirical risk minimization. In: *6th International Conference on Learning Representations, ICLR 2018 – Conference Track Proceedings.* 2018.

55. Cubuk ED, Zoph B, Mane D, Vasudevan V, Le QV. Autoaugment: learning augmentation strategies from data. In: *Proceedings of the IEEE Computer Society Conference on Computer Vision and Pattern Recognition.* 2019. https://doi.org/10.1109/CVPR.2019.00020

56. Beers A, Brown J, Chang K, et al. High-resolution medical image synthesis using progressively grown generative adversarial networks. May 2018. http://arxiv.org/abs/1805.03144. Accessed May 23, 2018.

57. Moradi M, Madani A, Karargyris A, Syeda-Mahmood TF. Chest x-ray generation and data augmentation for cardiovascular abnormality classification. In: Angelini ED, Landman BA, eds. *Medical Imaging 2018: Image Processing.* Vol 10574. SPIE; 2018:57. https://doi.org/10.1117/12.2293971

58. Jackson PT, Atapour-Abarghouei A, Bonner S, Breckon T, Obara B. Style augmentation: data augmentation via style randomization. September 2018. http://arxiv.org/abs/1809.05375. Accessed May 24, 2020.

59. Rocher L, Hendrickx JM, de Montjoye Y-A. Estimating the success of re-identifications in incomplete datasets using generative models. *Nat Commun.* 2019. https://doi.org/10.1038/s41467-019-10933-3

60. Schwarz CG, Kremers WK, Therneau TM, et al. Identification of anonymous MRI research participants with face-recognition software. *N Engl J Med.* 2019;381(17):1684–1686. https://doi.org/10.1056/NEJMc1908881

61. Mazura JC, Juluru K, Chen JJ, Morgan TA, John M, Siegel EL. Facial recognition software success rates for the identification of 3D surface reconstructed facial images: implications for patient privacy and security. *J Digit Imaging.* 2012;25(3):347–351. https://doi.org/10.1007/s10278-011-9429-3

62. Norgeot B, Muenzen K, Peterson TA, et al. Protected Health Information filter (Philter): accurately and securely de-identifying free-text clinical notes. *NPJ Digit Med.* 2020;3(1):1–8. https://doi.org/10.1038/s41746-020-0258-y

63. Collins SA, Wu J, Bai HX. Facial de-identification of head CT scans. *Radiology.* April 2020:192617. https://doi.org/10.1148/radiol.2020192617

64. Monteiro E, Costa C, Oliveira JL. A de-identification pipeline for ultrasound medical images in DICOM format. *J Med Syst.* 2017. https://doi.org/10.1007/s10916-017-0736-1

65. Larson DB, Magnus DC, Lungren MP, Shah NH, Langlotz CP. Ethics of using and sharing clinical imaging data for artificial intelligence: a proposed framework. *Radiology.* 2020;295(3):192536. https://doi.org/10.1148/radiol.2020192536

66. Smith-Bindman R, Kwan ML, Marlow EC, et al. Trends in use of medical imaging in US health care systems and in Ontario, Canada, 2000–2016. *JAMA.* 2019. https://doi.org/10.1001/jama.2019.11456

67. Chang K, Balachandar N, Lam C, et al. Distributed deep learning networks among institutions for medical imaging. *J Am Med Inform Assoc* March 2018. https://doi.org/10.1093/jamia/ocy017

68. Balachandar N, Chang K, Kalpathy-Cramer J, Rubin DL. Accounting for data variability in multi-institutional distributed deep learning for medical imaging. *J Am Med Inform Assoc.* 2020. https://doi.org/10.1093/jamia/ocaa017

69. Shokri R, Shmatikov V. Privacy-preserving deep learning. In: *2015 53rd Annual Allerton Conference on Communication, Control, and Computing, Allerton 2015.* 2016. https://doi.org/10.1109/ALLERTON.2015.7447103

70. Brendan McMahan H, Moore E, Ramage D, Hampson S, Agüera y Arcas B. Communication-efficient learning of deep networks from decentralized data. In: *Proceedings of the 20th International Conference on Artificial Intelligence and Statistics, AISTATS 2017.* 2017.

71. Gupta O, Raskar R. Distributed learning of deep neural network over multiple agents. *J Netw Comput Appl.* 2018. https://doi.org/10.1016/j.jnca.2018.05.003

72. Vepakomma P, Gupta O, Swedish T, Raskar R. Split learning for health: distributed deep learning without sharing raw patient data. December 2018. http://arxiv.org/abs/1812.00564. Accessed July 22, 2019.

73. Sheller MJ, Reina GA, Edwards B, Martin J, Bakas S. Multi-institutional deep learning modeling without sharing patient data: a feasibility study on brain tumor segmentation. In: *Lecture Notes in Computer Science (Including Subseries Lecture Notes in Artificial Intelligence and Lecture Notes in Bioinformatics).* Vol 11383 LNCS. Springer Verlag; 2019:92–104. https://doi.org/10.1007/978-3-030-11723-8_9

74. Li W, Milletarì F, Xu D, et al. Privacy-preserving federated brain tumour segmentation. *Lect Notes Comput Sci (including Subser Lect Notes Artif Intell Lect Notes Bioinformatics).* 2019;11861 LNCS:133–141. http://arxiv.org/abs/1910.00962. Accessed March 19, 2020.

75. Bartlett E, DeLorenzo C, Parsey R, Huang C. Noise contamination from PET blood sampling pump: effects on structural MRI image quality in simultaneous PET/MR studies: effects. *Med Phys.* 2018;45(2):678–686. https://doi.org/10.1002/mp.12715

76. Briggs R, Bailey JE, Eddy C, Sun I. A methodologic issue for ophthalmic telemedicine: image quality and its effect on diagnostic accuracy and confidence. *J Am Optom Assoc.* 1998;69(9):601–605.

77. Patel T, Peppard H, Williams MB. Effects on image quality of a 2D antiscatter grid in x-ray digital breast tomosynthesis: initial experience using the dual modality (x-ray and molecular) breast tomosynthesis scanner. *Med Phys.* 2016;43(4). https://doi.org/10.1118/1.4943632

78. Smet MH, Breysem L, Mussen E, Bosmans H, Marshall NW, Cockmartin L. Visual grading analysis of digital neonatal chest phantom X-ray images: impact of detector type, dose and image processing on image quality. *Eur Radiol.* 2018;28(7):2951–2959. https://doi.org/10.1007/s00330-017-5301-2

79. Goga R, Chandler NP, Love RM. Clarity and diagnostic quality of digitized conventional intraoral radiographs. *Dento Maxillo Facial Radiol.* 2004. https://doi.org/10.1259/dmfr/13010370

80. Lakshminarayanan B, Pritzel A, Blundell C. Simple and scalable predictive uncertainty estimation using deep ensembles. In: *Advances in Neural Information Processing Systems.* 2017.

81. Coyner AS, Swan R, Campbell JP, et al. Automated fundus image quality assessment in retinopathy of prematurity using deep convolutional neural networks. *Ophthalmol Retin.* 2019;3(5):444–450. https://doi.org/10.1016/j.oret.2019.01.015

82. Oksuz I, Ruijsink B, Puyol-Antón E, et al. Automatic CNN-based detection of cardiac MR motion artefacts using k-space data augmentation and curriculum learning. *Med Image Anal.* 2019;55:136–147. https://doi.org/10.1016/j.media.2019.04.009

83. Hoebel K, Chang K, Patel J, Singh P, Kalpathy-Cramer J. Give me (un)certainty – an exploration of parameters that affect segmentation uncertainty. November 2019. http://arxiv.org/abs/1911.06357. Accessed December 18, 2019.

84. Hoebel K, Andrearczyk V, Beers AL, et al. An exploration of uncertainty information for segmentation quality assessment. In: Landman BA, Išgum I, eds. *Medical Imaging 2020: Image Processing.* Vol 11313. SPIE; 2020:55. https://doi.org/10.1117/12.2548722

85. Piccini D, Demesmaeker R, Heerfordt J, et al. Deep learning to automate reference-free image quality assessment of whole-heart MR images. *Radiol Artif Intell.* 2020;2(3):e190123. https://doi.org/10.1148/ryai.2020190123

86. Sert E, Özyurt F, Doğantekin A. A new approach for brain tumor diagnosis system: single image super resolution based maximum fuzzy entropy segmentation and convolutional neural network. *Med Hypotheses.* 2019;133:109413. https://doi.org/10.1016/j.mehy.2019.109413

87. Haskell MW, Cauley SF, Bilgic B, et al. Network Accelerated Motion Estimation and Reduction (NAMER): convolutional neural network guided retrospective motion correction using a separable motion model. *Magn Reson Med.* 2019;82(4):1452–1461. https://doi.org/10.1002/mrm.27771

88. Masutani EM, Bahrami N, Hsiao A. Deep learning single-frame and multiframe super-resolution for cardiac MRI. *Radiology.* April 2020:192173. https://doi.org/10.1148/radiol.2020192173

89. Rauschecker AM, Rudie JD, Xie L, et al. Artificial intelligence system approaching neuroradiologist-level differential diagnosis accuracy at brain MRI. *Radiology.* 2020;295(3):626–637. https://doi.org/10.1148/radiol.2020190283

90. Zhang K. Clinically applicable AI system for accurate diagnosis, quantitative measurements and prognosis of COVID-19 pneumonia using computed tomography. *Cell.* 2020. https://doi.org/10.1016/j.chom.2020.04.004

91. De Fauw J, Ledsam JR, Romera-Paredes B, et al. Clinically applicable deep learning for diagnosis and referral in retinal disease. *Nat Med.* 2018;24(9):1342–1350. https://doi.org/10.1038/s41591-018-0107-6

92. Dreyer KJ, Kalra MK, Maher MM, et al. Application of recently developed computer algorithm for automatic classification of unstructured radiology reports: validation study. *Radiology.* 2005;234(2):323–329. https://doi.org/10.1148/radiol.2341040049

93. Hripcsak G, Austin JHM, Alderson PO, Friedman C. Use of natural language processing to translate clinical information from a database of 889,921 chest radiographic reports. *Radiology.* 2002;224(1):157–163. https://doi.org/10.1148/radiol.2241011118

94. Zech J, Pain M, Titano J, et al. Natural language-based machine learning models for the annotation of clinical radiology reports. *Radiology.* 2018. https://doi.org/10.1148/radiol.2018171093

95. Smit A, Jain S, Rajpurkar P, Pareek A, Ng AY, Lungren MP. CheXbert: combining automatic labelers and expert annotations for accurate radiology report labeling using BERT. April 2020. http://arxiv.org/abs/2004.09167. Accessed May 24, 2020.

96. Shih G, Wu CC, Halabi SS, et al. Augmenting the national institutes of health chest radiograph dataset with expert annotations of possible pneumonia. *Radiol Artif Intell.* 2019;1(1):e180041. https://doi.org/10.1148/ryai.2019180041

97. Filice RW, Stein A, Wu CC, et al. Crowdsourcing pneumothorax annotations using machine learning annotations on the NIH chest X-ray dataset. *J Digit Imaging.* 2019. https://doi.org/10.1007/s10278-019-00299-9

98. Irshad H, Montaser-Kouhsari L, Waltz G, et al. Crowdsourcing image annotation for nucleus detection and segmentation in computational pathology: evaluating experts, automated methods, and the crowd. In: *Pacific Symposium on Biocomputing.* 2015:294–305.

99. Candido dos Reis FJ, Lynn S, Ali HR, et al. Crowdsourcing the general public for large scale molecular pathology studies in cancer. *EBioMedicine.* 2015. https://doi.org/10.1016/j.ebiom.2015.05.009

100. Su H, Deng J, Fei-Fei L. Crowdsourcing annotations for visual object detection. In: *AAAI Workshop – Technical Report.* 2012.

101. Zhou ZH. A brief introduction to weakly supervised learning. *Natl Sci Rev.* 2018. https://doi.org/10.1093/nsr/nwx106

102. Yang L, Zhang Y, Chen J, Zhang S, Chen DZ. Suggestive annotation: a deep active learning framework for biomedical image segmentation. In: *Lecture Notes in Computer Science (Including Subseries Lecture Notes in Artificial Intelligence and Lecture Notes in Bioinformatics).* 2017. https://doi.org/10.1007/978-3-319-66179-7_46

103. Lecouat B, Chang K, Foo C-S, et al. Semi-supervised deep learning for abnormality classification in retinal images. December 2018. http://arxiv.org/abs/1812.07832. Accessed December 15, 2019.

104. Shen Y, Wu N, Phang J, et al. An interpretable classifier for high-resolution breast cancer screening images utilizing weakly supervised localization. February 2020. http://arxiv.org/abs/2002.07613. Accessed May 24, 2020.

105. Rolnick D, Veit A, Belongie S, Shavit N. Deep learning is robust to massive label noise. May 2017. http://arxiv.org/abs/1705.10694. Accessed June 22, 2019.

106. Ratner A, Bach SH, Ehrenberg H, Fries J, Wu S, Ré C. Snorkel: rapid training data creation with weak supervision. In: *Proceedings of the VLDB Endowment.* 2017. https://doi.org/10.14778/3157794.3157797

107. Wang X, Peng Y, Lu L, Lu Z, Bagheri M, Summers RM. ChestX-ray8: hospital-scale chest X-ray database and benchmarks on weakly-supervised classification and localization of common thorax diseases. In: *Proceedings – 30th IEEE Conference on Computer Vision and Pattern Recognition, CVPR 2017.* Vol 2017-Janua. 2017:3462–3471. https://doi.org/10.1109/CVPR.2017.369

108. Flanders AE, Prevedello LM, Shih G, et al. Construction of a machine learning dataset through collaboration: the RSNA 2019 brain CT hemorrhage challenge. *Radiol Artif Intell.* 2020;2(3):e190211. https://doi.org/10.1148/ryai.2020190211

109. Halabi SS, Prevedello LM, Kalpathy-Cramer J, et al. The RSNA pediatric bone age machine learning challenge. *Radiology.* 2019;290(2):498–503. https://doi.org/10.1148/radiol.2018180736

110. Cichonska A, Ravikumar B, Allaway RJ, et al. Crowdsourced mapping of unexplored target space of kinase inhibitors. *bioRxiv.* 2020. https://doi.org/10.1101/2019.12.31.891812

111. Ellrott K, Buchanan A, Creason A, et al. Reproducible biomedical benchmarking in the cloud: lessons from crowd-sourced data challenges. *Genome Biol.* 2019. https://doi.org/10.1186/s13059-019-1794-0

112. Allen GI, Amoroso N, Anghel C, et al. Crowdsourced estimation of cognitive decline and resilience in Alzheimer's disease. *Alzheimer's Dement.* 2016. https://doi.org/10.1016/j.jalz.2016.02.006

113. Saez-Rodriguez J, Costello JC, Friend SH, et al. Crowdsourcing biomedical research: leveraging communities as innovation engines. *Nat Rev Genet.* 2016. https://doi.org/10.1038/nrg.2016.69

114. Guinney J, Saez-Rodriguez J. Alternative models for sharing confidential biomedical data. *Nat Biotechnol.* 2018. https://doi.org/10.1038/nbt.4128

115. Kim JS, Greene MJ, Zlateski A, et al. Space-time wiring specificity supports direction selectivity in the retina. *Nature*. 2014. https://doi.org/10.1038/nature13240

116. Simpson R, Page KR, De Roure D. Zooniverse: observing the world's largest citizen science platform. In: *WWW 2014 Companion – Proceedings of the 23rd International Conference on World Wide Web*. 2014. https://doi.org/10.1145/2567948.2579215

117. Murthy VH, Krumholz HM, Gross CP. Participation in cancer clinical trials: race-, sex-, and age-based disparities. *JAMA*. 2004. https://doi.org/10.1001/jama.291.22.2720

118. Popejoy AB, Fullerton SM. Genomics is failing on diversity. *Nature*. 2016. https://doi.org/10.1038/538161a

119. Larrazabal AJ, Nieto N, Peterson V, Milone DH, Ferrante E. Gender imbalance in medical imaging datasets produces biased classifiers for computer-aided diagnosis. *Proc Natl Acad Sci U S A*. May 2020:201919012. https://doi.org/10.1073/pnas.1919012117

120. Crawford K, Dobbe R, Dryer T, et al. AI now 2019 report. *AI Now Inst*. 2019. https://ainowinstitute.org/AI_Now:2019_Report.html.

121. Wiens J, Saria S, Sendak M, et al. Do no harm: a roadmap for responsible machine learning for health care. *Nat Med*. 2019. https://doi.org/10.1038/s41591-019-0548-6

122. Pisano ED, Gatsonis C, Hendrick E, et al. Diagnostic performance of digital versus film mammography for breast-cancer screening. *N Engl J Med*. 2005;353(17):1773–1783. https://doi.org/10.1056/NEJMoa052911

15 On the Evaluation of Auto-Contouring in Radiotherapy

Mark J. Gooding

CONTENTS

15.1   INTRODUCTION

A first glance the evaluation of auto-contouring performance for radiotherapy seems straightforward: there is a wide range of quantitative measures that can be used to assess the correctness of an automatically generated contour against the ground truth. However, scratching the surface of this topic reveals numerous challenges that lead to a range of assessments being performed in practice. Even where essentially the same assessment has been performed the results may not be directly comparable as a result of implementation differences.

Table 15.1 reviews publications considering the assessment of auto-contouring in the context of radiotherapy. This is not intended as a comprehensive review, and it should be noted that auto-contouring is used in other contexts, such as neurology, and alternative assessments may have been performed. Therefore, Table 15.1 serves to review what has been applied in the domain of radiotherapy, not what could be applied, with a view to illustrating current practices in this field. Even at the high level of the information provided in the table, it can be observed that there is significant variation in the assessment of auto-contouring both in the approaches taken and in the scale of the evaluation performed. Nevertheless, there are commonalities between the assessments allowing broad categorization both by the purpose of the evaluation being performed and the type of evaluation being performed.

The purpose of the evaluation can be considered as falling into three classes: evaluation/demonstration of the benefits of a technical method being developed (Development), comparison of two or more methods to ascertain relative performance (Comparison), or the assessment of auto-contouring performance with the intent of clinical use (Commissioning). Some evaluations fall into more than one of these categories, for example it is common to show the benefits of a new method (Development) with respect to an existing benchmark method (Comparison). Although the purpose of evaluation may vary, there are common approaches to assessment used in each.

As noted in Gooding et al. [1] the types of evaluation can be grouped into quantitative, subjective, and clinical. While subjective and clinical assessments are often quantified in some way, e.g. time saving in minutes is a quantity, *quantitative assessment* can be defined as: *the calculation of the similarity or difference of a test contour with respect to a defined "ground truth"*. Notwithstanding the "ground truth" used for quantitative assessment may itself be a person's subjective opinion as to the correct contour for an organ or region, *subjective assessment* can be defined as *the evaluation made by an observer expressing their opinion of a contour's quality*. Finally, *clinical assessment* can be defined as *the evaluation or measurement of the impact the use of auto-contouring has on the clinical workflow*, although such clinical assessment may be both quantified and/or dependent on subjective opinion.

In this chapter, each of these categories of evaluation method will be explored in more detail. The strengths and limitations of each method will be considered, and implementational challenges and the corresponding impact will be discussed.

15.2   QUANTITATIVE EVALUATION

Quantitative evaluation is by far the most popular approach to auto-contouring assessment in published papers. A likely reason for this is that publications around the development of auto-contouring precede those related to clinical implementation. For those developing auto-contouring methods having a quantified measure of performance enables improvement in a system to be objectively monitored as research progresses. Thus, it could be expected that research focused on development will use quantitative evaluation, and subsequent research considering clinical implementation will derive its methodology from what has gone before in addition to any subjective or clinical evaluation.

TABLE 15.1

Summary of Papers Assessing Auto-Contouring in Radiotherapy. Papers are Categorized by the Purpose of Evaluation and the Type of Evaluation

Reference	Purpose of Evaluation	Type(s) of Evaluation	Number of Test Subjects	Summary of Evaluation Methods
Bondiau et al. 2005 [19]	Development	Quantitative	6	Measurement of volume, and sensitivity and specificity against average contour of seven experts. Comparison to inter-observer variability of these experts.
Levendag et al. 2008 [33]	Commissioning	Quantitative Clinical	10 (Leave-one-out)	Leave-one-out evaluation. 3D Dice similarity coefficient and 3D Average Distance comparison with left-out case. Measurement of editing time for clinical correction.
Commowick et al. 2008 [42]	Development	Subjective Quantitative	12	Visual inspection, sensitivity and specificity against manual contour.
Isambert et al. 2008 [5]	Commissioning	Quantitative	11	Relative volume difference, absolute centroid difference, 3D Dice similarity coefficient, and sensitivity and specificity against manual contour.
Gorthi et al. 2009 [43]	Development	Subjective Quantitative	10	Visual inspection by oncologist. Sensitivity and specificity, Dice similarity coefficient, and Hausdorff distance (3D) against manual contour.
Pekar et al. 2009 [15]	Comparison	Quantitative	7 + 8	Mean and median 2D Hausdorff distance, % of slices with 2D Hausdorff distance >3 mm, Mean & median 2D Dice similarity coefficient per slice, and Total Dice (3D) similarity coefficient against manual contour.
Reed et al. 2009 [36]	Development Commissioning	Clinical Quantitative	1 10	Editing time recorded for eight observers with comparison to de novo contouring time. 3D Average distance and 3D coefficient against single manual contour.
Sims et al. 2009 [44]	Commissioning	Quantitative	6 + 7	Relative volume, Sensitivity and specificity and 3D Dice similarity coefficient against manual contours from one institution and against the edited contour from a second institution.
Pekar et al. 2010 [2]	Comparison	Quantitative	7 + 8	Mean and median 2D Hausdorff distance, % of slices with 2D Hausdorff distance >3 mm, mean and median 2D Dice similarity coefficient per slice, and total Dice (3D) similarity coefficient against manual contour.
Ramus and Malandain 2010 [45]	Development	Quantitative	8	Evaluation performed using challenge in Pekar et al. [2].

(Continued)

TABLE 15.1 (CONTINUED)

Summary of Papers Assessing Auto-Contouring in Radiotherapy. Papers are Categorized by the Purpose of Evaluation and the Type of Evaluation

Reference	Purpose of Evaluation	Type(s) of Evaluation	Number of Test Subjects	Summary of Evaluation Methods
Stapleford et al. 2010 [6]	Commissioning	Quantitative Clinical	5	Absolute volume, 3D Dice similarity coefficient, 3D average distance and 3D Hausdorff distance, sensitivity and % false positive against average contour of five observers. Editing time compared to de novo contouring time.
Hwee et al. 2011 [27]	Commissioning	Quantitative Clinical Subjective	5	3D Dice similarity coefficient against average contour of five observers. Editing time compared to de novo contouring time. Visual assessment of acceptability and source of contours.
Teguh et al. 2011 [14]	Commissioning Comparison	Quantitative Clinical Subjective	10 (leave-one-out)	Leave-one-out evaluation. 3D Dice similarity coefficient and 3D average distance comparison with left-out patient. Measurement of editing time for clinical correction. Three-point scoring (Good, Moderate, Bad) of clinical contours. Four-point scoring (Poor, Major deviation, Minor deviation, Perfect) of auto-contours.
Granberg 2011 [16]	Commissioning	Quantitative Clinical Subjective	10	3D Dice similarity coefficient, 2D Hausdorff distance against manual contour. Volume. Editing time compared to de novo contouring time. Number of mouse-clicks to edit or draw from scratch. Visual inspection. Grading on three-point scale (not helpful, a little helpful, very helpful) of the contours.
Acosta et al. 2011 [25]	Development Comparison	Quantitative	24 (leave-one-out)	3D Dice similarity coefficient.
Young et al. 2011 [32]	Commissioning	Clinical Quantitative	10	Editing time compared to de novo contouring time. 3D Dice similarity coefficient between multiple observers.
Voet et al. 2011 [37]	Commissioning	Clinical Quantitative	9	Dose comparison between edited and unedited contours. 3D Dice similarity coefficient.
Chen et al. 2012 [46]	Development Comparison	Quantitative Subjective	20 (leave-one-out)	3D Dice similarity coefficient, 2D Hausdorff distance, % of slices with 2D Hausdorff distance >3 mm against manual contour. Visual inspection of regions of edit.

(Continued)

TABLE 15.1 (CONTINUED)
Summary of Papers Assessing Auto-Contouring in Radiotherapy. Papers are Categorized by the Purpose of Evaluation and the Type of Evaluation

Reference	Purpose of Evaluation	Type(s) of Evaluation	Number of Test Subjects	Summary of Evaluation Methods
La Macchia et al. 2012 [7]	Comparison	Clinical Quantitative	15	Editing time compared to de novo contouring time. Relative volume, 3D Dice similarity coefficient, sensitivity and specificity, and centroid displacement in each axis against manual contour.
Anders et al. 2012 [9]	Commissioning Comparison	Quantitative	9	3D Dice similarity coefficient and % overlap against manual contour.
Mattiucci et al. 2013 [47]	Commissioning	Quantitative Clinical	10	3D Dice similarity coefficient. Editing time compared to de novo contouring time.
Gambacorta et al. 2013 [48]	Commissioning	Clinical Quantitative	10	Editing time compared to de novo contouring time. 3D Dice similarity coefficient.
Yang et al. 2013 [10]	Commissioning	Clinical Quantitative	80	Editing time compared to de novo contouring time. 2D and 3D Dice similarity coefficients, 2D Hausdorff distance, mean 3D distance and correlation of DVH between edited and auto-contours.
Sjöberg el al. 2013 [49]	Commissioning Comparison	Clinical Quantitative	10+10	Editing time compared to de novo contouring time. 3D Dice similarity coefficient and volume.
Gooding et al. 2013a [50]	Commissioning	Subjective	8	Six observers. Four-point scale ("no useable contours, no time saving expected" to "most contours useful, significant timesaving expected").
Gooding et al. 2013b [29]	Commissioning	Subjective	18	Six observers. Four-point scale ("no useable contours, no time saving expected" to "most contours useful, significant timesaving expected").
Acosta et al. 2014 [51]	Development Comparison	Quantitative	19 (leave-one-out)	3D Dice similarity coefficient
Thomson et al. 2014 [52]	Commissioning	Clinical Quantitative	10	Five observers. Editing time compared to de novo contouring time. 3D Dice similarity coefficient, mean 3D distance, and 3D Hausdorff distance against manual contour.
Duc et al. 2015 [53]	Development Comparison	Quantitative Subjective Clinical	100 (leave-one-out) 15 for clinical	3D Dice similarity coefficient and 3D Hausdorff distance against manual contour. Three-point scale (clinically acceptable, reasonably acceptable, not acceptable). Editing time for "reasonably acceptable".

(Continued)

TABLE 15.1 (CONTINUED)

Summary of Papers Assessing Auto-Contouring in Radiotherapy. Papers are Categorized by the Purpose of Evaluation and the Type of Evaluation

Reference	Purpose of Evaluation	Type(s) of Evaluation	Number of Test Subjects	Summary of Evaluation Methods
Van Dijk-Peters et al. 2015 [54]	Commissioning	Quantitative Subjective Clinical	10	3D Dice similarity coefficient of auto-contours and four manual observers against manual reference. Three-point scale (no editing, minor editing, major editing). Editing time compared to manual contouring time.
Liu et al. 2016 [55]	Commissioning	Quantitative	6	3D mean surface distance and 3D Dice similarity coefficient against manual contours.
Delpon et al. 2016 [56]	Comparison	Quantitative	10	Volume ratio, 3D Dice similarity coefficient, and 3D 95% Hausdorff distance against manual ground truth.
Wardman et al. 2016 [57]	Development Commissioning	Quantitative	14 (leave-on-out)	3D Dice similarity coefficient, Mean surface distance, sensitivity index, inclusion index against manual delineation.
Dipasquale et al. 2016 [34]	Commissioning	Quantitative Clinical	27	3D Dice similarity coefficient, sensitivity, inclusiveness index, center of mass difference, percentage volume difference. Editing time compared to de novo contouring time.
Raudaschi et al. 2017 [3]	Comparison	Quantitative	10+5	3D Dice similarity coefficient, 3D 95% Hausdorff distance, 3D Hausdorff distance, and 3D mean surface distance against manual contour. DVH parameter difference between contour sets.
Fast et al. 2017 [58]	Development Comparison	Quantitative	6	3D Dice similarity coefficient, 3D Hausdorff distance, and 3D centroid difference against manual contour.
Ibragimov and Xing 2017 [59]	Development Comparison	Quantitative	50	Dice similarity coefficient against manual contour.
Men et al. 2017 [60]	Development Comparison	Quantitative	60	Dice similarity coefficient against manual contour.
Ibragimov et al. 2017 [61]	Development	Quantitative	n/a	Dice similarity coefficient against manual contour.
Men et al. 2017a [62]	Development Comparison	Quantitative	46	Dice similarity coefficient against manual contour.
Ciardo et al. 2017 [20]	Development Comparison	Quantitative	47	Centroid distance, Dice similarity coefficient, and mean surface distance against manual contour.
Yang et al. 2018 [4]	Comparison	Quantitative	12+12	3D Dice similarity coefficient, 3D 95% Hausdorff distance, and 3D mean surface distance against manual contour.

(Continued)

TABLE 15.1 (CONTINUED)

Summary of Papers Assessing Auto-Contouring in Radiotherapy. Papers are Categorized by the Purpose of Evaluation and the Type of Evaluation

Reference	Purpose of Evaluation	Type(s) of Evaluation	Number of Test Subjects	Summary of Evaluation Methods
Gooding et al. 2018 [28]	Commissioning Comparison	Subjective	60	Blinded questions on source of contour, acceptability (four-point scale), and contour preference.
Nikolov et al. 2018 [21]	Development	Quantitative	24	3D Dice similarity coefficient and 3D surface Dice against manual contour.
Kazemifar et al. 2018 [63]	Development	Quantitative	25	3D Dice similarity coefficient, 3D Hausdorff distance, 3D mean surface distance and positive predictive value against manual contour.
Astaraki et al. 2018 [64]	Development Comparison	Quantitative	20	3D Dice similarity coefficient, 3D Hausdorff distance, 3D mean surface distance, Percentage volume difference, and absolute volume difference against manual contour.
Kieselmann et al. 2018 [38]	Development Comparison	Quantitative Clinical	12 (leave-one-out)	3D Dice similarity coefficient, 3D 95% Hausdorff distance, and 3D mean surface distance against manual contour. Compared with inter-observer variability for three observers. Dose difference between plan on manual and plan on auto contours evaluated with manual contours.
Lustberg et al. 2018 [30]	Commissioning Comparison	Quantitative Subjective Clinical	20	3D Dice similarity coefficient and 3D Hausdorff distance against manual contour. Four-point scoring (no results useful, no time saving expected; some results useful, little time saving expected; many results useful, moderate time saving expected; most results useful, significant time saving expected). Editing time compared to de novo contouring time.
Tong et al. 2018 [65]	Development	Quantitative	10	3D Dice similarity coefficient, positive predictive value, sensitivity, 3D average surface distance, and 95% 3D Hausdorff distance against manual contour.
McCarroll et al. 2018 [11]	Commissioning	Subjective Quantitative	128 + 166	Grading on three-point scale (no edits, minor edits, major edits) by one observer. Subset of ten graded by five observers. 3D Dice similarity coefficient, 3D average distance, and 3D Hausdorff distance against manual contour. Same measures for second cohort against edited auto-contour.

(Continued)

TABLE 15.1 (CONTINUED)

Summary of Papers Assessing Auto-Contouring in Radiotherapy. Papers are Categorized by the Purpose of Evaluation and the Type of Evaluation

Reference	Purpose of Evaluation	Type(s) of Evaluation	Number of Test Subjects	Summary of Evaluation Methods
Cardenas et al. 2018 [66]	Development	Quantitative	10 (leave-one-out)	3D Dice similarity coefficient, 3D 95% Hausdorff distance, 3D mean surface distance, false negative Dice, false positive Dice, and relative volume difference against manual contour.
Simões et al. 2019 [39]	Commissioning	Quantitative Clinical	87	3D Dice similarity coefficient, 3D 95% Hausdorff distance and volume against manual contour. DVH parameter difference between plans from each set of contours.
Elmahdy et al. 2019 [67]	Development	Quantitative	6	3D 95% Hausdorff distance and 3D mean surface distance against manual contour.
Ahn et al. 2019 [68]	Development	Quantitative	10	3D Dice similarity coefficient, Hausdorff distance, volume overlap error, relative volume difference.
van Rooij et al. 2019 [69]	Development Commissioning	Quantitative Clinical	15	Sensitivity. Proportion of false positives, 3D Dice similarity coefficient, dose difference.
Kaderka et al. 2019 [39]	Commissioning	Quantitative Clinical	27	3D Dice similarity coefficient. Absolute difference in dose. Correlation of DVH parameters.
Zhu et al. 2019 [70]	Development Comparison	Quantitative	10	3D Dice similarity coefficient against manual contour.
Lee et al. 2019 [71]	Commissioning Comparison	Quantitative	10	3D Dice similarity coefficient and 3D Hausdorff distance against manual contour.
Van Dijk et al. 2020 [35]	Commissioning Comparison	Quantitative Subjective Clinical	104 20 14	3D Dice similarity coefficient, 95% 3D Hausdorff. Absolute dose difference between manual and auto-contour. Editing time compared to de novo contouring time. Three different questions for subjective evaluation; a single contour: "How was this contour drawn?", "Which contour do you prefer?", and "You have been asked to Quality Assure this contour. Would you…" with a four-point scale of quality.

15.2.1 STRENGTHS AND LIMITATIONS

Prior to considering specific methods or implementational details, the strengths and limitations of quantitative assessment are first considered given the general scope definition of *quantitative assessment as the calculation of the similarity or difference of a test contour with respect to a defined "ground truth"*.

The strengths of quantitative assessment come primarily from the *calculation of the similarity or difference*. In performing a calculation against a fixed "ground truth", the derived score is both objective and deterministic. As noted already, these properties mean that quantitative assessment approaches lend themselves to the development framework, whereby one auto-contouring method can to compared against another to demonstrate improvement resulting from research. To be able to measure this improvement, an assessment method is required that does not depend on the person running the evaluation (objective) and gives the same answer for the same experiment every time (deterministic). Without these properties a measure could suggest improvement where there has been no change. Furthermore, once the "ground truth" is available against which evaluation can be performed, the calculation itself requires minimal human effort to perform, an additional property which makes the use of quantitative assessment highly suited to the development framework. This is reflected in the use of quantitative scoring extensively in "grand challenges" [2–4].

In contrast, the limitations of quantitative assessment stem from the *similarity or difference ... with respect to a defined "ground truth"*. First, there is a challenge as to what "ground truth" is. While a region of an image is either part of a particular object or not – at least in an ontological sense, even if its anatomical boundary is hard to define – determining the boundary on the image is a challenge not only for the computer but also for the expert human observer. Consequently, a single observer may draw a different boundary on a different occasion for the same image, whether this is from random fluctuations in the accuracy/precision of their drawing or from a change in perception as to what the image shows. This is known as intra-observer variation. Furthermore, different observers may also draw regions differently, again as a result of random variation in contouring or differences in perception, or additionally resulting from differences in their definitions of the region. This is known as inter-observer variation. Therefore, the notion of "ground truth" is misleading, as it suggests that the contour being used as the reference is in some way more correct than any other expert contour and does not allow for the possibility that the test contour has the potential to be *more correct* than the reference in some places.

Second, *the similarity or difference* is defined in a way that makes it calculable and expresses something about the similarity or otherwise of the test contour with respect to the reference, however such a calculation does not necessarily relate to its suitability for clinical use. For example, the difference in volume of an auto-contoured region may be calculated compared to the reference; a difference shows that the auto-contour region has some error in size, yet no observed difference does not demonstrate that the auto-contoured region is accurate as it may not be in the right location or the correct shape.

These limitations of a reference that is the subject of inter-observer variation and the clinical meaning of the similarity must be borne in mind as the implementation of quantitative assessment is discussed.

15.2.2 IMPLEMENTATION

There are many choices to make in the implementation of quantitative evaluation measures. Some of these are dependent on the measure being used and some will be dependent on the biological region being segmented, while other choices are independent of both.

The first choice to be made is the region representation to use for the assessment. A binary 3D voxel-based representation is most commonplace in research, as this representation lends itself to simple implementation. However, regions are normally saved in the Digital Imaging and Communications in Medicine (DICOM) format as radiotherapy structure sets (RTSS) in radiotherapy clinical practice. Typically, radiotherapy imaging is stored in the DICOM image format as individual 2D slice instances, and the RTSS format reflects this. The segmentation is stored as a series of ordered points per 2D image in real-world units. The neighboring points in the list are assumed to form a line segment. Thus, the RTSS implies a region representation that is a 2D irregular polygon. While other DICOM objects exist that represent segmentations as voxel masks, e.g. a DICOM segmentation object, these are not the standard format for radiotherapy.

In many implementations, including the 2017 AAPM Thoracic Auto-segmentation Challenge, the 2D polygon representation is converted to a voxel array for ease of implementation. This conversion process itself can be subject to implementation variations between programs and therefore introduces uncertainty in comparing results from different implementations. Thus, it would be better to define any measures in real-world space using a 2D polygon region representation to reduce the potential for discrepancies.

A second choice required is whether measures are implemented in 2D or 3D. While RTSS contour representation is 2D, the biological objects are defined as 3D objects, therefore it is of interest to know the discrepancies in 3D. For example, volume differences are likely to be more meaningful than cross-sectional area ones. However, this choice depends both on the measure being used and on the biological structure being contoured. Some objects are well-defined in 3D, e.g. a lung has a clear superior/inferior extent in addition to inferior/anterior and left/right, while others, particularly tubular structures such as the spinal cord, are defined over the treatment region but poorly defined in superior/inferior extent. Therefore, surface distance measures may be meaningful for the lung in 3D, but less meaningful for the spinal cord where a difference in contoured extent may be treated differently to inaccuracy in an individual slice. Yet, a 3D assessment is only meaningful when considering volume differences, regardless of the organ.

While not able to provide a simple number, a visual slice-by-slice assessment, as shown in Figure 15.1, can be used to present multiple 2D measures in an easy-to-interpret form. This facilitates the assessment of extent differences, as can be seen in the figure where the area for the comparison contour exceeds the zero area of the reference at the top middle plot, showing that the comparison contour over-segments the base of the heart.

How measurements can be combined should also be considered. Assessment is more helpful if it is performed for multiple patient cases, to evaluate the performance over a range of variations in

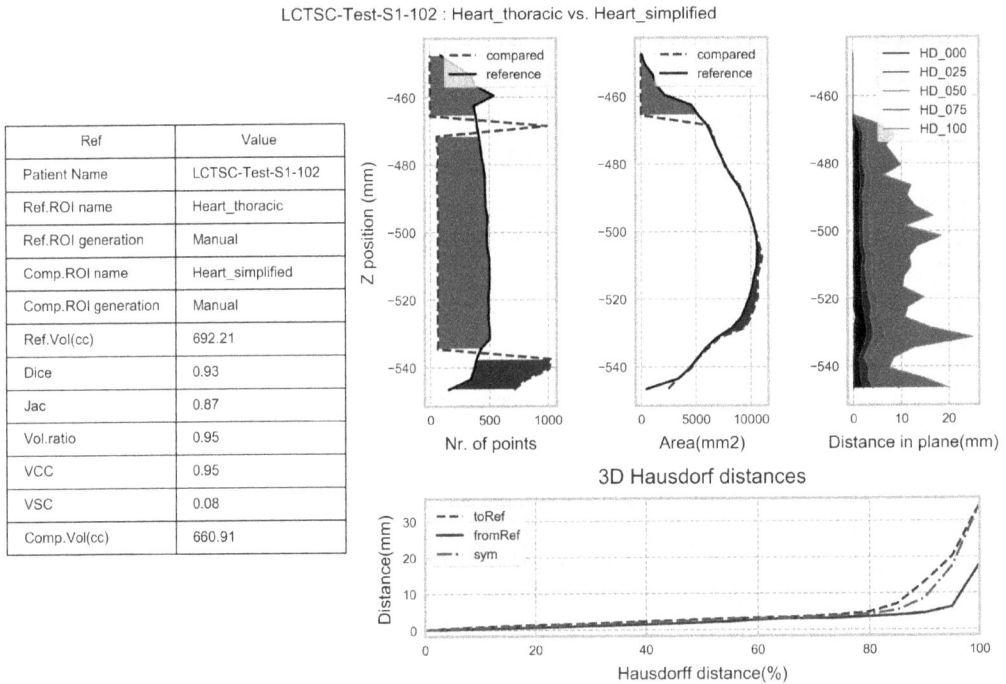

FIGURE 15.1 Example contour evaluation report, showing contour point density, area, and banded 2D Hausdorff distances over the range of the contours. It can be easily observed that will slice-by-slice area agreement is good over the majority of the organ, the comparison contour over segments at the base of the heart. Example courtesy of Akos Gulyban.

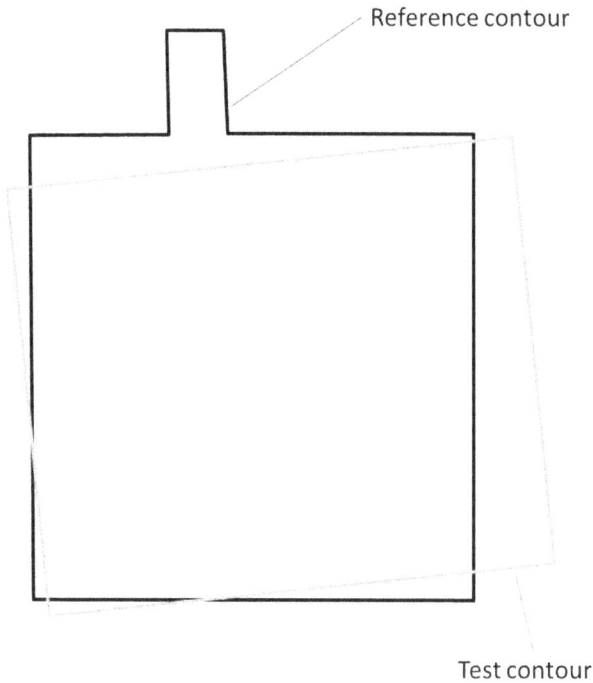

FIGURE 15.2 Example contours used to illustrate contour evaluation measures. The dark contour indicates the reference contour. The light contour indicates the test "auto-contour".

patient appearance. Where a mean and standard deviation can tell the average performance and its variation, for some measures it may be appropriate to combine results in a different way, for example the Hausdorff distance reports a maximum distance error, thus it might be appropriate to report the maximum Hausdorff distance across cases rather than the mean. It may also be desirable to implement quantitative measurements to allow comparison of auto-contouring methods for multiple organs and accounting for a range for measures. For example, for the 2017 AAPM Thoracic Auto-segmentation Challenge [4] a single score per method was required for all cases. The approach taken is discussed later.

A range of quantitative assessment methods is now described, considering their clinical meaning, method/structure-specific implementation details together, and how the assessment approach can be adapted for inter-observer variation. The measures selected are by no means comprehensive but are those that have been popular in the past or have recently been proposed and offer new potential.

To work through each of the measures, a toy example will be considered. This is shown in Figure 15.2 – while not anatomically realistic, it is much easier to draw!

15.2.3 CLASSIFICATION ACCURACY

In earlier assessments of auto-contouring, measures relating to classification accuracy were popular. Such measures historically had been used in computer science where pixel-wise classification was considered, and clinically in computer-aided detection where the correct detection of lesions is important. Classification accuracy is normally evaluated by counting the proportion of correctly labeled pixels, true positive (TP) and true negative (TN), and the incorrectly labeled pixels, false positive (FP) and false negative (FN). In Figure 15.3, regions have been labeled for TP, FN, and FP. However, the evaluation of true negative pixels becomes a challenge with auto-contouring, as the extent of the background region outside a structure (i.e. the negative classification) is unclear. Assuming the background is the rest of the image, the true negative proportion is always (assuming

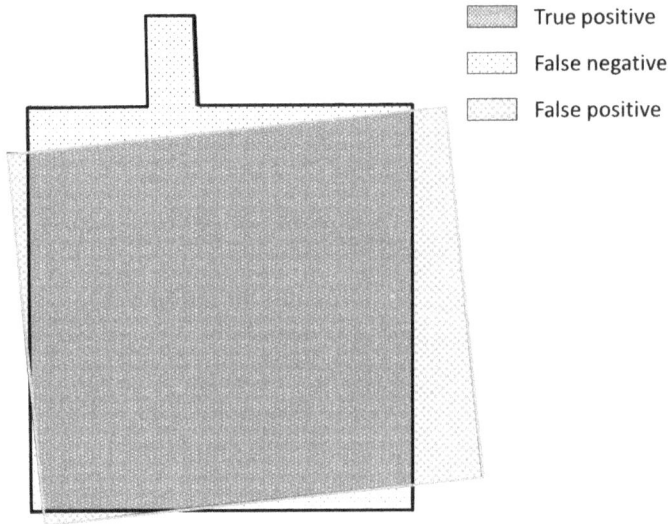

FIGURE 15.3 Various measures are based on pixel/voxel classification accuracy. True positive, False negative, and False positive are well defined for anatomical structures. True negative cannot easily be defined as the space outside the anatomical structure has no bounds.

a competent system worthy of testing) high and will be dependent on the image size. Thus, true negative is not a helpful measure when evaluating auto-contouring unless a region is defined over which it will be measured (as in Isambert et al. [5]). However, the choice of definition of such a region makes the comparison of systems by different researchers/clinics potentially inconsistent.

Within a clinical context, such as Computer-Aided Detection (CAD), the performance of a system is often cast into sensitivity and specificity. Sensitivity, defined as TP/(TP + FN), measures the number of correct foreground labels as a proportion of the possible number of correct foreground labels. Thus, sensitivity will measure the ratio of the correctly identified area of a structure as a proportion of the total expected structure area. Specificity, on the other hand, defined as TN/(TN + FP), measures the correctly identified area of background as a proportion to the total expected background. However, limitations of measuring TN in the context of auto-contouring, specificity also becomes a difficult measure to compare between implementations. Other alternatives that have been proposed are the percentage of FP [6], defined as FP/(TP + FN), and the inclusiveness index [7], defined as TP/(TP + FP).

15.2.3.1 Implementation

True positive, false positive, etc. are defined as the number of pixels/voxels – thus a correct implementation would be to convert RTSS to a voxel grid. However, the absolute values of TP counts will depend on the resolution at which this is done (which can be assumed to be the image resolution), therefore may vary between patients and clinical centers.

A more consistent and natural implementation may be to compute these measures in terms of volume or area. These values can be directly computed from RTSS using simple Boolean operations: the true positive region is the intersection of the reference and test contours, and the FP and FN regions can be calculated as the difference between the intersection and the test contour and reference contour respectively. A 2D implementation may give information about performance of an auto-contouring system over a range of slices and indicate slices with better or worse classification performance. However, given the inherent assumption of spatial independence of the observations, a volumetric implementation is more appropriate. A simple interpretation for RTSS is to convert the area to volume by multiplication by the slice thickness. While sensitivity is a ratio and is not dependent on this calculation, it is necessary if reporting TP, FP, or FN volume.

15.2.3.2 Advantages and Limitations

Such classification approaches have the benefit that measures such as sensitivity and specificity are reasonably well understood both within the clinical domain and within the computer science community. However, the approach has limited usefulness in evaluation of auto-contouring since each pixel is treated as an independent observation. The importance of spatial location and the concept of the structure are not considered by these measures; thus, the measures do not provide any information about the utility of the contours – for example the same results can be measured by a random spattering of incorrectly classified pixels and the same number of incorrectly classified pixels systematically located.

15.2.4 DICE SIMILARITY COEFFICIENT

Closely related to the classification accuracy measures previously described, the Dice similarity coefficient, often abbreviated as just Dice or DSC, is a measure of overlap of two structures. It was first proposed as a measure looking at species overlap in ecological studies and is named after the author [8]. The measure evaluates the intersection of the two regions as a ratio to the total area of them both, such that where there is total agreement the measure is one and where there is no agreement the value is zero.

$$DSC = A \cap B / (|A| + |B|) = 2 \times TP / (2 \times TP + FN + FP)$$

The Jaccard index, also known as a percent overlap [9], which is the intersection divided by the union, can be directly calculated from the DSC.

$$JI = A \cap B / (A \cup B) = TP / (TP + FN + FP) = DSC / (2 - DSC)$$

While these measures have the same bounds, and can be calculated from one another directly, care must be taken not to confuse or compare them directly on account of the non-linear relationship between them.

15.2.4.1 Implementation

It is relatively commonplace to voxelize the contour representation prior to calculating Dice. However, Dice can be easily calculated in the same way that the classification measures directly from RTSS contours using Boolean operations.

Although a 2D implementation may enable slice-by-slice assessment – under the assumption of slice by slice contouring – as with classification measures, 3D implementation is intuitive since the measure does not really consider spatial context. The choice of 3D or 2D implementation may depend on the organ for which Dice is being calculated. For some structures, the extent in one axis may be operator dependent, such as the spinal cord, yet variation in this direction may be considered insignificant. However, in this context steps can be taken to standardize the axial extent over which accuracy is assessed, mitigating this issue [10–11].

15.2.4.2 Advantages and Limitations

This measure has been the most popular method to date for the assessment of auto-contouring, perhaps because it is a simple single value with a bounded range that can be more easily interpreted than sensitivity/specificity. Nevertheless, the DSC suffers from the same main limitation as the classification accuracy measures from which it is derived. Namely, the measure is ambivalent to spatial context and therefore does not give an indication of clinical utility of a contour. While its clinical interpretation is unclear, a consequence of its historical popularity for contour evaluation is that it is reasonably well accepted and understood. Success breeds success, and thus it has become the current de facto standard.

15.2.5 DISTANCE MEASURES

Measures of the distance between contour measures have also been very popular to date. These measures are typically based on the principle that the distance from a point on one structure to the other structure is measured as the shortest distance between them. Given two structures X and Y, the distance from point x to structure Y is given as:

$$d(x,Y) = \min_{y \in Y} d(x,y)$$

where $d(x,y)$ is the Euclidean distance between points x and y.

15.2.5.1 Hausdorff Distance

The Hausdorff distance (HD) is defined as the maximum distance between these structures.

$$d_H(X,Y) = \max\left(\max_{x \in X} d(x,Y), \max_{y \in Y} d(X,y) \right)$$

Figure 15.4 shows the maximum distance between the reference contour and test contours and vice versa. It should be noted that the Hausdorff distance is symmetric in taking the maximum of these two possible distances, therefore it is independent of the measure that is defined as the reference. There are a range of variants that have been proposed taking direction into account [12].

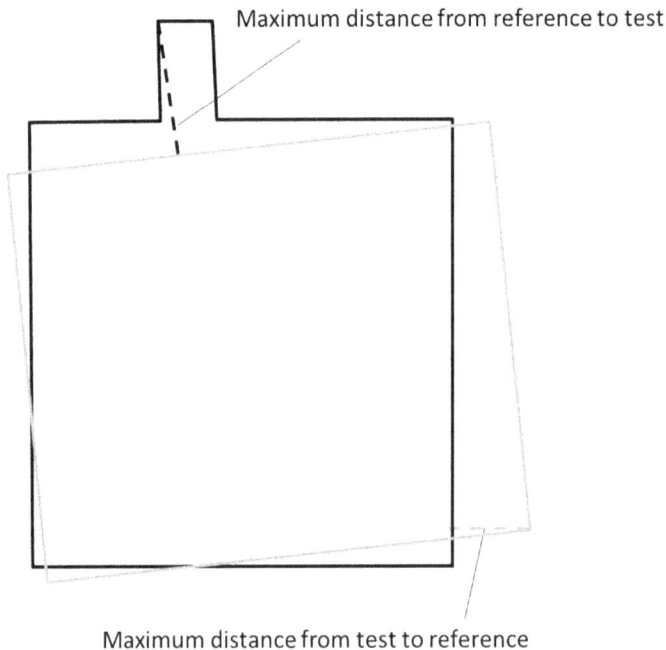

FIGURE 15.4 Distance between contours. The distance from one contour to another at a point is represented as the shortest distance from that point to the other contour. This is not symmetrical. The largest distances between the contours are shown by the dashed lines and are in very different locations. The maximum of these distances in the symmetric Hausdorff distance.

15.2.5.2 95% Hausdorff Distance

While the Hausdorff distance gives the true maximum distance between two structures, in finding the extreme it is very sensitive to small outliers [3]. The 95% Hausdorff distance [13] has been used as a more robust estimate of the maximum error. The definition of the 95% Hausdorff distance is less well defined, but is widely considered as the 95th percentile of the ordered distance measures:

$$d_k(X,Y) = K_{x \in X}^{th} \left(\min_{y \in Y} d(x,y) \right)$$

where $K_{x \in X}^{th}$ gives the kth percentile of the order distances over $x \in X$. However, this is unidirectional. Thus, following [13], a symmetric 95% Hausdorff distance can be defined as:

$$d_{H95}(X,Y) = \max \left(d_{95}(X,Y), d_{95}(Y,X) \right)$$

noting that the two sets may have a differing number of surface points included. Again, alternatives have been proposed and used, for example the average 95% distance, defined as $\left(d_{95}(X,Y) + d_{95}(Y,X) \right)/2$, was used in the 2017 AAPM Thoracic Auto-segmentation Challenge [4].

15.2.5.3 Average Distance

Rather than looking at the extreme deviations, it is also helpful to know the average difference between the structures. Noting that structures X and Y may have differing length/surface area, the average distance can be defined as:

$$d_{ave}(X,Y) = \left(\frac{1}{|X|} \sum_{x \in X} d(x,Y) + \frac{1}{|Y|} \sum_{y \in Y} d(y,X) \right)/2$$

This is also referred to as the average Hausdorff distance, although once again there is variation between authors, with the average Hausdorff distance also being defined as:

$$\max \left(\frac{1}{|X|} \sum_{x \in X} d(x,Y), \frac{1}{|Y|} \sum_{y \in Y} d(y,X) \right).$$

15.2.5.4 Implementation

While distance measures may be easy to define mathematically – even if these definitions are inconsistent between publications – their implementation in a discrete manner is not. The range of choices that can be made can lead both to quite substantial variation in the value reported and the speed of calculation. Mathematical definitions often consider a continuous contour or surface from which distance can be calculated. While it is possible to analytically calculate the distribution of distances between surface patches or line segments, extending such an approach to multiple elements quickly becomes non-trivial. Therefore, it is more convenient to discretize their calculation to implement these measures in practice, yet the resolution at which discretization takes place will affect the distances calculated and their distribution.

While the RTSS format is defined in terms of points and thus provides an obvious set of points from which distance can be calculated, these do not necessarily need to be evenly spaced to represent a contour and therefore do not form a good basis for calculation. The resolution of the image on which the contours are defined provides a natural discretization under the assumption of voxelized implementation, yet this makes the quantization of the measurement potentially vary between test cases on which the calculation is performed.

The choice of the dimensionality in which measurements are to be made affects both the result and the method of implementation. RTSS is an inherently 2D representation, thus calculation in 3D requires reconstruction of the contours into a volumetric surface. The most common approaches to this are voxelization into a binary mask (e.g. [2]) or creation of a surface mesh (e.g. [14]). As noted already, the former carries difficulties with regards to quantization error but carries the benefit of an inherent discretization. Construction of a surface measure can lend itself to a discretized approach using the nodes of the mesh (as used in Teguh et al. [14]), and precision can be increased by using a fine mesh. However, meshing is normally achieved by converting to a voxel representation first to enable the surfaces to be found, thus potentially introducing the error.

Calculation of distance measures in 2D enables distance measures to be calculated from the RTSS polygon, albeit with some additional discretization. However, this risks ignoring errors whereby slices have or have not been contoured. This can lead to reporting of reduced values where the errors are mainly in the extent over which a structure should be contours, e.g. for the spinal cord. Conversely, it may lead to increased errors where the surface is a single slice-out in some locations. Figure 15.5 illustrates these sensitivities of using Hausdorff distance using data from the 2017 AAPM Thoracic Auto-segmentation Challenge. In the left-hand example, a small difference in the axial extent of a small region of lung present on the auto-contour in the neighboring slice, results in a large 2D Hausdorff distance. In the other example, the difference in extent contoured for the spinal cord results in a large 3D Hausdorff distance for an otherwise well-contoured structure. Various approaches have been taken to overcome some of these limitations, such as limiting the extent or

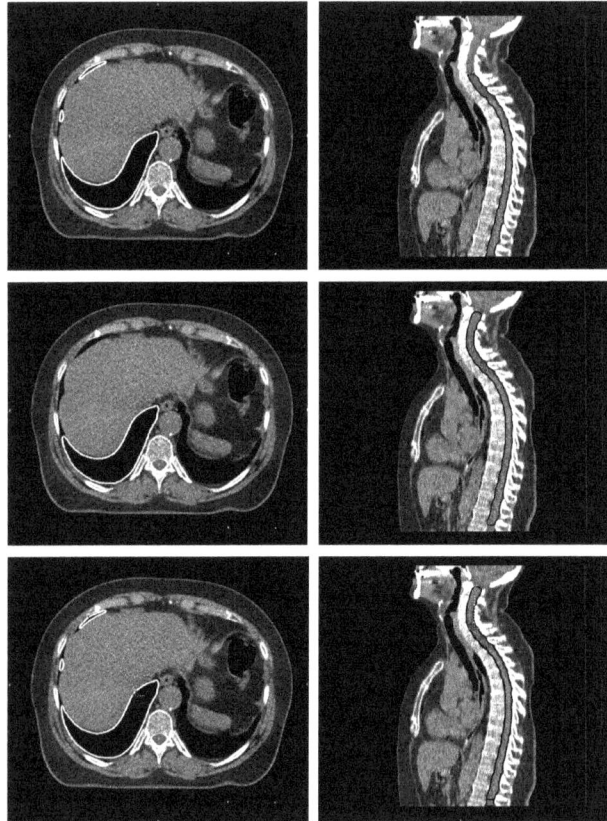

FIGURE 15.5   Example of sensitivities of Hausdorff distance. The top row shows reference contour for case LSTSC-Test-102. The middle row shows auto-contour for evaluation. The bottom row illustrates the Hausdorff distance. The left-hand column illustrates 2D Hausdorff. The difference in a small part of the lung results in an 11 cm 2D Hausdorff distance. The right-hand column shows 3D Hausdorff distance, a difference in the extent contoured for the spinal cord results in a 3D Hausdorff of about 8 cm. In both instances the contouring might otherwise be considered good.

slices in which calculation is performed (e.g. [11, 15, 16]). However, these methods are effectively seeking the definition of the Hausdorff distance; the maximum distance between structures.

15.2.5.5 Advantages and Limitations

Distance measures are generally welcomed within radiation oncology, as location error is a meaningful concept in treatment planning. A treatment volume will have margins applied to account for clinical and operational uncertainty, and therefore it is appropriate that auto-contouring errors are put into this context. Nevertheless, there is a risk that the interpretation is incorrect; Hausdorff distance (and its 95% variant) represents a single large error. While this shows inaccuracy in a contour, it may not represent a bad volume in terms of clinical utility – the outlier may be clinically insignificant in location or volume. The average distance may be small, suggesting good agreement between the contours, but does not indicate bias which could lead to volume distance. Taking additional measures, such as volume difference, into account may overcome some of these limitations.

In the presence of contouring errors, distance measures also reveal little about the effort required to correct an erroneous contour to make it suitable for clinical use. A large Hausdorff distance may represent a large volume error requiring substantial recontouring, or a tiny outlier that could be easily deleted. A small average distance may represent an accurate unbiased contour with random error needing minimal editing, or a small but systematic error needing correction over a wide area. Thus, distance measures do not correlate well with clinical opinion [17] or with the time required to edit contours to a clinically acceptable standard [18].

From the perspective of comparing auto-contouring methods, the greatest challenge lies in the inconsistency between definitions and approaches to implementation. The different definitions can lead to substantial differences in the output value, making comparison between publications meaningless unless full implementation details are known.

15.2.6 GEOMETRIC PROPERTIES

The focus on distance measures has appeal in radiation oncology, as these measurements bear some resemblance to a concept that is already used: treatment margins. Simple geometric properties, namely volume and centroid location, have been used from some of the earliest evaluations of auto-contouring [5, 19] following similar reasoning. In terms of treatment delivery, accurate dose delivery requires accurate positioning. Thus, it seems natural to measure the error in position of a centroid of a structure. Similarly, the interpretation of dose to a structure is tied to its volume through the dose volume histogram (DVH), thus the correct estimation of volume is required for a correct interpretation of dose. Therefore, these measures are familiar to those in radiation oncology, leading to their use.

15.2.6.1 Centroid Location Comparison

The center of mass/gravity, C, of a structure is defined as:

$$C = \frac{1}{M} \iiint \rho(r) r dV$$

where M is the mass of the structure, and $\rho(r)$ is the density of the object at a point, r, within the volume, V. However, for any structure it is normal to assume constant density throughout, thus the center of mass is equivalent to the centroid, defined as:

$$C = \frac{1}{V} \iiint r dV$$

removing the need to know the density or mass.

In assessing auto-contouring, it is normal to consider either the difference in centroid location between the reference and the test contour in each direction (e.g. [5]):

$$\Delta x = \left| cx_{test} - cx_{ref} \right|, \Delta y = \left| cy_{test} - cy_{ref} \right|, \Delta z = \left| cz_{test} - cz_{ref} \right|$$

or the magnitude of the centroid error (e.g. [20]):

$$d = \sqrt{\Delta x^2 + \Delta y^2 + \Delta z^2}$$

15.2.6.2 Volume Comparison

Various ways to compare volume have been used. The volume ratio, given by:

$$VR = \frac{V_{test}}{V_{ref}}$$

assesses the relative volumes of the reference and test contours. The volume difference between the reference and test contour can be expressed as:

$$\Delta V = \left| V_{test} - V_{ref} \right|$$

A signed difference enables the identification of bias in the volume over multiple test subjects, while the absolute value enables the average magnitude of volumetric error to be determined.

However, the volume difference is dependent on the size of the organ being contoured, thus comparison between organs is unhelpful. To overcome this the relative volume difference can be used (e.g. [5]):

$$\Delta V \left(\% \right) = \frac{\left| V_{test} - V_{ref} \right|}{V_{ref}} \times 100$$

15.2.6.3 Implementation

The centroid and volume of a structure are inherently 3D properties, removing the decision as to whether to calculate these in 2D or 3D.

Volume and volume ratio can easily be calculated using the classification parameters, as illustrated in Figure 15.2. The volume difference between the reference and test is given by $FP - FN$ multiplied by a scaling factor. The scaling factor is the volume of a voxel if using a mask-based representation. If calculating directly from the RTSS contours, the total FP area and FN area can be calculated over all slices and multiplied by the slice thickness. The relative volume difference can be calculated using $(FP - FN)/(TP + FN)$, while the volume ratio can be expressed as $(TP + FN)/(TP + FN)$.

Using a voxel mask approach, the centroid can be calculated as the average voxel location given by:

$$C = \frac{1}{V} \sum_{i \in V} \left[x_i, y_i, z_i \right]$$

To calculate the centroid directly from RTSS, break the contours into simple primitives and use the area-weighted average of their centroids to find the object centroid. This can be performed slice-by-slice from the contours, then the overall centroid can be subsequently calculated using the area-weighted average of these slices.

15.2.6.4 Advantages and Limitations

Simple geometric measures are appealing since they give the appearance of relating to other measures used in radiotherapy. However, this presents a risk in reality; a difference in centroid position, might imply that a shift, as would be performed for the patient position on the day of treatment, would be sufficient to correct the auto-contour. Yet, this is very unlikely to be the case. Similarly, having no difference in volume, or a volume ratio of one, implies nothing about the spatial correctness of the object indicating only that the test contour is of the same size as the reference. Taken together these measures demonstrate an object of the same size at the correct location, yet still they are uninformative as to the accuracy of the contour itself. For example, the test contour could be a rotation of the reference contour about its centroid. Thus, their familiarity as concepts in radiation oncology is misleading, giving the impression of types of errors found elsewhere in the radiotherapy process while at the same time providing very limited information as to the quality of the auto-contouring.

15.2.7 MEASURES OF ESTIMATED EDITING

More recently, two approaches have been proposed that try to take into account the end use of the contours that are being generated. For use in a clinical context it is reasonable to assume that the contours will be reviewed, and that any errors in the contours will be corrected before they are used in treatment planning. Thus, a good measure of contouring performance will be one that reflects how much editing a contour requires. The *Surface Dice* [21] and *Added Path Length* [18] are measures that seek to achieve this.

Surface Dice [21] assumes that any surface area outside of an acceptable tolerance, τ, will require editing. Therefore, the surface Dice duplicates the Dice measure but using the surface of the structure rather than the volume, with the intersection of areas as those within the tolerance. The standard Dice measure has no tolerance resulting in twice the intersection of the structures in the numerator. Two possible intersections can be defined when the tolerance is applied to each structure, as shown in Figure 15.6 and Figure 15.7. The surface Dice is then defined as:

$$DSC_{surface} = \frac{\left|R \cap T^\tau\right| + \left|T \cap R^\tau\right|}{\left|R\right| + \left|T\right|}$$

where $|R|$ and $|T|$ represent the surface area of the structures, and $|R \cap T^\tau|$ and $|T \cap R^\tau|$ represent the surface intersections as illustrated in the figures.

The added path length (APL) measure [18] builds on this, observing that most manual editing is performed on a slice-by-slice basis and that the total amount of editing is likely to relate to the time required for editing. It is also noted that in most treatment planning systems and contouring workstations it is much easier and quicker to delete a contour than to draw an additional contour. Therefore, the measure is designed only to evaluate the amount of contour to be added, rather than the length of contour to remove.

No tolerance measure was applied in Femke et al. [18] as the authors were comparing the measure to timings of actual edits made to a contour. Therefore, the contours were in perfect agreement where no edit had been made. However, a tolerance must be applied where auto-contours are being compared to a reference with the assumption that editing will be performed in the future. In this case, it is assumed that the edit will be made to the path to match the reference, but that there is no cost to removing contour where the test contour needs deleting. Figure 15.8 illustrates this, shown the contour that would need to be added to correct the test contour to match the reference for sections outside the acceptance tolerance. Thus, the APL can be defined as:

$$APL = \left|R\right| - \left|R \cap T^\tau\right|$$

Sections inside tolerance considered surface intersection $R \cap T^\tau$

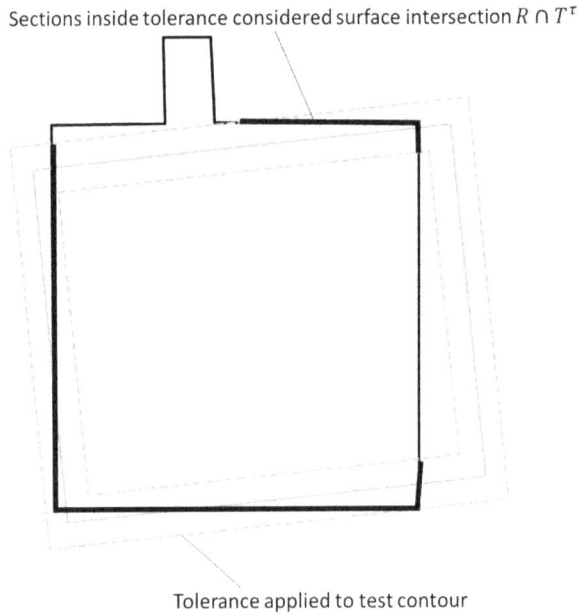

Tolerance applied to test contour

FIGURE 15.6 Definition of the intersection of the surfaces used for surface Dice with a tolerance applied to the test contour.

Tolerance applied to reference contour

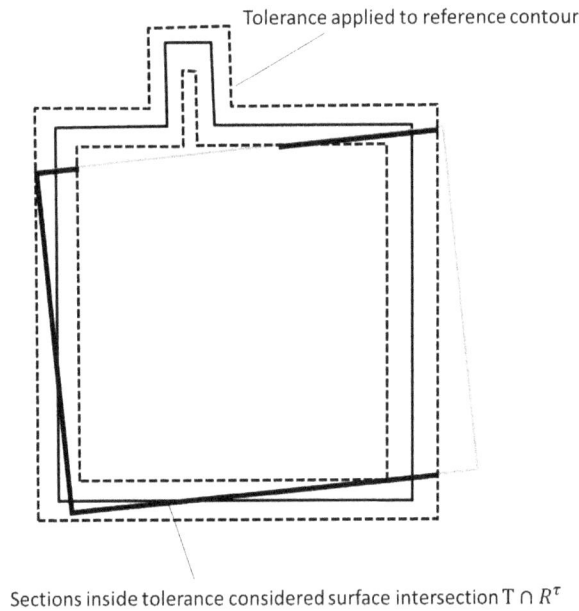

Sections inside tolerance considered surface intersection $T \cap R^\tau$

FIGURE 15.7 Definition of the intersection of the surfaces used for surface Dice with a tolerance applied to the reference contour.

Three important differences between these measures can be noted; surface DSC is a 3D measure whereas APL is a 2D measure. Although it measures surface, it is surface in 3D, and must be implemented in 3D to address differences in contours at the superior/inferior extents of a structure. APL on the other hand assumes slice-by-slice editing and is therefore calculated slice by slice. Second, APL is not symmetric in its evaluation unlike surface DSC. The APL measure only considers the

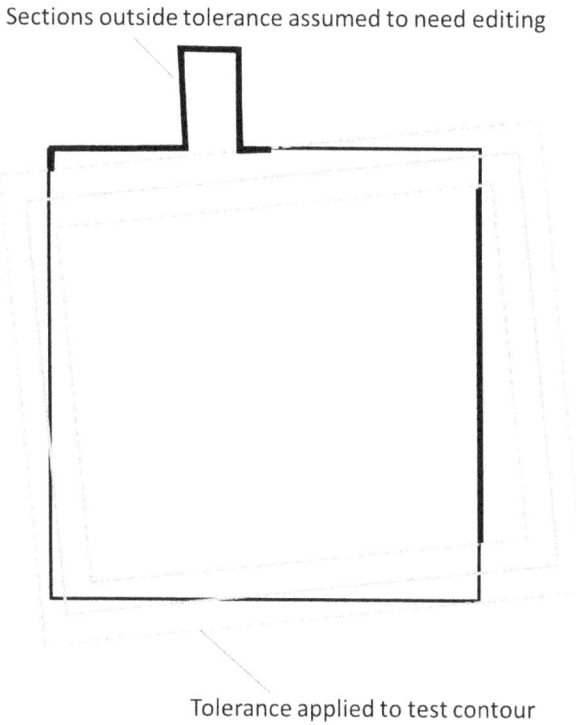

Sections outside tolerance assumed to need editing

Tolerance applied to test contour

FIGURE 15.8 The boundary of the reference contour outside the acceptance tolerance of the test contour represents the length of contour that would need adding, i.e. the "added path length"

path length added and applies a tolerance only to the test contour, whereas the surface DSC considers both addition and removal of surface by applying the tolerance to each surface in turn. Finally, surface DSC is a ratio of the editing to the total surface, whereas APL is an absolute measure. It might be assumed therefore that surface DSC aims to represent the proportional editing time, while APL seeks to represent the absolute editing time.

12.2.7.1 Implementation

Surface DSC is an inherently 3D measure, requiring a 3D representation of the structure for evaluation. Nikolov et al. [21] implemented this measure using a mesh-based representation, with the mesh created using the marching cubes algorithm [22] from an initial mask-based representation. While a mask-based implementation would be easier to implement directly, it would quantize the tolerances that could be applied to the size the voxel. The use of a mesh allows for sub-voxel tolerances to be applied. With surface DSC being defined on a 3D surface, it is not obvious how this measure could be implemented from RTSS, without reconstructing a volumetric representation.

Conversely, APL lends itself to an implementation that works directly on the RTSS. Although Vaassen et al. [18] opted for a 3D mask-based implementation the measure itself is naturally 2D. While it is non-trivial to apply margins and perform Boolean operations on a polygon, many software libraries exist which enable this. The APL can be thought of as the sum of the segments of the reference contour that fall either fully inside the contracted test contour or fully outside of the expanded test contour. Having identified the intersection of the refence contour with each of the expanded and contracted test contours, the mid-point along each line segment can be tested to identify if is outside or inside. This is illustrated in Figure 15.9, which shows the segments that would be generated by performing the contour intersection and their mid-points. The dark segments make up the APL.

Mid-point of segment

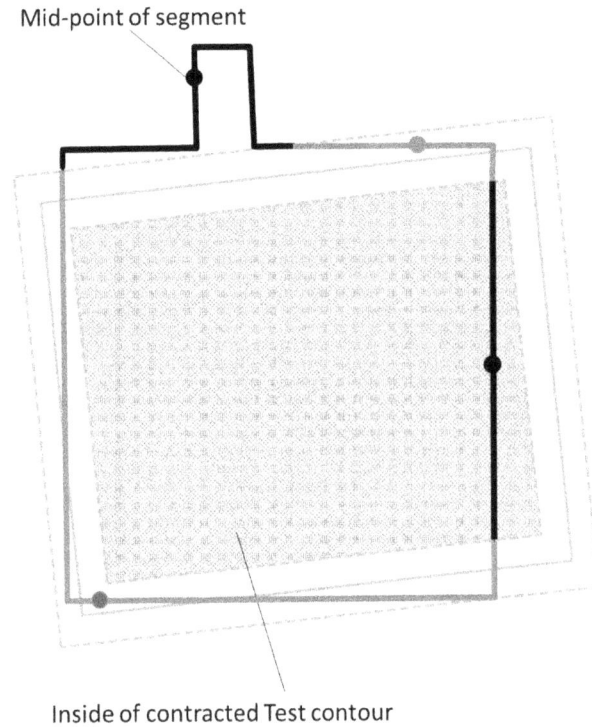

Inside of contracted Test contour

FIGURE 15.9 Added path length can be calculated from the segments formed from the intersection between Reference and each of the expanded and contracted Test contours. A test on the mid-point to determine if it is fully inside or fully outside these contours reveals that the dark segments would constitute the APL.

15.2.7.2 Advantages and Limitations

Both the surface DSC and APL are encouraging new measures that have shown better correlation with the editing time for contouring compared to more conventional measures such as DSC and 95% Hausdorff distance [18], with APL showing a tighter correlation than surface DSC. Both measures however carry similar limitations; first, the tolerance to use is unknown. The idea that a clinician would only adjust a contour that is sufficiently incorrect seems a reasonable assumption. However, it is unknown what "sufficiently incorrect" is. Inter-observer variation in contouring could be used to determine this, although it is not necessarily the case that one clinician would accept another clinician's contours, or even their own – if blinded to its origin [1]. Experiments would be needed to calibrate the tolerance on these measures to maximize their correlation with editing time and to be able to predict the required editing time given a measurement. The tolerance might be expected to vary structure by structure, and therefore such experiments would need doing for all structures. Second, both measures assume a fixed tolerance for a structure. However, inter-observer variation in contouring is known to vary across a structure [23], therefore it might be reasonable to expect differing degrees of acceptable tolerance across a structure. This would be challenging to implement in practice for either method.

Despite these limitations, these measures represent the state-of-the-art when it comes to quantitative assessment of auto-contouring. If the goal of auto-contouring is to save clinical contouring time, then these measures are closer to measuring performance in terms of that goal than others previously used.

15.2.8 HANDLING INTER-OBSERVER VARIATION IN QUANTITATIVE ASSESSMENT

All of the quantitative measures discussed here rely on having a reference contour for comparison. This reference is assumed to be correct. However, this is perhaps an unreasonable assumption, since there is variation in manual contouring. Thus, is could be considered that a manually drawn reference contour is only one interpretation of correct. The quantitative assessment between two manual contours may not result in a perfect score as a result of inter- and intra-observer variation. In this, a suboptimal measurement reported for an auto-contour may not reflect an incorrect contour necessarily, merely a different contour. Although a very bad score would indicate a bad contour, a good score does not necessarily imply good contouring, while a perfect score is not achievable as a result of the variability affecting the reference. From the earliest publications accessing auto-contouring, attempts have been made to place quantitative measures in the context of this inter-observer variability [19].

A common approach is to generate a consensus contour, representing the average opinion of multiple experts. The popular Simultaneous truth and performance level estimation (STAPLE) method [24] seeks to achieve this while at the same time estimating the confidence that can be placed on any particular contour contributing to the consensus, down-weighting the contribution of experts who do not appear to agree with the rest. If the auto-contour is used within this STAPLE process, together with manual estimates then it is possible to get this performance (confidence) level for the auto-contour at the same time as the experts. In practice, however, the method is commonly used either to generate the consensus of multiple auto-contours (e.g. [25]), as a way of reducing random error, or to generate a single reference from multiple experts (e.g. [6]). This single reference is then used as a "hard" ground truth for quantitative assessment. Despite benefits in reducing random error, the use of a consensus contour as a reference does not overcome the challenges of inter-observer variability. The final contour is still a fixed reference and does not acknowledge that experts had variation between them and may even have disagreed as to what the correct boundary definition should be.

The introduction of a tolerance parameter in surface DSC [21] and APL [18] goes some way towards addressing this issue, acknowledging that a certain degree of variation in the contour position should be acceptable. This approach is beneficial in allowing a perfect score where there may be minor differences and could potentially be applied to other quantitative measures. However, as noted previously, this tolerance is fixed (per structure), whereas inter-observer variation may be spatially varying.

Various approaches have been used within "Grand Challenges" to address the problem of inter-observer variation, seeking not to discriminant between participants on the basis of small error from an arbitrary reference rather to penalize significant errors. The 2010 Head & Neck Contouring Challenge [2] used a tolerance to allow for small errors, with the percentage of slices with a 2D Hausdorff exceeding 3 mm being evaluated among other measures. The observation that *deviations below 3 mm are often considered acceptable by the clinicians* was used to explain this scoring. The 2017 AAPM Thoracic Auto-segmentation Challenge [4] applied a linear function to scores within twice the inter-observer variation for each structure, as shown in Figure 15.10. Scores showing perfect adherence to the reference scored 100, while scores twice, or more, the inter-observer variation away from perfect, scored zero. Although this approach still rewards adherence to the reference contour, the deviations from this reference are scaled according to expected human variation. The approach also has the added benefit of normalizing scores between measures and organs enabling an overall average score to be computed.

The tolerance-based methods represent the state-of-the-art in handling inter-observer variation, however this approach does not fully allow quantitative assessment of what is acceptable or not. This is the greatest challenge for handling inter-observer variability using quantitative measures. Such measures are blunt tools incapable of distinguishing different types of error. Some larger errors (compared to the reference) may be considered reasonable interpretation of the underlying image data with which another observer may have agreed, while other small errors, although close to the reference, would not be considered a reasonable interpretation by any observer. This motivates the qualitative subjective assessment discussed in the next section.

15.2.9 SUMMARY OF QUANTITATIVE EVALUATION

In this section, both popular and recent quantitative measures used for assessing auto-contouring in radiotherapy have been reviewed. It has been seen that these measures all have limitations in

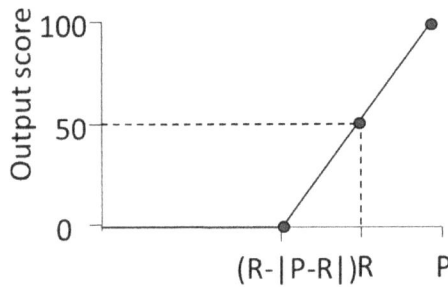

FIGURE 15.10 Normalization of scores based on the inter-observer variation as used in the 2017 AAPM Thoracic Auto-segmentation Challenge. Measures within twice the inter-observer variation of the reference score in the range 0–100, with 100 representing perfect adherence to the reference. P represents the perfect measure score with respect to the reference. R represents the average score of 3 independent observers with respect to the reference contour.

what they reveal about the clinical benefit of any particular contouring approach. The recently proposed surface Dice and APL measures come closest to showing clinical relevance, with some degree of correlation to the time required to fix any errors. Nevertheless, all measures are limited by their reliance on a "true" reference contour and understanding contouring accuracy with respect to inter-observer variation remains a challenge. Of the approaches discussed, the tolerance threshold appears to be the most useful, yet requires further research in the area of spatially varying tolerance to fully represent contouring quality compared to the human reference.

15.2.10   Example Implementation

In this section, an attempt has been made to consider implementation choices that might impact the results of quantitative evaluation. While others have sought to develop libraries which would standardize measurements if used widely (e.g. [12, 26]), these all rely on converting from the RTSS format, either as part of the measurement calculation or to make use of the library. Throughout this section, thought has been given to how these measures might be implemented directly from the contours native RTSS format to avoid issues of conversion to other representations such as masks or meshes. Therefore, to facilitate use of these measures an open source implementation is provided at: https://github.com/Auto-segmentation-in-Radiation-Oncology/Chapter-15

15.3   SUBJECTIVE EVALUATION

Quantitative evaluation tries to measure the error in contouring, but until recently has failed to properly consider the endpoint of auto-contouring – the clinical use of those contours. Clinical use will depend on the acceptance or editing of those contours by a human, and this is inherently subjective. Subjective assessment of auto-contouring in radiotherapy has largely followed the same format since 2011, when a number of studies were published using this approach [14, 16, 27], but can be broadly broken down into considering the acceptance of contouring, the source of contouring, and the preference for contouring

15.3.1   The Acceptance of Contours

The majority of investigations have posed the question as to whether the contours being evaluated are suitable for clinical use, although the exact framing of the question has varied. Hwee et al. [27] asked the question whether the contours were clinically acceptable, offering the binary choice of "yes" or "no". In contrast, Teguh et al. [14] anticipated editing of contours asking experts to rate contours on a four-point scale as: 0 poor, 1 major deviation (editable), 2 minor deviation (editable), 3 perfect. Curiously, in the same study, a three-point scale was used to assess manual clinical contours and edited auto-contours making direct comparison difficult. In the same year, Granberg [16] asked clinicians to rate contours as to how helpful they perceived them to be on a three-point scale of: not at all, a little, very much.

Table 15.2 summaries the studies in the domain of radiotherapy that have used a subjective approach to evaluate contour acceptability. In most papers the question asked is not explicitly stated but can be inferred from the answers available, thus the table may not reflect the exact wording used by the investigators.

Since Hwee et al. [27], few papers have explicitly stated the score that is considered clinically acceptable, with most opting to allow major editing in some way. This implicit expectation of editing could introduce bias in the interpretation of results, with contours being considered useful. Furthermore, most investigations since Hwee et al. have not hidden the source of the contours from the evaluators, also potentially introducing bias. Gooding et al. [28],* recognizing that there may be bias if the evaluator knows the source of the contouring, implemented their assessment in a blinded manner but also expressed the question so as to place the assessment within a clinical context. Peer review of contouring is considered best practice, yet to reject another person's contours requires confidence in the significance of the error. Thus, the question was phrased to express a peer review context, with a balanced scoring scale between acceptance and rejection of the contours. It is interesting to note that both Hwee et al. [27] and Gooding et al. [28] found that clinical observers will reject around 25% of clinical contours (ones that have been approved for clinical use previously) when blinded as to the source of contouring.

However, the acceptance of contours still requires agreement as to what the correct contouring standard is. As a consequence, the assessment of acceptance of contours from an auto-contouring system may reflect the style of contouring with which the system has been developed/trained, rather than the efficacy of the system itself. In Gooding et al. [29], wide variability between the acceptance of contours was found between institutions, despite the same contours being shown.

15.3.2   THE SOURCE OF CONTOURING

This blinding to the source of the contour, together with the use of AI (deep learning) in their auto-contouring research [30], inspired Gooding et al. to frame the question not as one of acceptance but as one of identification [1, 28] following the idea of the Imitation Game proposed by Turing [31] and asking the question: *Was this contour drawn by a human or a computer?* While the question of the source of contours was also used by Hwee et al. [27], their purpose was to assess whether blinding observers to the source of contouring reduced potential bias in asking about contour acceptability. Gooding et al. propose that the inability to determine the source of the contours could itself be used as performance criterion [1].

They pose the argument that where quantitative measures are blind to the type of error, as discussed above, the human observer is not. If it is assumed that a clinically drawn contour is acceptable, then the inability to distinguish the clinical contour from the auto-contour would suggest that the auto-contour is equally good. This is not to say that either are correct or incorrect, overcoming the issue of inter-observer variability or disagreement, but that the types of error being made by the auto-contouring system are similar in nature to those made by a human expert. Thus, the auto-contouring requires no more editing than those of another clinical expert. In their initial study, they provide indicative results that the misclassification rate of the source of contouring is a better surrogate of editing time than quantitative measures such as Dice. One objection that can be leveled at this approach is that it does not easily allow for the auto-contouring to outperform manual

---

* The published abstract does not contain this detail. However, the survey questions are still available online at www.auto-contouring.com/

TABLE 15.2

Summary of Studies Using Subjective Evaluation to Ask about the Clinical Acceptability of Auto-Contours in the Domain of Radiotherapy. Question Phrasing Has Been Inferred Where Not Explicitly Stated in the Chapter

Study	Question	0	1	2	3
Hwee et al. [27]	Are the contours acceptable?	No	Yes		
Teguh et al. [14]	How well do the contours agree with published guidelines?	Poor	Major deviation, editable	Minor deviation, editable	Perfect
Granberg [16]	How useful are the segmentation proposals?	Not at all	A little	Very much	
Gooding et al. [50, 29]	How useful are the contours, how much time would you expect to save?	None of the results would form a useful basis for further editing, no time is expected to be saved compared to manual contouring	Some of the results form a useful basis for further editing, little time would be saved compared to manual contouring	Many of the results form a useful basis for further editing, a moderate time saving is expected compared to manual contouring	Most of the results form a useful basis for further editing, a significant time saving is expected compared to manual contouring
Hoang Duc et al. [53]	How well do the contours adhere to guidelines for clinical use?	The segmentation does not meet universal guidelines. Some slices show gross mis-delineation that cannot be attributed to segmentation variability	The segmentation is reasonably acceptable but needs some manual editing. Some contour lines need to be corrected to meet universal guidelines	The segmentation is clinically acceptable and satisfies universal OAR delineation guidelines and can be used as created for radiotherapy planning	
Van Dijk-Peters et al. [54]	How much editing is required prior to clinical use?	Major editing required	Minor editing required	No editing required	
Lustberg et al. [30]	How useful are the contours, how much time would you expect to save?	None of the results would form a useful basis for further editing, no time is expected to be saved compared to manual contouring	Some of the results form a useful basis for further editing, little time would be saved compared to manual contouring	Many of the results form a useful basis for further editing, a moderate time saving is expected compared to manual contouring	Most of the results form a useful basis for further editing, a significant time saving is expected compared to manual contouring
McCarroll et al. [11]	How much editing is required for use in dose-volume-histogram (DVH)-based planning?	Major editing needed	Minor editing needed	No editing needed	
Gooding et al. 2018 [28]	You have been asked to QA these contours for clinical use by a colleague. Would you...	Require them to be corrected; there are large, obvious, errors	Require them to be corrected; There are minor errors that need a small amount of editing	Accept them as they are; there are minor errors, but these are clinically not significant	Accept them as they are; the contours are very precise

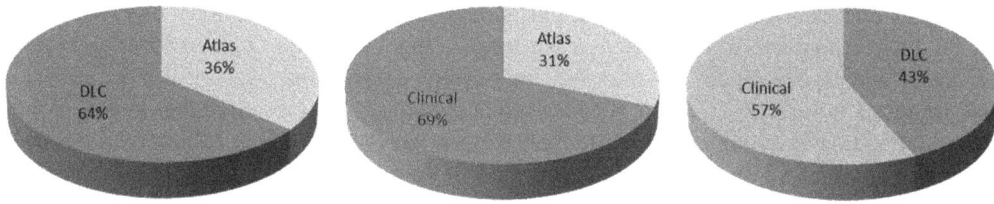

FIGURE 15.11 Preference for contouring for three different methods (clinical contours, atlas-based auto-contouring, and deep learning contouring) after blinded side-by-side assessment, as reported in Gooding et al. [28].

contouring with the aim being to imitate human contouring behavior. To achieve immitigability, the auto-contour must be equally as good, but also equally as bad.

15.3.3 The Preference for Contouring

The remaining type of subjective assessment used to date was also posed in the discussion by Gooding et al. [1] and used in practice in Gooding et al. [28]; this is to ask which contour is preferred when two contours are shown side by side in a blinded fashion. This approach again sidesteps the question of contour correctness, allowing contours to be incorrect and/or subject to inter-observer variability. Such a question allows the comparison of multiple contouring methods. A reference contour (or manual clinical contour) can then be used to benchmark performance yet allowing auto-contouring to outperform this benchmark, as illustrated in Figure 15.11.

15.3.4 Challenges of Subjective Assessment

Subjective assessment is very helpful in understanding the acceptance and utility of auto-contouring. However, there are several limitations and challenges to using this approach.

As has already been noted when considering Table 15.2, there is a substantial risk in introducing bias into any subjective study since human observers are involved. Consequently, any study must be carefully designed. Much research has been conducted in polling,* but as professionals in radiation oncology, medical physics, medical devices, computer science, etc., the authors are no experts in the psychology behind the design of such studies. Therefore, caution must be exercised to avoid introducing bias in how questions are phrased, and the choice of answers given. As is commonplace in medicine, blinding, and controls are important to reduce bias and should be used for any subjective assessment.

An additional factor that needs consideration is whether the assessment should be performed in 2D or 3D. 2D assessment reflects slice-by-slice contouring and offers more data for the same number of patients yet does not allow the assessment of correctness of the axial extent of contouring. However, 3D assessment will require more cases to be meaningful, and human slice-by-slice contouring results in telltale jaggedness on a coronal or sagittal view giving away the source of contouring. Similarly, efforts must be made to avoid clear differences resulting from the contouring method, but not related to the quality of the contouring. Some treatment planning systems and auto-contouring solutions produce pixelized RTSS, while other produce smooth contours. Such artifacts, if different between methods, can provide evidence for assessment that divert the observer from the purpose of the assessment.

The qualitative nature of these assessments introduces challenges in the development of auto-contouring methods. While results should be reasonably reproducible for large studies,

---

* As an outsider to the field of opinion polling, the most suitable review paper to cite is unknown to the author. Therefore, no reference is offered.

a degree of variation would be expected, particularly for smaller evaluations. Consequently, it may be difficult to perform repeated assessments to demonstrate the improvement of a system over time as it is developed, or to assess the current performance during the development. This limitation is further impacted because subjective assessment requires the time of clinical experts. This time is valuable, and while their input is worthwhile where subjective assessment is useful in commissioning a clinical system, their involvement will be limited in the development or comparison of systems. Consequently, studies focusing on the development or comparison of methods, such as the grand challenges [2–4], will resort to quantitative assessment over subjective assessment.

15.3.5 SUMMARY OF SUBJECTIVE EVALUATION

In this section the use of subjective assessment for evaluation of auto-contouring in radiation oncology is considered. Three types of question have been considered relating to the acceptance of contouring, the source of contouring, and the preference for contouring. It has been highlighted that study design is an important factor in such investigations on account of the human element involved. Despite the challenges of implementing subjective evaluation well, this approach to assessing auto-contouring can overcome some of the limitations of quantitative assessment in determining the suitability of a system for clinical use, particularly those related to inter- and intra-observer variability.

15.4 ASSESSMENT BASED ON INTENDED CLINICAL USE

The end purpose of auto-contouring in radiation oncology is that those auto-contours are used in the clinic for treatment planning. To this end, the best form of assessment is to understand the impact that their use would have on clinical practice. Ideally, one would study the impact on patient outcomes since this is the endpoint for assessment of any intervention. It may be argued that such an assessment would be pointless anyway since auto-contours should be checked and approved before use in treatment planning. However, it is known that having an auto-contour as a starting point improves consistency between manual observers [32], therefore auto-contouring will be having some impact on the treatment plan and thus the patient outcome. However, contouring is very early in the treatment planning processes, and other steps are likely to obscure any meaningful observation of differences in patient outcome that can be attributed only to the difference in contouring methodology.

15.4.1 EVALUATION OF TIME SAVING

If contours are to be edited following auto-contouring, as some have deemed inevitable [33], then the benefit of auto-contouring is in reducing the contouring time required as far as possible. Thus, it follows that evaluating the time saving to get to an acceptable clinical contour is a good way to quantify the clinical impact of using auto-contouring. Correspondingly, this is the most common approach used for clinical benefit assessment, as observed in Table 15.1.

Most studies follow the same pattern; the structures for a set of patients are first manually contoured, with the time taken for manual contouring being recorded. Subsequent to this, an auto-contouring method is applied to the same set of patients, and the auto-contours are edited to an acceptable standard, with the time taken for this step also recorded.

15.4.1.1 Challenges for Study Design

To evaluate contouring time inherently takes time. As mentioned previously, the clinicians time is valuable, therefore, to ask them to contour the same cases twice only for evaluation purposes represents a barrier to performing a large assessment. Consequently, the size of study tends to be small, with the largest study listed in Table 15.1 only considering evaluation on 27 patients [34].

Since the sample sizes are small, all investigations have used the same set of patients for manual contouring and editing of auto-contours. This introduces a risk that the observers become familiar with the case, and thus familiarity influences the time required for contouring or editing. Steps can be taken to mitigate this risk, such as randomizing the order in which contouring is performed between cases or ensuring a large gap between de novo contouring and editing [35]. The scenario of repeated contouring is artificial and introduces the risk that the time recorded is not reflective of clinical practice – since observers know that they are being monitored for this evaluation. This could influence the results (either positively or negative depending on the observers general feeling towards auto-contouring). Mitigation of this effect can be achieved to some extent by evaluating how similar contours are following manual contouring and after-editing, with comparison to inter-observer variation.

A better approach to overcome these risks could be to perform evaluation on a large number of consecutive patients, studying the time taken for contouring within the clinical workflow, before and after introducing an auto-contouring system. While the patient cases assessed will be different, and the structures required to be contoured for each patient will vary according to their treatment, such timing will reflect an average clinical workflow. Such a study does not appear to have been performed to date. This perhaps is unsurprising as it is preferable to conduct a small cohort study for commissioning a system prior to clinical use, to avoid disrupting clinical workflow with an unproven system. Nevertheless, such an impact analysis study would be very beneficial in demonstrating the true clinical impact of auto-contouring.

A further consideration relates to what is being compared. Some studies (e.g. [7, 32]) compared to manual contouring, while other investigators have considered the impact to their current clinical practice which may already include a different form of auto-contouring (e.g. [35, 36]) or semi-automatic tools (e.g. [30]). The study suggested above would also reflect the impact of auto-contouring in clinical practice, rather than an artificial use of manual contouring. However, changing existing clinical practice will also influence the time taken. Where clinical staff are familiar with using a set of tools for manual clinical contouring, editing of auto-contouring may be best achieved with a different set of tools. Such a change in the suitability of the contouring/editing tools should be expected, and alongside that it would be anticipated that there would be a learning curve. Thus, the impact on clinical practice would need to be assessed after a period of adoption.

15.4.2 IMPACT OF AUTO-CONTOURING ON PLANNING

Beyond contouring, the contours themselves are an input to treatment planning. Thus, to get an indication of whether the use of auto-contouring is impacting the patient, it is necessary to look at the dosimetric impact in planning. To discuss what has been done more easily, what could be done will be considered first. Figure 15.12 illustrates the combinations of assessments that could be calculated. Three possible contouring approaches can be considered: manual clinical contouring, unedited auto-contouring, and editing auto-contouring. A treatment plan can be created from each one of these contour sets, assuming these were the contours available to the planner. These plans will be referred to as A, B, and C. However, if one assumes that the true structure boundary could be any one of the contour sets, it is possible to calculate the DVH from any plan with any contour set, leading to nine possible DVHs. These combinations will be referred to in the form A1, A2, etc. where A1 would represent the DVH assessment of the plan made using contour set 1 with the contours of set 1, A2 would be the DVH of the plan created using the contour set 1 evaluated with the contour set 2.

Voet et al. [37] asked the question as to whether auto-contours need editing by first evaluating B3 for the elective nodal volumes in the head and neck. B3 shows the impact of dose to structures should a plan be generated with unedited contours, on the assumption that the adjusted auto-contours are correct and represent the true anatomy. Therefore, this could be used to assess potential under-dosage to the Planning Treatment Volume (PTV) resulting from not editing contours. Subsequently

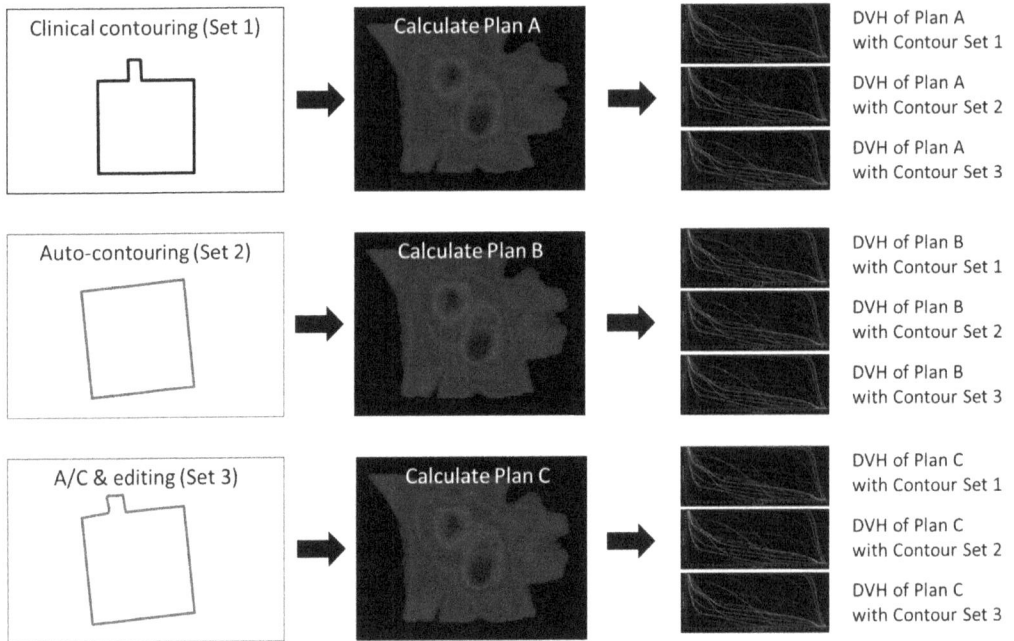

FIGURE 15.12 Possible investigations into the impact on treatment planning. Three possible contour sets could lead to three different treatment plans. The dose to structures can be investigated assuming each contour set is correct, leading to nine possible DVH curves. The most common approach so far is to compare the dose from Plan A using contour sets 1 and 3.

they compared C2 to C3. C3 represents the DVH for OAR doses that would be performed during clinical planning using edited auto-contours, whereas C2 gives the DVH that would be found if the contours were not edited. However, this comparison assumes that the OAR dose is not critical to the plan generated. Thus, they suggest that auto-contours of OARs only need editing if they are approaching a critical planning dose.

Yang et al. [10] performed a comparison using A2 and A3 to assess the impact of editing on the reported OAR dose. Similarly, Van Dijk et al. [35] used A2 and A3 to study difference in the impact on OAR dose of two different auto-contouring approaches. Kieselmann et al. [38] compared the OAR dose using B1 and B2 for head and neck contouring on magnetic resonance imaging (MRI). However, the PTV was manually drawn. Thus, while the dose was optimized on unedited OAR and the DVH compared with those using the clinical contours, only clinical PTV were used in planning. Kaderka et al. [39] similarly compared OAR dose, using A1 and A2, for breast OAR contouring.

For target contouring, Dipasquale et al. [34] made a target dose assessment using C3 to C1 for the whole breast PTV. Here, the aim was to assess whether the planning on edited contours would provide appropriate dose coverage, assuming manually drawn contours represent the true anatomy. In contrast, Simões et al. [40] compared dose using B3 and C3 to suggest that planning on unedited contours was acceptable for the whole breast PTV.

In all cases, comparison between different doses was done measuring the difference in selected DVH parameters, typically those used for planning either as OAR constraint parameters, such as V20Gy for lungs, or as PTV coverage measures, such as the V95%.

15.4.2.1 Challenges for Study Design

As has been noted there are nine combinations of plans and contours sets that could be used to assess the dosimetric impact of planning. Table 15.3 summarizes these possibilities, showing that all the available options have been used in at least one research study.

TABLE 15.3

Use of Plan and Contour Combinations in Studies of the Impact of Auto-Contouring

Dose plan on \ Contour set	1. Manual	2. Automatic	3. Adjusted Automatic
A. Manual	[39, 69]	[10, 35, 39]	[10, 35]
B. Automatic	[37, 38]	[38, 69]	[37, 40]
C. Adjusted Automatic	[34]	[37]	[34, 37, 40]

While there appears to be a broad choice of contour/plan combinations, the purpose of evaluation is similar for all studies; can auto-contours be used clinically, and do they require editing? Thus, most studies either hold the plan constant and vary the contours for assessment, or use the same contour set for evaluation but vary the plan. There is an underlying assumption in these assessments that the clinical contours, whether manual (1) or adjusted (3), and the plans based on these (A or C) represent the reference contours and the "correct" plan. Therefore, the choice becomes contour set 1 or 3 vs 2 on a constant plan, or plan A or C vs B on a constant contour set.

Only two studies fall outside of this. Dipasquale et al. [34] seeking to show that the plans on edited auto-contours are equivalent to those on manual contours – not considering the possibility that the contours would be use unedited. Voet et al. [37] evaluates B3 in isolation of a comparator for target volumes, since the plan has been optimized assuming the unedited contours are correct, and the "true" contours can be used to assess what the under-dosage would be.

This assumption that the clinical contour is truth, raises the same concern as that of quantitative assessment: inter-observer variability. However, in addition to inter-observer variability in the contours, the potential for inter-observer variability in the planning may come into play. Several studies take steps to address this through the use of auto-planning systems to produce plans from human subjectivity [38, 40]. The same studies seek to measure the impact of inter-observer variation in contours on the planning, so as to place the impact of the auto-contouring into context.

While the need to edit contours and perform planning generally keeps studies small (<30 cases). The use of auto-planning reduces the demand for human effort, thus facilitating slightly larger studies, with Simões et al. [40] performing an evaluation on 87 patient cases.

15.4.3 SUMMARY OF CLINICAL IMPACT EVALUATION

The main purpose of auto-contouring is to reduce the burden of contouring. Therefore, to really evaluate the clinical impact of auto-contouring, it is natural to measure the time saving by using it. In this chapter, it has been considered that there are a number of challenges with implementing such experiments in practice – particularly with regard to inter-observer variability. Ultimately, auto-contouring would be of such a standard as to not require editing. Yet, human nature appears to be to correct within the margins of inter-observer variation, as was evidenced by subjective tests where blinded observers would edit original clinical contours. Therefore, a better clinical test of auto-contouring, or human editing, is to evaluate the dosimetric impact that editing, or the absence of editing, will have on the patient. An alternative approach would be to introduce clinical contours from another observer into the assessment blindly. This would demonstrate the potential impact of an auto-contouring method at human expert level. Nevertheless, all clinical impact studies are time consuming and challenging to implement for larger numbers of cases, and therefore might be best reserved for clinical commissioning rather than during algorithm development.

15.5   DISCUSSION AND RECOMMENDATIONS

In this chapter the main approaches to assessing auto-contouring have been reviewed. Three main approaches were discussed: quantitative assessment, subjective studies, and the evaluation of clinical impact. It was noted that all of these approaches have limitations and challenges, with the greatest of these challenges being that of inter-observer variation. If experts cannot agree or be relied on to delineate contours in the same location, then auto-contouring methods cannot be expected to be able to do so.

Recommendations as to how clinical commissioning and validation of auto-contouring should be performed have been proposed [41]. While many studies have been performed since these recommendations were made, they are still valid. However, these recommendations do not separate the various purposes of evidence gathering, and the nature of the evaluation required. Therefore, Table 15.4 adds to the recommendations of Valentini et al. [41], by proposing the type of evaluation that should be performed depending on the context.

Noting the challenge inter-observer variation brings, the following statement from the previous recommendation is highlighted: "...must still be considered along with what we refer to as the 'benchmark trap': are we confident that the daily 'human' inter-observer variability could show better performances, in terms of dose and volumes, when considering a comparison with the same manual benchmark?"

Any evaluation must consider the variability of the benchmark. Inter-observer variation should be considered in all studies to provide context for the results presented.

TABLE 15.4

Recommended Approaches for Validation of Auto-Contouring for the Different Purposes of Evaluation

Purpose of Study	Validation Approach(es) Suggested
Development: The validation of a newly proposed method of auto-contouring	Quantitative measures are the primary method since these do not require clinical time. Current state-of-the-art measures, added path length, or surface Dice should be used to place a clinical interpretation on results. Further measures should be reported for context.
	Inter-observer variation should be reported using the same measures.
	If expert opinion is available, blinded subjective evaluation to assess potential clinical performance is highly beneficial. Clinical contours should be assessed alongside auto-contours as a benchmark.
Comparison: The validation of two or more auto-contouring methods for the purpose of determining their relative strengths	For challenge type evaluation, only quantitative assessment can be performed effectively for multiple methods. Inter-observer variation must be considered prior to drawing strong conclusions as to a "best" method. Methods using tolerances, such as added path length or surface Dice, should be considered.
Clinical commissioning: The validation of an auto-contouring method for the purpose of evaluating its suitability for clinical use	Methods assessing clinical impact should be the primary mode of validation. Care should be taken in implementation of such evaluations to avoid bias. Consideration should be given to blinding to measure the impact that using an alternative clinical contour would have in the same setting. Evaluation should be performed against current clinical practice rather than artificial contouring scenarios.
	Blind subjective assessment is recommended in addition to enable larger dataset and multiple observers to participate in validation.
	Quantitative evaluation could be performed to demonstrate similar performance in the clinic as was observed in development studies. However, quantitative measures cannot demonstrate clinical acceptability.

REFERENCES

1. MJ Gooding et al., "Comparative evaluation of autocontouring in clinical practice: a practical method using the turing test," *Med. Phys.*, vol. 45, no. 11, pp. 5105–5115, 2018.

2. V Pekar, S Allaire, and A Qazi, "Head and neck auto-segmentation challenge: segmentation of the parotid glands," *MICCAI 2010 A Gd. Chall. Clin.*, no. October 2015, pp. 273–280, 2010.

3. PF Raudaschl et al., "Evaluation of segmentation methods on head and neck CT: auto-segmentation challenge 2015," *Med. Phys.*, vol. 44, no. 5, pp. 2020–2036, 2017.

4. J Yang et al., "Autosegmentation for thoracic radiation treatment planning: a grand challenge at AAPM 2017," *Med. Phys.*, vol. 45, no. 10, pp. 4568–4581, Oct. 2018.

5. A Isambert et al., "Evaluation of an atlas-based automatic segmentation software for the delineation of brain organs at risk in a radiation therapy clinical context," *Radiother. Oncol.*, vol. 87, no. 1, pp. 93–99, 2008.

6. LJ Stapleford et al., "Evaluation of automatic atlas-based lymph node segmentation for head-and-neck cancer," *Int. J. Radiat. Oncol. Biol. Phys.*, vol. 77, no. 3, pp. 959–966, 2010.

7. M La Macchia et al., "Systematic evaluation of three different commercial software solutions for automatic segmentation for adaptive therapy in head-and-neck, prostate and pleural cancer," *Rad. Oncol.*, vol. 7, no. 1, pp. 1–16, 2012.

8. LR Dice, "Measures of the amount of ecologic association between species," *Ecology*, vol. 26, no. 3, pp. 297–302, 1945.

9. LC Anders, F Stieler, K Siebenlist, J Schäfer, F Lohr, and F Wenz, "Performance of an atlas-based autosegmentation software for delineation of target volumes for radiotherapy of breast and anorectal cancer," *Radiother. Oncol.*, vol. 102, no. 1, pp. 68–73, 2012.

10. J Yang et al., "Automatic contouring of brachial plexus using a multi-atlas approach for lung cancer radiation therapy," *Pract. Radiat. Oncol.*, vol. 3, no. 4, pp. 1–16, 2013.

11. RE McCarroll et al., "Retrospective validation and clinical implementation of automated contouring of organs at risk in the head and neck: a step toward automated radiation treatment planning for low- and middle-income countries," *J. Glob. Oncol.*, no. 4, pp. 1–11, 2018.

12. AA Taha, and A Hanbury, "Metrics for evaluating 3D medical image segmentation: analysis, selection, and tool," *BMC Med. Imaging*, vol. 15, no. 29, 2015. https://doi.org/10.1186/s12880-015-0068-x, https://link.springer.com/article/10.1186/s12880-015-0068-x#citeas

13. DP Huttenlocher, GA Klanderman, and WJ Rucklidge, "Comparing images using the hausdorff distance," *IEEE Trans. Pattern Anal. Mach. Intell.*, vol. 15, no. 9, pp. 850–863, 1993.

14. DN Teguh et al., "Clinical validation of atlas-based auto-segmentation of multiple target volumes and normal tissue (swallowing/mastication) structures in the head and neck," *Int. J. Radiat. Oncol. Biol. Phys.*, vol. 81, no. 4, pp. 950–957, 2011.

15. V Pekar, J Kim, and DA Jaffray, "Head and neck auto-segmentation challenge," *MIDAS J.*, vol. 11, no. November 2009. www.midasjournal.org/browse/publication/703

16. C Granberg, "Clinical evaluation of atlas based segmentation for radiotherapy of prostate tumours," MSc Thesis, pp. 1–66, 2011.

17. MJ Gooding, L Durrant, KY Chu, CL Eccles, F Kaster, and T Kadir, "Methods for assessing atlas-based contouring in head and neck cancer," *Radiother. Oncol.*, vol. 1, no. 106, pp. S365–S366, 2013.

18. F Vaassen et al., "Evaluation of measures for assessing time-saving of automatic organ-at-risk segmentation in radiotherapy," *Phys. Im. Radiother. Oncol.*, vol. 13, pp. 1–6, 2020.

19. PY Bondiau et al., "Atlas-based automatic segmentation of MR images: validation study on the brainstem in radiotherapy context," *Int. J. Radiat. Oncol. Biol. Phys.*, vol. 61, no. 1, pp. 289–298, 2005.

20. D Ciardo et al., "Atlas-based segmentation in breast cancer radiotherapy: evaluation of specific and generic-purpose atlases," *Breast*, vol. 32, pp. 44–52, 2017.

21. S Nikolov et al., "Deep learning to achieve clinically applicable segmentation of head and neck anatomy for radiotherapy," arXiv:1809.04430, 2018.

22. WE Lorensen, and HE Cline, "Marching cubes: a high resolution 3D surface construction algorithm," *ACM Transactions on Graphics*, vol. 21, no. 4. pp. 163–169, 1987.

23. EL Lorenzen et al., "Inter-observer variation in delineation of the heart and left anterior descending coronary artery in radiotherapy for breast cancer: a multi-centre study from Denmark and the UK," *Radiother. Oncol.*, vol. 108, no. 2, pp. 254–258, 2013.

24. SK Warfield, KH Zou, and WM Wells, "Simultaneous Truth and Performance Level Estimation (STAPLE): an algorithm for the validation of image segmentation," *IEEE Trans. Med. Imaging*, vol. 23, no. 7, pp. 903–921, Jul. 2004.

25. O Acosta et al., "Evaluation of multi-atlas-based segmentation of CT scans in prostate cancer radiotherapy," *Proc. – Int. Symp. Biomed. Imaging*, pp. 1966–1969, 2011.

26. G Sharp et al., "Plastimatch: an open source software suite for radiotherapy image processing," *Proc. XVI'th Int. Conf. use Comput. Radiother.*, no. October 2017, 2010.

27. J Hwee et al., "Technology assessment of automated atlas based segmentation in prostate bed contouring," *Rad. Oncol.*, vol. 6, no. 1, pp. 1–9, 2011.

28. M Gooding et al., "PV-0531: multi-centre evaluation of atlas-based and deep learning contouring using a modified turing test," *Radiother. Oncol.*, vol. 127, pp. S282–S283, 2018.

29. MJ Gooding et al., "Multicenter clinical assessment of DIR atlas-based autocontouring," *Int. J. Radiat. Oncol.*, vol. 87, no. 2, pp. S714–S715, 2013.

30. T Lustberg et al., "Clinical evaluation of atlas and deep learning based automatic contouring for lung cancer," *Radiother. Oncol.*, vol. 126, no. 2, pp. 312–317, 2018.

31. AM TURING, "Computing machinary and intelligence," *Mind*, vol. LIX, no. 236, pp. 433–460, Oct. 1950.

32. AV Young, A Wortham, I Wernick, A Evans, and RD Ennis, "Atlas-based segmentation improves consistency and decreases time required for contouring postoperative endometrial cancer nodal volumes," *Int. J. Radiat. Oncol. Biol. Phys.*, vol. 79, no. 3, pp. 943–947, 2011.

33. PC Levendag et al., "Atlas based auto-segmentation of CT images: clinical evaluation of using auto-contouring in high-dose, high-precision radiotherapy of cancer in the head and neck," *Int. J. Radiat. Oncol. Biol. Phys.*, vol. 72, no. 1, p. S401, 2008.

34. G Dipasquale, X Wang, V Chatelain-Fontanella, V Vinh-Hung, and R Miralbell, "Automatic segmentation of breast in prone position: correlation of similarity indexes and breast pendulousness with dose/volume parameters," *Radiother. Oncol.*, vol. 120, no. 1, pp. 124–127, 2016.

35. LV van Dijk et al., "Improving automatic delineation for head and neck organs at risk by deep learning contouring," *Radiother. Oncol.*, vol. 142, pp. 115–123, 2020.

36. VK Reed et al., "Automatic segmentation of whole breast using atlas approach and deformable image registration," *Int. J. Radiat. Oncol. Biol. Phys.*, vol. 73, no. 5, pp. 1493–1500, 2009.

37. PWJ Voet, MLP Dirkx, DN Teguh, MS Hoogeman, PC Levendag, and BJM Heijmen, "Does atlas-based autosegmentation of neck levels require subsequent manual contour editing to avoid risk of severe target underdosage? A dosimetric analysis," *Radiother. Oncol.*, vol. 98, no. 3, pp. 373–377, 2011.

38. JP Kieselmann et al., "Geometric and dosimetric evaluations of atlas-based segmentation methods of MR images in the head and neck region," *Phys. Med. Biol.*, vol. 63, no. 14, pp. 145007, 2018.

39. R Kaderka et al., "Geometric and dosimetric evaluation of atlas based auto-segmentation of cardiac structures in breast cancer patients," *Radiother. Oncol.*, vol. 131, pp. 215–220, 2019.

40. R Simões, G Wortel, TG Wiersma, TM Janssen, UA van der Heide, and P Remeijer, "Geometrical and dosimetric evaluation of breast target volume auto-contouring," *Phys. Imaging Radiat. Oncol.*, vol. 12, no. September, pp. 38–43, 2019.

41. V Valentini, L Boldrini, A Damiani, and LP Muren, "Recommendations on how to establish evidence from auto-segmentation software in radiotherapy," *Radiother. Oncol.*, vol. 112, no. 3, pp. 317–320, 2014.

42. O Commowick, V Grégoire, and G Malandain, "Atlas-based delineation of lymph node levels in head and neck computed tomography images," *Radiother. Oncol.*, vol. 87, no. 2, pp. 281–289, 2008.

43. S Gorthi et al., "Segmentation of head and neck lymph node regions for radiotherapy planning using active contour-based atlas registration," *IEEE J. Sel. Top. Signal Process.*, vol. 3, no. 1, pp. 135–147, 2009.

44. R Sims et al., "A pre-clinical assessment of an atlas-based automatic segmentation tool for the head and neck," *Radiother. Oncol.*, vol. 93, no. 3, pp. 474–478, 2009.

45. L Ramus, and G Malandain, "Multi-atlas based segmentation: application to the head and neck region for radiotherapy planning," *Med. Image Anal. Clin.*, pp. 281–288, 2010.

46. A Chen, KJ Niermann, MA Deeley, and BM Dawant, "Evaluation of multiple-atlas-based strategies for the segmentation of the thyroid gland in head and neck CT images for IMRT," *Phys. Med. Biol.*, vol. 57, no. 1, pp. 93–111, 2012.

47. GC Mattiucci et al., "Automatic delineation for replanning in nasopharynx radiotherapy: what is the agreement among experts to be considered as benchmark?" *Acta Oncol. (Madr).*, vol. 52, no. 7, pp. 1417–1422, 2013.

48. MA Gambacorta et al., "Clinical validation of atlas-based auto-segmentation of pelvic volumes and normal tissue in rectal tumors using auto-segmentation computed system," *Acta Oncol. (Madr).*, vol. 52, no. 8, pp. 1676–1681, 2013.

49. C Sjöberg, M Lundmark, C Granberg, S Johansson, A Ahnesjö, and A Montelius, "Clinical evaluation of multi-atlas based segmentation of lymph node regions in head and neck and prostate cancer patients," *Rad. Oncol.*, vol. 8, no. 1, pp. 1–7, 2013.

50. MJ Gooding et al., "Multicenter assessment of autocontouring using deformable image registration for adaptive therapy," *Int. J. Radiat. Oncol.*, vol. 87, no. 2, p. S717, 2013.

51. O Acosta, J Dowling, G Drean, A Simon, R De Crevoisier, and P Haigron, *Multi-Atlas-Based Segmentation of Pelvic Structures from CT Scans for Planning in Prostate Cancer Radiotherapy.* 2014.

52. D Thomson et al., "Evaluation of an automatic segmentation algorithm for definition of head and neck organs at risk," *Rad. Oncol.*, vol. 9, no. 1, pp. 1–12, 2014.

53. AKHoang Duc et al., "Validation of clinical acceptability of an atlas-based segmentation algorithm for the delineation of organs at risk in head and neck cancer," *Med. Phys.*, vol. 42, no. 9, pp. 5027–5034, 2015.

54. FBJ Van Dijk-Peters et al., "OC-0259: validation of a multi-atlas based auto-segmentation of the heart in breast cancer patients," *Radiother. Oncol.*, vol. 115, no. April, pp. S132–S133, 2015.

55. Q Liu, A Qin, J Liang, and D Yan, "Evaluation of atlas-based auto-segmentation and deformable propagation of organs-at-risk for head-and-neck adaptive radiotherapy," *Recent Patents Top. Imaging*, vol. 5, no. 2, pp. 79–87, 2016.

56. G Delpon et al., "Comparison of automated atlas-based segmentation software for postoperative prostate cancer radiotherapy," *Front. Oncol.*, vol. 6, no. August, pp. 1–6, 2016.

57. K Wardman, RJD Prestwich, MJ Gooding, and RJ Speight, "The feasibility of atlas-based automatic segmentation of MRI for H&N radiotherapy planning," *J. Appl. Clin. Med. Phys.*, vol. 17, no. 4, pp. 146–154, 2016.

58. MF Fast et al., "Tumour auto-contouring on 2d cine MRI for locally advanced lung cancer: a comparative study," *Radiother. Oncol.*, vol. 125, no. 3, pp. 485–491, 2017.

59. B Ibragimov, and L Xing, "Segmentation of organs-at-risks in head and neck CT images using convolutional neural networks," *Med. Phys.*, vol. 44, no. 2, pp. 547–557, 2017.

60. K Men, J Dai, and Y Li, "Automatic segmentation of the clinical target volume and organs at risk in the planning CT for rectal cancer using deep dilated convolutional neural networks," *Med. Phys.*, vol. 44, no. 12, pp. 6377–6389, Dec. 2017.

61. B Ibragimov, DAS Toesca, DT Chang, AC Koong, and L Xing, "Deep learning-based autosegmentation of portal vein for prediction of central liver toxicity after SBRT," *Int. J. Radiat. Oncol.*, vol. 99, no. 2, p. E672, 2017.

62. K Men et al., "Deep deconvolutional neural network for target segmentation of nasopharyngeal cancer in planning computed tomography images," *Front. Oncol.*, vol. 7, no. December, pp. 1–9, 2017.

63. S Kazemifar et al., "Segmentation of the prostate and organs at risk in male pelvic CT images using deep learning," *Biomed. Phys. Eng. Express*, vol. 4, no. 5, pp. 055003, 2018.

64. M Astaraki et al., "Evaluation of localized region-based segmentation algorithms for CT-based delineation of organs at risk in radiotherapy," *Phys. Imaging Radiat. Oncol.*, vol. 5, no. February, pp. 52–57, 2018.

65. N Tong, S Gou, S Yang, D Ruan, and K Sheng, "Fully automatic multi-organ segmentation for head and neck cancer radiotherapy using shape representation model constrained fully convolutional neural networks," *Med. Phys.*, vol. 45, no. 10, pp. 4558–4567, 2018.

66. CE Cardenas et al., "Deep learning algorithm for auto-delineation of high-risk oropharyngeal clinical target volumes with built-in dice similarity coefficient parameter optimization function," *Int. J. Radiat. Oncol. Biol. Phys.*, vol. 101, no. 2, pp. 468–478, 2018.

67. MS Elmahdy, JM Wolterink, H Sokooti, I Išgum, and M Staring, "Adversarial optimization for joint registration and segmentation in prostate CT radiotherapy," in *International Conference on Medical Image Computing and Computer-Assisted Intervention 2019*, Cham: Springer, pp. 366–374, 2019.

68. SH Ahn et al., "Comparative clinical evaluation of atlas and deep-learning-based auto-segmentation of organ structures in liver cancer," *Rad. Oncol.*, vol. 14, no. 1, pp. 1–13, 2019.

69. W van Rooij, M Dahele, H Ribeiro Brandao, AR Delaney, BJ Slotman, and WF Verbakel, "Deep learning-based delineation of head and neck organs at risk: geometric and dosimetric evaluation," *Int. J. Radiat. Oncol.*, vol. 104, no. 3, pp. 677–684, Jul. 2019.

70. W Zhu et al., "AnatomyNet: deep learning for fast and fully automated whole-volume segmentation of head and neck anatomy," *Med. Phys.*, vol. 46, no. 2, pp. 576–589, 2019.

71. H Lee et al., "Clinical evaluation of commercial atlas-based auto-segmentation in the head and neck region," *Front. Oncol.*, vol. 9, no. April, pp. 1–9, 2019.

Index

A

Abdomen, 3, 85, 154, 167, 177, 181, 184
Abnormal, 5, 14, 85–87, 190, 193, 196, 201, 205
Absolute volume, 220, 223
Abstraction, 201
Acceptance, 6, 192, 233, 238, 240–244, 248
Accuracy, *see* Evaluation measures
Activation function, 73, 87, 126, 170
Active contours, *see* Low-level segmentation
Adam, 88, 128, 156, 171
Anatomist, 196
Anatomy, 4, 19–21, 65, 76, 174, 183, 191
    aorta, 51, 63, 89
    bladder, 102
    bowels, 182
    brachial plexus, 2, 49, 249
    brainstem, 175
    breast, 89, 177–180, 210, 246
    cervical esophagus, 50
    clavicle, 177
    cricoid, 50, 174
    duodenum, 154
    esophagus, 3–5, 55–65, 102–104, 174
    gallbladder, 154
    gastroesophageal junction, 177
    head and neck, 55, 56, 175, 193
    heart, 4, 20, 63, 174–177, 179–183, 206
    liver, 85, 87, 92, 177, 182
    lungs, 4, 5
    lymph node, 101, 154
    mandible, 35, 193
    optic nerves, 89
    pancreas, 88, 92, 154
    parotid gland, 89, 187, 195
    pelvis, 2, 3, 92, 102
    pericardium, 172, 180
    pituitary, 159
    prostate, 103, 166, 197
    rectum, 102, 166
    ribcage, 20
    skull, 134, 174
    spinal cord, 4, 42, 58, 63, 103, 116, 119, 125, 168, 174, 175, 177, 226, 229, 232
    stomach, 154
    submandibular glands, 194
    thyroid gland, 89
    trachea, 89, 90, 174
    vertebral bodies, 51, 174
    vessels, 174
Annotation, 92, 97, 101, 103, 205, 207–210
Anonymization, 205–207, 209, 210
Architecture, 82–85, 87–97, 99–101, 114–123
    auto-encoder, 83–85, 87
    cascaded convolutional neural networks, 89
    cascaded fully convolutional network, 91–93, 96
    CFUN, 101
    convolutional SDAE, 86, 87

    cycle GAN, 102, 154
    deep convolutional neural network, 88, 125
    denoising autoencoder, 83, 85
    DenseNet, 88, 93
    Dense-U-net, 92
    DSAE, 86
    faster R-CNN, 100–102
    fast R-CNN, 99, 100
    generative adversarial network, 96, 97, 99, 102, 154
    hybrid methods, 102–104
    mask r-CNN, 100–102
    regional proposal network, 101
    region-CNN, 99–102
    residual network, 88, 89
    stacked autoencoders, 83, 85
    stacked denoising autoencoders, 85
    stacked SAE, 86, 87
    structure correcting adversarial network, 97
    U-net, 90–92, 95–97, 114–122, 127, 168–172
    VGG, 88
    V-net, 91, 92, 96
    WGAN, 154
Artificial neural network, 3, 72, 73
Artificial neuron, 72, 73
Assessment, 6, 30, 190, 192, 194, 218–227, 229, 238–241, 243–248
Atlas clustering, 15
Atlas selection, 14, 15, 21–27, 29–36, 42, 50–53, 62, 63, 65
    age-based, 26–28
    image-based, 26–29, 31, 32
    manifold-based, 28–30
    offline, 14, 26, 49
    online, 15, 50, 55, 62, 65
    pre-selection, 14
    random selection, 19, 26, 27, 29, 31
Attention gate, 91
Augmentation, 78, 128, 151–155, 157, 158, 206
    cropping, 152, 154, 155, 158
    elastic, 153, 206
    geometric transformation, 152, 153, 155
    intensity transformation, 14, 152, 153, 206
    jittering, 206
    mirroring, 152, 206
    rotation, 128, 152, 155, 157, 158, 206
    scaling, 128, 152, 153, 206
    shearing, 152
    translation, 152, 157, 171
    zoom, 152, 171
Auto-planning, 247

B

Backpropagation, 88, 91, 134, 136, 137, 146, 148
Batch normalization, 87, 102, 114
Batch size, 122
Bayesian classifiers, *see* Low-level segmentation
Benchmark, 4–6, 15, 55, 63–66, 103, 114, 191, 248
Bias, 142, 159, 209, 210, 233, 241, 243

For Product Safety Concerns and Information please contact our EU
representative GPSR@taylorandfrancis.com
Taylor & Francis Verlag GmbH, Kaufingerstraße 24, 80331 München, Germany

www.ingramcontent.com/pod-product-compliance
Lightning Source LLC
Chambersburg PA
CBHW061353210326
41598CB00035B/5973